Pocket Guide to Rhododendron Species

based on the descriptions by H.H. Davidian

Kew Publishing
Royal Botanic Gardens, Kew

Pocket Guide to Rhododendron Species

based on the descriptions by H.H. Davidian

J.F.J. McQuire and M.L.A. Robinson

PLANTS PEOPLE
POSSIBILITIES

First published in 2009 by
Royal Botanic Gardens, Kew
Richmond, Surrey, TW9 3AB, UK.
www.kew.org

ISBN 978 1 84246 148 8

British Library Cataloguing in Publication Data
A catalogue record for this book is available from the British Library.

Production Editor: Sharon Whitehead.
Copy Editor: John Sanders
Typesetting and page layout: Margaret Newman.
Design by Publishing, Design and Photography Department,
Royal Botanic Gardens, Kew.

Cover design by Jeff Eden.

Printed in the UK by Cambrian Printers Ltd.
ISO-14001 accredited with award winning Environmental Management Systems.

For information or to purchase all Kew titles please visit
www.kewbooks.com or email publishing@kew.org

Kew's mission is to inspire and deliver science-based plant conservation worldwide,
enhancing the quality of life.

Mixed Sources
Product group from well-managed
forests and other controlled sources
www.fsc.org Cert no. TT-COC-2200
© 1996 Forest Stewardship Council

The paper used in this book contains wood from well-managed forests, certified in
accordance with the strict environmental, social and economic standards of the Forest
Stewardship Council (FSC).

Contents

'Azaleas'

Lepidote Rhododendrons

Foreword

The genus *Rhododendron* is an incredibly large and diverse group of woody plants. Some estimates place the number of species at more than 1000, and newly discovered taxa are being described on a regular basis. When dealing with such a large group of often very similar plants, it can be difficult and take many years of effort and experience to become comfortable with identifying rhododendrons in a garden or in their wild habitats. Even those of us who have considerable experience with rhododendrons can embarrass ourselves by erring or suffering a lapse in memory when confronted with an unidentified pink-flowered rhododendron.

I am sure there are people who manage to keep straight, and at ready recall, the myriad of more or less minor botanical details used to separate the hundreds of *Rhododendron* species from each other. I, however, am not one of these people. For people like me, then, with a strong interest in rhododendrons and their identification, here at last is a reference book actually designed for use in the field. It provides a ready and accessible resource for both amateur gardener and professional horticulturist. And, it is suitable and relevant while either touring a botanical garden where labels often disappear, usually on the plants of particular interest to you, or trekking in the remote mountains of China where there are no labels at all. John McQuire and Mike Robinson, two of the world's foremost experts on the genus, have produced a handy, concise and well-illustrated pocket guide highlighting the most important morphological details relevant to the recognition and identification of virtually every species currently in cultivation. While there have been many fine books on rhododendrons published in the past few years, this is the first practical guidebook for the field.

I can assure you that this book will be on my list of travel necessities for future plant-hunting expeditions in the Sino-Himalaya.

Steve Hootman
Co-Director
Rhododendron Species Botanical Garden,
Seattle, Washington, USA

Acknowledgements

This book would never have been written but for the meticulous research of Pam Hayward, who informed us that H.H. Davidian was intending to publish a handbook of abbreviated texts of *Rhododendron* species but sadly died before full completion. Through the courtesy of his loyal secretary – Eileen Wood – the typescript of the work (which represented all of the then current species arranged in alphabetical order) was made available to us. This triggered the impulse to rearrange the species into their revised Sections, updating them, and adding new species and comments where appropriate together with illustrations.

We have received generous assistance from enthusiasts throughout the world. Tony Cox spent considerable time assisting our attempt to make sense of the confusing *Lapponica* Subsection. Mike Creel introduced us to *R. eastmanii*, used his unique experience of American azaleas to edit our text, and generously gave us permission to use his photographs. Yuji Kurashige advised us on the difficult task of classifying taxa related to *R. reticulatum*, and made his photographs available. The late Ambrose Bristow gave of his expertise in botanical Latin. Keith Rushforth, Tom Hudson and the late Peter Wharton provided useful information about species collected by them and others.

Unless otherwise stated, all the photographs in this book were taken by J.F.J. McQuire, mostly in his garden at Deer Dell, Farnham, Surrey.

In addition, Steve Hootman supplied many photographs of newly introduced and rare species – many flowering for the first time in cultivation in the superb collection at the Rhododendron Species Botanic Garden, Seattle; Steve also provided photographs by the late Art Dome. Jens Birck provided photographs of more unusual species from his connoisseur's collection in Denmark. We are also grateful to Philip Evans for his slides taken in the Arunachal Pradesh, to Barry Starling for yet more gap-filling, and to Everard Daniel for his excellent close-ups of rhododendron flowers. Special thanks go to the Rhododendron Society Gardens at Blackheath, NSW, Australia for allowing a photograph to be taken of the woodcut depicting the geographical extent of rhododendron species.

The photographs that we have used from the archives of the late H.H. Davidian have been loaned to us by his Executor, Eileen Wood. Some of the photographers cannot be credited as no source is indicated on several of the slides, but we should like to acknowledge the authorship of Sydney J. Clark, R. Eudall, Debbie White (all former official photographers at the Royal Botanic Garden Edinburgh), Hamish Gunn and Britt M. Smith.

Finally, we must acknowledge the advice and guidance of our publishers for their care in the preparation of this book, together with the reproduction of diverse types of photographs.

Naturally the identification of rhododendrons is a dynamic process, subject to review, opinions, and differences of opinion. The authors would be grateful to hear of new developments, the opinions of others, and any errors or misconceptions in this book.

Preface

There has been much controversy over the revision of the species of *Rhododendron*. The publication of the *RHS Rhododendron Handbook* (1998) finally replaced the Balfourian classification by one based on the Edinburgh revision, which was originally published in 1982.

Unfortunately, this has led to the discrediting of the work of H.H. Davidian. This is a great pity, as detailed reading of both Davidian's prefaces and those of subsequent writers shows that there is no conflict at all between the old and newer approaches. Davidian, with his usual clarity, describes the reasons for dividing rhododendrons into Series in his preface. Suffice to say that the system was designed as an aid to identifying **species in cultivation** by arranging them in groups of **superficially similar appearance**, one Series necessarily overlapping with another, with no definite demarcation intended but some separation being required in order to help users to memorise a remarkably complex genus. It is a great compliment to the Series system that so much of it has fitted in with later botanical studies.

More modern approaches have been able to take into account the evolution of rhododendrons and a greater awareness of their geographical distributions than was available 150 years ago. Subsequently, the division into Subgenera, Sections and Subsections has been possible on scientific and genetic grounds, and it is only reasonable that this book should follow the newer classification system based on molecular evidence as well as morphology.

For emphasis, however, the lack of conflict between the most up-to-date classification and the Balfourian system is demonstrated in this book by including the older plant names and Series (which are often found on labels in gardens). The current accepted name or, in a few cases, the one the authors believe is more apt, is given in bold. Synonyms of long-submerged species are in smaller type.

Revisions continue as DNA, leaf wax and other laboratory studies become more widespread, and as more species, clines and hybrid swarms are documented. There is still a great deal to be investigated and major revisions cannot be ruled out.

The authors make no apologies for publishing H.H. Davidian's descriptions of the species known to him. For painstaking regard to detail and accuracy, these descriptions are still unequalled. Each was based on exhaustive comparisons between herbarium specimens and, always, on more than one flowering plant grown from wild collected seed – at Edinburgh and in other gardens throughout the UK. A further mark of his scholarship is his acknowledgement of the variation within a species, even uncommon forms being carefully noted.

As the descriptions were written under a system designed primarily to assist identification, they form an ideal basis for a guide designed to assist in the identification of rhododendron species in cultivation.

Introduction

The majestic genus *Rhododendron* boasts both numerous and varied species, together with a very large number of additional 'natural' and 'man-made' hybrids.

This photographic pocket guide has been developed to aid the recognition and identification of those species that have been fully described and are currently in cultivation. In a publication of this size, a few rarities or species of doubtful provenance are not included; those omitted, together with others not yet in cultivation, are listed as an appendix. Descriptions of many of these can be found in less portable books, which are listed in the bibliography.

Interpreting the species descriptions

With a few exceptions, designed to assist in comparing two species of very similar appearance, the species are placed in the Subsections of the Edinburgh Review, as listed in *The Rhododendron Handbook 1998* (RHS, 1997). With a few exceptions, where they are misleading, the names used in that classification are given first, followed by Davidian's name if this differs.

Close attention to Davidian's species descriptions will usually lead to the identification of a rhododendron. To help speed up this process, the more significant features are **highlighted next to the photographs, with the key points being given double emphasis in bold type**.

Additional brief comments, where thought helpful, have been given at the end of Davidian's description.

Where a taxon described as a full species by Davidian has been reclassified as a subspecies or variety, or submerged altogether, the description is **indented** under the species into which it has been incorporated.

A full glossary and explanation of terms used can be found on page 668.

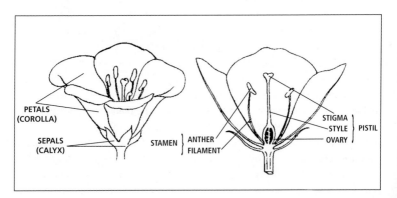

Basic rhododendron anatomy

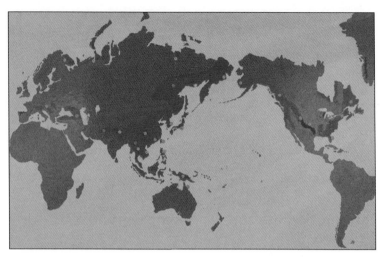

Distribution of Rhododendrons

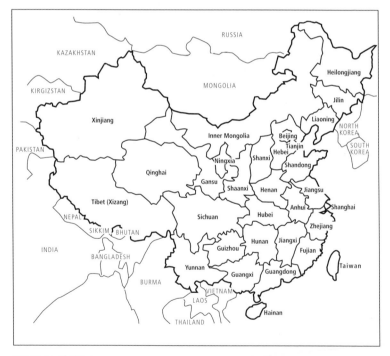

Provinces of China

Locations of the photographs

Arduaine, Loch Melfort, Oban, Scotland
Barnhourie Mill, Colvend, Dumfriesshire, Scotland
Baravalla, West Loch Tarbert, Argyll, Scotland
Benmore, Dunoon, Argyll, Scotland
Beresford Gardens, Trinity, Scotland
Blackhills, Elgin, Scotland
Borde Hill Gardens, Haywards Heath, West Sussex, England
Botallick, Lanreath, Cornwall, England
Brodick, Isle of Arran, Scotland
Caerhays Castle, Gorran, St. Austell, Cornwall, England
Clyne Castle, Swansea, South Wales
Crarae, Inverary, Argyll, Scotland
Darts Hill Garden, 16th Ave and 170 Street, Vancouver, British Columbia, Canada
Dawyck, Stobo, Peebles, Scotland
Deer Dell, The Sands, Farnham, Surrey, England
Eckford, Dunoon, Argyll, Scotland
Elizabeth Miller Botanical Garden, PO Box 77377, Seattle, WA 98177, USA
Exbury Gardens, Southampton, Hampshire, England
Glenarn, Rhu, Helensburgh, Scotland
Glendoick, Perth, Scotland
Hawkwood West, Crowborough, East Sussex
Hergest Croft, Kington, Herefordshire, England
Hillier Arboretum, Romsey, Hampshire, England
Hindleap Lodge, Forest Row, East Sussex, England
Kilbryde, Northumberland, England
Knightshayes Court, Tiverton, Devon, England
Lochinch, Stranraer, Scotland
Loder Plants, Leonardslee, Lower Beeding, West Sussex, England
Meerkerk Garden, 3531 Meerkerk Lane, Greenbank, WA 98253, USA
Millais Nurseries, Crosswater Farm, Farnham, Surrey, England
Mount Tomah, NSW, Australia
Muncaster Castle, Ravenglass, Cumbria, England
Nymans, Handcross, West Sussex, England
Pukeiti Rhododendron Trust, Taranaki, New Zealand
Quarryhill Botanical Garden, PO Box 232, Glen Ellen, CA 95442, USA
Royal Botanic Garden (RBG) Edinburgh, Scotland
Rhododendron Species Botanic Garden, Seattle, WA 98063, USA
Sarum, Worplesdon, Surrey
Savill Garden, Windsor Great Park, England
Stonefield Castle, Tarbert, Argyll, Scotland
Tregrehan, Par, Cornwall, England
Valley Gardens, Windsor Great Park, England
Inverewe, Achnasheen, Ross-shire, Scotland
University of British Columbia (UBC) Botanical Gardens, Vancouver V6T 1Z4, Canada
Wakehurst Place, Ardingly, West Sussex, England
Warren Berg Garden, 91 Wren Court, Port Ludlow, WA 98365, USA
Werrington, Launceston, Cornwall, England

Key to species descriptions

Authority

Group (if applicable)
used as an aid to
identification

4 Pocket guide to Rhododendrons

Name — *R. fortunei* ssp. *discolor* D.F.Chamb.

Name used by
Davidian (if different) — *R. discolor* Franch.
(*R. kwangfuense* Cun & Fang)

Fortunea Gp A

Names 'submerged'
into this species

Shrub or tree 1.2–8 m (4–26 ft) high, **branchlets** pale green, glabrous.
Leaves oblanceolate, oblong or oblong-obovate, **lamina** 10–20 cm long, 2.4–7 cm broad;
upper surface dark or pale green; **underside** pale glaucous green, glabrous.
Petiole glabrous.
Inflorescence in trusses of 6–10 flowers.

Description
(yellow background
indicates that a
description is not the
work of Davidian)

Pedicel glandular or eglandular.
Calyx margin glandular.
Corolla funnel-campanulate, 4.6–8 cm long, white, pink or rosy-pink, with or
without a yellowish-green blotch, lobes 7.
Stamens 12–14.
Style glandular throughout to the tip with white or yellowish glands.

Distribution and
altitude in the wild — E Sichuan, W Hubei. 1,000–2,300 m (3,500–7,500 ft).

RHS awards — AM 1921, flowers white tinged pink.
AM 1922 (Bodnant), flowers pale pink, with crimson blotch.
FCC 1922 (RBG Kew), flowers white, tinged pink.

Normal
flowering
time in
the UK

Derivation of
the plant name

AGM (1969) 2002.
Epithet: Of various colours.　Hardiness 3.　June–July.

Hardiness scale 1–4

Additional
comments

*Some doubt exists as to the exact taxonomic placing of this species: in cultivation
many hybrids shelter under this name.*

Hardiness ratings

H4 Hardy anywhere
in the UK (USDA zone
7 or below).

H3 Hardy in the UK in
sheltered gardens and
on favourable sites
inland (approximately
USDA zone 8).

H2 Hardy in sheltered
gardens on the
western coasts of the
UK (approximately
USDA zone 9).

H1 Requires cool
greenhouse
conditions in the UK
(some parts of USDA
zone 9 plus zones 10
and above).

R. maoerense W.P.Fang & G.Z.Li

Fortunea Gp A

Tree 8–12 m (25–40 ft); **branchlets** purplish at first, maturing to grey, glabrous.
Leaves oblanceolate, rarely obovate, apex nearly obtuse and cuspidate, base
cuneate, 14–16 cm long, 4–5 cm broad, with 19–21 pairs of veins; **underside**
pale green, glabrous.
Petiole pale purple, glabrous.
Inflorescence 9–12-flowered.
Pedicel 3–4.5 cm long; purplish, markedly glandular.
Corolla campanulate, 6–6.5 cm long, pale pink fading to white, lobes 7, reflexed.
Stamens 14–17, filaments glabrous.
Calyx persistent, glandular.
Ovary densely glandular; style sparsely glandular.
NE Guangxi. 1,800–1,900 m (5,700–6,200 ft).
Epithet: From Maoershan, Guangxi.　Hardiness 3?　May.

*Recently introduced from Maoershan by Alan Clark. On plants in cultivation, the
young growth is maroon or deep red, but this may well be a juvenile plant characteristic.*

Similar to
R. fortunei, but
with narrower
often recurved
leaves.

In cultivation, it is
late flowering.

Stamens 12–14.

Corolla white, pink
or rosy-pink.

Pedicel glandular
or eglandular.

Style glandular
throughout to the
tip with white
or yellowish glands.

**Principal
defining
characteristics**

**Secondary
defining
characteristics**

Foliage of *R. fortunei* ssp. *discolor* PW38 at the Millais Nursery

Leaves similar to
R. glanduliferum,
but more
noticeably
oblanceolate.

Petiole pale
purple.

Pedicel noticeably
glandular.

Corolla pale pink.

Style sparsely
glandular.

R. maoerense AC4207 at Deer Dell

Pontica
(Fortunea Group A)

**Section
(Subsection
and group)**

How to use this book

It is, of course, far easier to identify a plant when it is in flower, but the authors believe that it is often necessary to make the attempt when no flowers are visible, and have therefore put much emphasis on the foliage. This makes the use of the published botanists' keys impossible at times, so a new and fresh approach to classification within a Subsection based on many years experience of the genus is often used in this book. The notes and photographs at the beginning of each Subsection are designed to help those with, as yet, little experience.

For beginners, a few major divisions need to be explained:

The genus is currently divided into nine Subgenera, which are here listed in the order in which they are dealt with in the main text. As can be seen below, the Subgenera *Hymenanthes*, *Rhododendron* and *Vireya* contain the great majority of species in cultivation. *Vireyas* are excluded from this book as the great majority are either not hardy in cool temperate climates or are very rare in cultivation. They are superbly described in George Argent's recent monograph.

Subgenus *Hymenanthes*
    Section *Pontica*   over 200 species in cultivation in 24 Subsections

Subgenus *Azaleastrum*
    Section *Azaleastrum*   4 species in cultivation
    Section *Choniastrum*   approx. 6 species in cultivation

Subgenus *Pentanthera*
    Section *Pentanthera*   18 species in cultivation in 2 Subsections
    Section *Rhodora*   2 species in cultivation
    Section *Sciadorhodion* 5 species in cultivation
    Section *Viscidula*   1 species

Subgenus *Tsutsusi*
    Section *Brachycalyx*   15 species in cultivation
    Section *Tsutsusi*   approx. 40 species in cultivation

Subgenus *Candidastrum*   1 species

Subgenus *Mumeazalea*   1 species

Subgenus *Therorhodion*   probably 1 species
Subgenus *Rhododendron*
    Section *Pogonanthum*   10 species in cultivation
    Section *Rhododendron* over 200 species in cultivation in 28 Subsections
Subgenus *Vireya*   approx. 150 species in cultivation (not included)

Identifying Subgenera

When identifying a rhododendron, the first step, which will rapidly become automatic with a little practice, is to decide to which Subgenus it belongs.

Representative examples of Subgenera *Hymenanthes* (a. *R. wallichii* [M.L.A. Robinson]), *Azaleastrum* (b. *R. latouchae* [M.L.A. Robinson]), *Pentanthera* (c. deciduous azalea – *R. occidentale* [M.L.A. Robinson]) and *Tsutsusi* (d. evergreen azalea – *R. kiusianum*).

Subgenus *Hymenanthes* contains **elepidote** (pronounced 'ee-lepidote' and meaning non-scaly) rhododendrons. These have no scales anywhere, most significantly not on the leaves, which are almost always evergreen and may be glabrous or covered with simple or compound hairs. The flowers are in a terminal inflorescence.

Subgenus *Azaleastrum* (elepidote) species were formerly not considered hardy outside in the UK. The inflorescence is lateral – below the terminal buds. The leaves have no scales, and the indumentum, when present, consists of simple hairs or glands.

Subgenus *Pentanthera* (elepidote) contains most of the deciduous azaleas.

Subgenus *Tsutsusi* (elepidote) contains one Section of deciduous azaleas and all the evergreen or 'Japanese' azaleas.

Subgenus *Candidastrum* (elepidote) contains one deciduous species. The flowers are lateral below vegetative buds. Stamens 10.

Subgenus *Mumeazalea* (elepidote) contains one deciduous species. The flowers are lateral below vegetative buds. Stamens 5. Indumentum of simple hairs.

Subgenus *Therorhodion* (elepidote) contains dwarf evergeeen or deciduous shrubs.

Subgenus *Rhododendron* contains **lepidote** rhododendrons that are usually evergreen. They have **scales**, most significantly on the leaves, the scales on the leaf underside usually being more obvious. There are sometimes hairs on the leaves as well. The inflorescences are terminal and occasionally also in the axils of the upper leaves. The differences between Section *Pogonanthum*, which has lacerate scales, and the other sections, are discernable only under a good microscope.

Identifying Sections and Subsections

Once the Subgenus is identified, the next step is to look for the Section or Subsection.

Although it would seem to be a formidable task to determine to which Subsection an unlabelled rhododendron belongs, after a short time studying the plants (and there is no substitute for looking at the plants rather than photographs or descriptions), the task is well within the reach of most amateurs.

Identification of a plant starts with the examination of:
    the general habit of growth;
    the leaf shape, size, and the nature of the upper surface;
    the leaf underside;
    the bark;
    the young growth;
and, if in flower:
    the inflorescence;
    the corolla;
and, if the flowers are past, the calyx is often useful.

A few typical characteristics of each Subsection are given below.

It must be noted that exceptions to the general characteristics given below will be found in many Subsections. What is intended here is an introductory guide for the non-specialist. The number of taxa quoted refers to those currently in cultivation.

A full glossary and explanation of terms used can be found on page 668.

Subgenus *Hymenanthes* (Elepidote)
Section *Ponticum*:
The elepidote rhododendrons are all members of the single Section *Ponticum* – the reasons for this are described in specialist botanical books on this genus.

There are 24 Subsections included within this Section.

The corolla is 5-lobed unless stated otherwise.

Subsections:
Arborea 12 taxa
Tree-like: leaves rugulose, with a shiny or matt indumentum below, leaving the veins visible; truss compact and rounded; corolla with nectar pouches.

Photo: *R. arboreum* f. *roseum* at Inverewe

Argyrophylla approx. 20 taxa
Large shrubs or small trees: leaves rugulose but usually less so than in *Arborea*; indumentum thin, shiny or matt; prominent tapering yellowish main vein on the underside; corolla without nectar pouches.

Photo: *R. argyrophyllum* W1210 at Bemore Botanic Gardens, Argyll

Auriculata 2 taxa
Small trees: young growth hairy with long sticky hairs/glands; leaf not shiny; underside with scattered hairs that may not persist; corolla 7-lobed.

Photo: *R. auriculatum* at Deer Dell

Barbata with **Fulgensia** probably 6 taxa
Shrubs or small trees: bark smooth, flaking, red to brown; young shoots and leaf petioles usually barbed; leaf rugulose and almost always strongly concave; inflorescence compact; corolla with nectar pouches, usually red.

Photo: *R. imberbe* in the Valley Gardens, Windsor

Campanulata 4 taxa
Shrubs: bark rough; leaves elliptic or ovate; underside with a brown felted indumentum, partly or wholly obscuring the veins; corolla mauvish or white without nectar pouches.

Photo: *R. campanulatum* ssp. *campanulatum* at Werrington, Cornwall

Campylocarpa 7 taxa
Shrubs: branchlets usually glandular; leaves not large, glabrous on both sides at maturity, narrowly rounded or round; corolla campanulate or saucer-shaped, white, pink or yellow.

Photo: *R. callimorphum* var. *callimorphum* at Deer Dell showing the campanulate corolla and round leaf

Falconera 14 taxa
Trees or (usually large) lax shrubs: big leaves, underside with a **thick** matt indumentum; inflorescence dense, many flowers; corolla ventricose, usually 7–10-lobed.

Photo: *R. rothschildii* F30528 at Deer Dell

Fortunea 27 taxa

A very diverse Subsection. Shrubs or trees: leaves oblanceolate or oblong to orbicular; essentially glabrous when mature, essentially smooth surfaces; corolla usually 5- or 7-lobed, usually pink or white, funnel or open-campanulate, sometimes fragrant.

Photo: The newly introduced *R. decorum* ssp. *cordatum* AC1043 at Hindleap Lodge showing the typical funnel campanulate 7-lobed corolla. (M.L.A. Robinson)

Fulva 5 taxa

Large shrubs or small trees: leaf underside with a dense matt grey to brown indumentum; corolla pink or white, usually with a basal blotch.

Photo: *R. fulvoides* R10931 in the Valley Gardens, Windsor

Glischra 11 taxa

Shrubs or small trees: bark rough; young shoots densely glandular; leaves at least slightly rugulose, undersides at maturity with bristles at least on the midrib; corolla almost always with a basal blotch.

Photo: *R. crinigerum* var. *euadenium* F25633 in the Valley Gardens

Grandia probably 15 taxa

Large lax shrubs or trees: bark rough; big leaves, underside **at maturity** with a whitish, fawn or brown indumentum; inflorescence dense, many flowers; corolla usually 6–10-lobed; often ventricose.

Photo: *R. montroseanum* KW6261A at Deer Dell

Griersoniana 1 taxon
Straggly shrub or small tree: young growth hairy with long sticky hairs/glands; the indumentum on the leaf underside can easily be rubbed off; corolla brick red.

Photo: *R. griersonianum* at Deer Dell

Irrorata about 16 taxa
Shrubs or small trees: leaves usually long compared to their breadth, apex often acuminate; corolla usually 5-lobed, with or without nectar pouches.

Photo: *R. annae* at Deer Dell

Lanata probably 6 taxa
Shrubs: branchlets densely tomentose; leaves comparatively small; underside with a thick brownish indumentum obscuring the veins.

Photo: *R. flinckii* (Glenarn form) at Deer Dell

Maculifera 14 taxa
Shrubs or small trees: young growth floccose or setulose-glandular; leaf underside with indumentum on the midrib at maturity, indumentum on the rest of the leaf underside sometimes present.

Photo: *R. longesquamatum* at Deer Dell

Neriiflora about 44 taxa
Prostrate shrubs to small trees: corolla with nectar pouches, and almost always fleshy, often red. A diverse Subsection best divided into groups: see the Species descriptions.

Photo: The growth habit of *R. forrestii* ssp. *forrestii* 'Repens' in the Valley Gardens

Photo: *R. horaeum* F21850 at Deer Dell

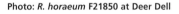

Parishia with **Venatora** about 5 taxa
Large shrubs or small trees: leaf underside with a detersile indumentum of stellate hairs, often persistent near the petiole, lamina usually shining beneath; corolla usually bright red with nectar pouches.

Photo: *R. facetum* F15917 at Deer Dell

Pontica about 19 taxa
Shrubs or small trees; leaf usually smooth, not rugulose; inflorescence candelabroid on a long rhachis; corolla lobes long.

Photo: *R. degronianun* ssp. *yakushimanum* at Deer Dell

Selensia about 8 taxa
Shrubs or small trees: young growth glandular; leaves rounded, underside glabrous or with a very thin indumentum; corolla without nectar pouches.

Photo: *R. eurysiphon* KW21557 at Deer Dell (no nectar pouches)

Sherriffii Series 3 taxa

Shrubs or small trees: leaves rounded; inflorescence lax; corolla crimson or carmine, with nectar pouches.

Photo: *R. sherriffii* **L&S2751 AM at Deer Dell**

Taliensia about 49 taxa

Dwarf to large shrubs: leaf underside with a thick woolly or thinner suede-like indumentum; corolla campanulate or funnel campanulate without nectar pouches. A diverse Subsection best divided: see the main text.

Photo: *R. iodes* **at RBG Edinburgh**

Thomsonia 15 taxa

Shrubs or small trees: bark smooth and peeling; leaves round, underside glabrous or with a thin indumentum; corolla with nectar pouches; calyx usually prominent. Similar to Subsection *Selensia*.

Photo: *R. cerasinum* **KW6923 'Cherry Brandy' at Deer Dell showing nectar pouches**

Williamsiana 1 taxon

Small spreading shrub: leaf small, round, often with a cordate or truncate base; inflorescence lax, pedicels long.

Photo: *R. williamsianum* **at Deer Dell**

Subgenus *Azaleastrum* (Elepidote)
Section *Azaleastrum* 4 taxa

Elepidote shrubs or trees: leaf underside glabrous, young leaves often coloured; inflorescence axillary in the uppermost 1–4 leaves, 1–2-flowered; ovary bristly; calyx lobes large, persistent; stamens 5.

Photo: New foliage of *R. viallii* at Deer Dell

Section *Choniastrum* about 6 taxa

Elepidote shrubs or trees: leaves usually glossy; young leaves often coloured; inflorescence axillary from the uppermost 1–3 leaves, 1–several-flowered; corolla narrowly funnel-shaped, often fragrant; calyx minute; stamens 10.

Photo: *R. stamineum* at Glendoick (M.L.A. Robinson)

Subgenus *Pentanthera* (Elepidote)
Section *Pentanthera*

Subsections:

Pentanthera 16 taxa
Deciduous shrubs: leaves alternate, though often close together; inflorescence terminal; corolla zygomorphic, **narrowly** funnel-shaped, outer surface pubescent; stamens 5, strongly exserted.

Photo: *R. atlanticum* (D. White) showing the narrowly tubular campanulate corolla

Sinensia 2 taxa
Deciduous shrubs: inflorescence terminal; corolla zygomorphic; **broadly** funnel-shaped, outer surface pubescent; stamens 5, not strongly exserted.

Photo: *R. molle* ssp. *japonicum* at Beresford Gardens (Mrs. E. Wood) showing the wider corolla

Section *Rhodora* 2 taxa

Deciduous shrubs: inflorescence terminal; corolla 2-lipped, the top 3 lobes fused together; stamens usually 7–10.

Photo: *R. canadense* (M. Creel)

Section *Sciadorhodion* 4 taxa

Deciduous shrubs: leaves usually in whorls or pseudowhorls of **five** at the end of the branches; more or less obovate; corolla zygomorphic, not 2-lipped, outer surface glabrous; stamens 10.

Photo: New foliage of *R. quinquefolium* at Deer Dell

Section *Viscidula* 1 taxon

Deciduous shrub: inflorescence often hidden by the leaves; corolla tubular campanulate, outer surface glabrous; stamens 10.

Photo: *R. nipponicum* (S. Hootman)

Subgenus *Tsutsusi* (Elepidote)

The leaves and the inflorescence emerge from the same bud scales.

Section *Brachycalyx* 14 taxa

Deciduous shrubs: rhombic to rhombic-ovate leaves in pseudowhorls usually of **three** at the ends of the branches; corolla funnel-shaped or funnel campanulate; stamens usually 5–10.

Photo: New foliage of *R. weyrichii* at Deer Dell

Section *Tsutsusi* about 40 taxa

Partly evergreen or evergreen shrubs: spring leaves usually deciduous, the summer leaves usually smaller and always evergreen; indumentum of simple hairs or bristles; corolla rotate to tubular-campanulate.

Photo: *R. kiusianum* (Davidian)

Subgenus *Candidastrum*

R. albiflorum
One deciduous species: flowers lateral below vegetative buds; stamens 10.

Photo: *R. albiflorum* (B. Starling)

Subgenus *Mumeazalea*

R. semibarbatum
One deciduous species: flowers lateral below vegetative buds; indumentum of simple hairs; stamens 5.

Photo: *R. semibarbatum* (S. Hootman)

Subgenus *Therorhodion*

Probably only *R. camtschaticum* in cultivation. Dwarf evergreen or decidous shrub.

Photo: *R. camtschaticum* at Deer Dell

Subgenus *Rhododendron* (Lepidote)
Section *Pogonanthum* 14 taxa

Small shrubs: leaf undersurface scaly with overlapping lacerate scales; corolla narrowly tubular with spreading lobes, resembling *Daphne*; stamens short, often concealed in the corolla tube.

Photo: *R. cephalanthum* at Eckford

Section *Rhododendron*

Subsections:

Afghanica 1 taxon
Small shrubs: leaf undersurface laxly scaly; inflorescence with a distinct elongated rhachis; corolla whitish.

Photo: Leaves of *R. afghanicum* (A. Dome)

Baileya 1 taxon
Small shrubs: leaf underside densely scaly with overlapping crenulate scales; inflorescence 5–12-flowered; corolla reddish purple with a large calyx.

Photo: *R. baileyi* LS&H17359 at Deer Dell

Boothia 8 taxa
Small or straggly shrubs: leaf underside pale and usually glaucous, densely scaly with saucer-shaped scales; corolla yellow.

Photo: *R. megeratum* at Glenarn

Camelliiflora 1 taxon

Small or medium shrub, often lax: leaves to 12 cm, underside densely scaly; inflorescence 1–2-flowered; corolla with spreading lobes.

Photo: *R. camelliiflorum* S&L5696 at Benmore (M.L.A. Robinson)

Campylogyna about 4 taxa

Dwarf, often prostrate shrubs: foliage small; leaf margin crenulate; corollas held above the foliage on long pedicels.

Photo: *R. campylogynum* (bush) Brodick form at Deer Dell

Caroliniana 2 taxa

Shrubs: leaf underside densely scaly; inflorescence terminal or terminal and axillary, 4–12-flowered, usually white or pale pink.

Photo: *R. minus* var. *minus* (pink form) in the Valley Gardens

Cinnabarina 5 taxa

Leaves often glaucous; flowers tubular to campanulate, usually pendulous.

Photo: Foliage of *R. cinnabarinum* BLM234 at Nymans (E. Daniel)

Edgeworthia 3 taxa

Leaves usually bullate, always with a dense indumentum; branchlets, petiole, pedicel and ovary densely woolly.

Photo: *R. pendulum* L&S6660 at Deer Dell

Fragariiflora 1or 2 taxa
Small shrublets: margins of young leaves
ciliate, upper surface with yellow or brown
scales; calyx conspicuous.

Photo: *R. fragariiflorum* at Tsari SE Tibet (P. Evans)

Genestieriana 1 taxon
Leaf underside almost uniformly white;
corolla small, bloomed plum-purple,
pendulous on a long pedicel.

Photo: *R. genestierianum* at Brodick

Glauca 10 taxa
Leaves aromatic, underside white or greyish
glaucous with scales of two kinds; style short
and sharply deflexed.

Photo: *R. charitopes* at Deer Dell

Heliolepida 4 taxa
Leaves very aromatic, underside with large
scales; corolla conspicuously scaly outside;
style slender and straight.

Photo: *R. heliolepis* var. *brevistylum* F26928 at
Deer Dell

Lapponica about 30 taxa
Low to small shrubs: leaves small, densely
scaly on both surfaces; corolla usually widely
funnel-shaped.

Photo: *R. russatum* F21932 at Deer Dell

Ledum 5 taxa
Branchlets lepidote and tomentose; leaf
undersides whitish, papillate or setulose and
often tomentose; inflorescence many
flowered; corolla small, white.

**Photo: *R. tomentosum* ssp. *tomentosum* at
Deer Dell**

Lepidota 5 taxa
Small to medium shrubs: leaves evergreen
or deciduous; corolla rotate or rotate-
campanulate; style short, stout and
sharply bent.

**Photo: *R. lepidotum* Obovatum Group at
Deer Dell**

Maddenia about 40 taxa
Except for the old ciliatum subseries, the
growth is often straggly; leaves and corolla
are often large (for lepidotes). Corolla funnel-
campanulate or tubular-campanulate, often
waxy; often fragrant.

Photo: *R. nuttallii* at Borde Hill

Micrantha 1 taxon
Leaves narrowed toward the base, underside
with conspicuous scales; inflorescence many
flowered (10–25); white.

Photo: *R. micranthum* at Deer Dell

Monantha Probably 4 taxa
Small shrubs: branchlets lepidote; inflorescence
1–3-flowered; autumn flowering.

**Photo: *R. kasoense* HECC10009 flowering in
November at Hindleap Lodge (M.L.A. Robinson)**

Moupinensia 3 taxa
Small to medium shrubs: often straggly; branchlets bristly; leaf thick and rigid; inflorescence 1–2-flowered, corolla widely campanulate.

Photo: *R. moupinense* W4256 at Deer Dell

Rhododendron 3 taxa
Compact shrubs: branchlets and leaf underside densely lepidote; leaf margin often crenulate; rhachis elongated; corolla tubular with spreading lobes.

Photo: *R. ferrugineum* in the Valley Gardens

Rhodorastra 5 taxa
Shrubs: leaves usually wholly or partly deciduous; young growth early, leaves thin; corolla widely funnel-shaped; style slender and straight; early flowering.

Photo: *R. mucronulatum* 'Cornell Pink' in the Valley Gardens

Saluenensia 6 taxa
Small shrubs: leaf undersides densely lepidote with crenulate scales; corolla widely funnel-shaped or rotate; calyx usually coloured.

Photo: *R. calostrotum* at Eckford

Scabrifolia 8 taxa
Shrubs: often straggly, branchlets often pubescent or bristly; leaf upper surface pubescent; inflorescence usually axillary.

Photo: *R. scabrifolium* var. *pauciflorum* **KR342 at Deer Dell. Note the flower buds in the leaf axils**

Tephropepla 5 taxa
Shrubs: leaves often narrow, undersides green or glaucous, but always pale; corolla tubular campanulate with a long, straight, slender style; calyx sometimes conspicuous.

Photo: *R. auritum* **KW6278 at Deer Dell**

Trichoclada about 8 taxa
Branchlets usually bristly; leaves usually deciduous; scales vesicular; corolla yellow or with yellowish shades.

Photo: *R. lepidostylum* **at Deer Dell**

Triflora 25 taxa
Usually upright spreading shrubs: corolla zygomorphic, usually widely funnel- or 'butterfly'-shaped, with exerted stamens and a long slender style; calyx small.

Photo: *R. augustinii* **ssp.** *augustinii* **at Deer Dell**

Uniflora 5 taxa

Prostrate to compact shrubs: inflorescence 1–2-flowered; corolla densely pilose outside; style slender and straight.

Photo: *R. ludlowii* at Bracken Hill (Davidian)

Virgata 2 taxa

Sprawling shrubs: inflorescence 1–2-flowered, always axillary in the upper 1–12 leaves.

Photo: *R. virgatum* ssp. *oleifolium* 'Penheale Pink' in the Valley Gardens

Species Descriptions
and Photographs

R. macabeanum KW7724 at Deer Dell

Subgenus *Hymenanthes*
Section *Ponticum* G.Don
Subsection *Arborea* Sleumer
Arboreum Series

Upright shrubs or trees.
Leaves rugose, mostly shiny above, mostly lanceolate, with
  prominent veins.
Inflorescence rounded, usually compact.
Corolla with nectar pouches.

*The classification of the various subspecies and varieties of R. arboreum is
confusing and needs to be related more closely to the geographical
distribution of the various forms.*

*The important distinguishing features are the presence of either a thin-
plastered or a bistrate indumentum, and the colours of the indumentum
and flowers.*

R. arboreum ssp. arboreum Sm.
R. arboreum Sm.

Shrub or **tree** 1.2–18 m (4–60 ft) high, **branchlets** densely tomentose.
Leaves lanceolate, oblong-lanceolate or oblanceolate; **lamina** 7.2–20 cm long,
  2–5.6 cm broad; **upper surface** dark green or olive-green, glabrous;
  underside with a thin-plastered or sometimes with a thin somewhat woolly,
  continuous, unistrate indumentum of rosulate hairs.
Inflorescence very compact truss of 15–20 flowers.
Corolla tubular-campanulate, 3–5 cm long, deep crimson, crimson, scarlet or red,
  5 nectar pouches.

Extends from Kashmir, Punjab, Uttar Pradesh, through Nepal, Sikkim, Bhutan to
  Assam. 1,200–3,350 m (4,000–11,000 ft).

AM 1964 (Brodick) 'Goat Fell', flowers cherry red.

AM 1968 (Exbury) 'Rubaiyat', flowers red.
Epithet: Tree-like.   Hardiness 1–3.   February–May.

Leaf undersides of Subsect. *Arborea* **at Deer Dell.**
Clockwise from top right: *R. arboreum* ssp. *delavayi, R. arboreum*
ssp. *arboreum, R. arboreum* ssp. *cinnamomeum* (as *R. campbelliae,*
R. arboreum ssp. *cinnamomeum, R. niveum, R. lanigerum,*
R. arboreum ssp. *albotomentosum* and *R. zeylanicum.* Bar = 5 cm

Plastered whitish
indumentum, usually
unistrate but may
be loosely felted.
Red or crimson
flowers.

R. arboreum ssp. *arboreum* at Brodick

Ponticum
(Arborea)

R. arboreum f. album Wall.

R. arboreum ssp. *cinnamomeum* var. *album*

Two growth forms are in cultivation.

Form 1. A broadly upright tree 10–15 ft high.

Form 2. A columnar tree 10–20 ft high.

Nepal. 2,900–3,050 m (9,500–10,000 ft).

Epithet: With white flowers.   Hardiness 2–3.   February–May.

The modern name is misleading so far as identification in cultivation is concerned. This applies also to R. arboreum *f.* roseum.

R. arboreum f. roseum (L.) Tagg

R. arboreum ssp. *cinnamomeum* var. *roseum*

Nepal. 2,600–3,050 m (8,500–10,000 ft).

FCC (Wakehurst) 1974 'Tony Schilling'.

Epithet: With rose flowers.   Hardiness 2–3.   February–May.

***R. arboreum* f. *roseum* at Baravalla, Argyll (M.L.A. Robinson)**

R. arboreum ssp. albotomentosum D.F.Chamb.

R. delavayi var. *albotomentosum*

Mount Victoria, West Central Burma. 3,000 m (10,000 ft).

Epithet: With white tomentum.   Hardiness 3.   April–May.

Rare in cultivation in the UK. Formerly classified as a subspecies of R. delavayi.

The indumentum may be thin and plastered but is usually bistrate. It is pale, **not cinnamon**. White flower with purple spotting.

R. arboreum f. *album* at Nymans, Sussex

Indumentum white or whitish, **not cinnamon**. It may be thin and plastered but is sometimes bistrate.
Pink flowers with deeper spotting, sometimes bi-coloured.

R. arboreum f. *roseum* at Inverewe

R. arboreum f. *roseum* at the Hillier Arboretum, Hampshire

White tomentum on the petiole and branchlets. Indumentum on leaf underside pure white.
Flowers intense scarlet without spots.

R. arboreum ssp. *albotomentosum* at the Oz Blumhardt nursery, N. Island, New Zealand (M.L.A. Robinson)

R. arboreum ssp. *albotomentosum* KW21976 at Deer Dell

We suggest that the least misleading name for this taxon is:

R. arboreum ssp. cinnamomeum (L.) Tagg

R. cinnamomeum Wall ex G.Don.
R. arboreum ssp. *cinnamomeum* var. *cinnamomeum*

Shrub or **tree** 3–7.5 m (10–25 ft) high, **branchlets** densely or moderately tomentose.

Leaves oblong-lanceolate, lanceolate or rarely oblong; **lamina** 8–14.8 cm long, 2–5 cm broad; **upper surface** dark green or olive-green; **underside** matt, with a thin woolly, continuous bistrate indumentum, upper layer cinnamon, of small dendroid hairs, which are more or less detersile on old leaves, under layer rosulate.

Inflorescence very compact or somewhat compact racemose umbel of 15–20 flowers.

Pedicel tomentose, glandular.

Calyx minute, 1 mm long, glandular.

Corolla tubular-campanulate, 2.6–3.8 cm long, white with purple spots, 5 nectar pouches.

Nepal. 2,750–3,650 m (9,000–12,000ft).

Epithet: Cinnamon-coloured.   Hardiness 2–3.   April–May.

R. arboreum ssp. cinnamomeum
R. campbelliae Hook.f.

Tree 3–12 m (10–40 ft) high, **branchlets** densely or moderately tomentose, glandular.

Leaves oblong-lanceolate or lanceolate; **lamina** 6.8–13.5 cm long, 2.1–3.8 cm broad; **upper surface** dark green or olive-green, glabrous; **underside** matt, with a thin, woolly, continuous, bistrate indumentum, upper layer cinnamon, rust-coloured or dark brown, small dendroid hairs.

Petiole tomentose.

Inflorescence very compact truss of 15–20 flowers.

Corolla tubular-campanulate, 2.6–3.8 cm long, purplish-rose, pink, purple, scarlet or crimson, 5 nectar pouches.

Sikkim, Nepal. 2,750–3,050 m (9,000–10,000 ft).

Epithet: After the wife of Dr. Campbell, co-traveller with Hooker in the Himalayas.   Hardiness 2–3.   April–May.

*Not recognised as different from the subspecies above by most recent authors: the indumentum differs only slightly but **note the difference in flower colour**.*

Underside matt, with a thin woolly, bistrate, brown, granular **cinnamon** indumentum, often at least partly detersile.
White flowers with purple spots.

R. arboreum ssp. *cinnamomeum* at Deer Dell

New foliage of *R. arboreum* ssp. *cinnamomeum* at Deer Dell

Indumentum matt, bistrate, **cinnamon**, rust-coloured or brown, persistent.
Flowers purplish-rose, pink, purple, scarlet or crimson.

R. arboreum ssp. *cinnamomeum* (*R. campbelliae*) at the Valley Gardens

Ponticum
(Arborea)

R. arboreum ssp. delavayi var. delavayi

D.F.Chamb.

R. delavayi Franch. & Chamb.

Shrub or **tree** 1–12 m (3–40 ft) high, **branchlets** with a fawn or whitish tomentum.

Leaves lanceolate, oblong-lanceolate or oblong; **lamina** coriaceous, 5–16.3 cm long, 1.6–5 cm broad; **upper surface** dark green, margin recurved; **underside** with a somewhat thin, woolly, spongy, or sometimes plastered, brown or fawn, continuous bistrate indumentum of hairs.

Inflorescence a very compact umbel of 10–30 flowers.

Calyx 1–3 mm long.

Corolla tubular-campanulate, 3–5 cm long, deep crimson, crimson or scarlet, 5 nectar pouches.

W and NW Yunnan, West Central and NE Upper Burma. 1,500–3,350 m (5,000–11,000ft).

FCC 1936.

Epithet: After L'Abbe J.M. Delavayi (1838–1895), an early collector in China. Hardiness 1–3.   March–May.

There appear to be at least two forms of this variety in cultivation. Form 1 has a globular inflorescence. Form 2 has a hemispherical truss with a thinner flatter leaf. These may be geographical variations. A related species that has thickly tomentose young growth and a unistrate dendroid indumentum has been found in N Vietnam. In the 'Flora of China' 2005 R. delavayi has been reinstated as a species.

R. arboreum ssp. delavayi Franch. var. album W.Watson

Corolla pure white, with a purple blotch and with small purple spots on the upper lobes.

Epithet: With white flowers.   Hardiness 1–2.   April–May.

This is doubtfully distinct from R. arboreum *f.* album. *The famous Hooker plant at Stonefield Castle is* R. arboreum *f.* album. *Strictly,* R. delavayi *var.* album *should be from China. Probably not in cultivation.*

Indumentum usually woolly, spongy, bistrate, fawn or brown. Flowers deep crimson, crimson or scarlet.

R. arboreum ssp. *delavayi* var. *delavayi* at Clyne Castle, Swansea

R. arboreum ssp. *delavayi* var. *delavayi* at Benmore

R. arboreum ssp. delavayi var. peramoenum (Balfour & Forrest) T.L.Ming

R. peramoenum Balf.f. & Forrest

Shrub or **tree** to 12 m (40 ft) high, **branchlets** tomentose, glandular.

Leaves linear-lanceolate or lanceolate; **lamina** 5.4–16 cm long, 1–3 cm broad, apex tapered, acuminate, base tapered; **upper surface** pale green, glabrous; **underside** matt, with a thin-plastered or somewhat plastered, continuous, bistrate indumentum of hairs.

Petiole tomentose, glandular.

Inflorescence in very compact trusses of 15–20 flowers.

Calyx minute, 1 mm long.

Corolla campanulate, 3–5 cm long, deep rose-crimson, bright cherry-scarlet, bright scarlet-crimson or black-crimson, 5 nectar pouches.

W and mid-W Yunnan, Assam. 2,500–3,350 m (8,000–11,000 ft).

Epithet: Very pleasing.   Hardiness 1–3.   March–May.

Extremely rare in cultivation. The variety has recently been found in North Vietnam, and a possible pink-flowered form is recorded from Bhutan.

R. arboreum ssp. nilagiricum D.F.Chamb.

R. nilagiricum Zenker

Shrub or **tree** 2.4–12 m (8–40 ft) high, **branchlets** moderately or densely tomentose, eglandular.

Leaves elliptic, oblong-elliptic or ovate; **lamina** 5–15 cm long, 2.5–6.5 cm broad; **upper surface** dark green or olive-green, convex, markedly bullate, glabrous; margin recurved; **underside** matt with a thick, somewhat woolly, continuous, bistrate indumentum of hairs, midrib eglandular.

Petiole densely tomentose, eglandular.

Inflorescence a very compact racemose umbel of 10–20 flowers.

Pedicel eglandular.

Calyx hairy, eglandular.

Corolla tubular-campanulate, 2.8–4 cm long, deep crimson, rose, crimson-rose or pink, 5 nectar pouches.

Nilgiris, Madras, India. 1,850–2,250 m (6,000–7,300 ft).

Epithet: From the Nilgiris, Madras, India.   Hardiness 1–3.   April–May.

Extremely rare in cultivation. Indumentum very similar to R. zeylanicum. *Reported occurences in Assam, previously called* R. kingianum, *are hybrids of* R. arboreum *and* R. macabeanum.

Leaves linear-lanceolate, narrow in comparison with their length.
Indumentum thin-plastered or somewhat plastered, continuous, bistrate.
Flowers red or crimson.

R. arboreum ssp. *delavayi* var. *peramoenum* (M. Shuttleworth)

Branchlets moderately or densely tomentose, eglandular.
Leaf upper surface convex, markedly bullate. Leaf margin recurved; underside with a thick somewhat woolly or thin-plastered, continuous bistrate coppery indumentum.
Petiole tomentose, eglandular.
Pedicel eglandular.
Calyx hairy, eglandular.

Indumentum similar to that of *R. zeylanicum*.
Corolla usually crimson.

Foliage of *R. arboreum* ssp. *nilagiricum* (possibly) at Arduaine, Argyll (M.L.A. Robinson)

R. arboreum ssp. zeylanicum (Booth) Tagg

R. zeylanicum Booth

Shrub or **tree** 1.8–11 m (6–35 ft) high or more, **branchlets** moderately or densely tomentose, glandular.

Leaves ovate, ovate-elliptic, elliptic or oblong-elliptic; **lamina** coriaceous, 5.6–17.5 cm long, 2.5–8 cm broad; **upper surface** dark green, convex, markedly bullate, glabrous; margin recurved; **underside** matt or shiny, with a thick somewhat woolly or thin-plastered, continuous, bistrate indumentum of hairs, midrib often glandular. Primary veins often sparsely glandular.

Petiole tomentose, glandular.

Inflorescence very compact trusses of 12–20 flowers.

Pedicel tomentose, glandular.

Corolla tubular-campanulate, 3.2–4.1 cm long, crimson or pink, 5 nectar pouches.

Sri Lanka, Manipur, Assam. 900–2,750 m (3,000–9,000 ft).

Epithet: From Ceylon.   Hardiness 1–3.   June–July.

Recent observations by S. Hootman indicate that R. arboreum *ssp.* zeylanicum *is variable in the wild, extreme forms having flatter leaves or a whitish indumentum. A recent introduction by T. Hudson is much earlier (April) flowering. The two relict populations of* R. nilagiricum *and* R. zeylanicum *are properly distinguished by* R. zeylanicum *being a glandular expression of* R. nilagiricum.

R. lanigerum Tagg

Shrub or **tree** 2.4–6 m (8–20 ft) high, **branchlets** densely fawn or whitish tomentose, eglandular.

Leaves oblong, oblong-lanceolate or oblanceolate; **lamina** 10–25 cm long, 3–7.5 cm broad; **upper surface** dark green, glabrous (in young leaves densely tomentose with a thick creamy or creamy-white tomentum); **underside** with a thick woolly, continuous, bistrate indumentum of brown or fawn hairs.

Petiole densely or moderately tomentose.

Inflorescence in very compact trusses of 16–35 or sometimes up to 50 flowers; flower-bud ovate or nearly orbicular, large, 3–3.4 cm long, densely tomentose.

Corolla tubular-campanulate, 3–5 cm long, rose-purple, 5 nectar pouches.

Assam. 3,000–3,350 m (10,000–11,000 ft).

AM 1949 (Trengwainton), AM 1961 (Stonehurst) 'Stonehurst',

AM 1961 and FCC 1967 (Windsor) 'Chapel Wood'.

Epithet: Woolly.   Hardiness 3.   February–April.

The red and crimson forms were known as R. lanigerum *var.* silvaticum *but are now regarded as insufficiently distinct to merit varietal status:*

AM 1951 (Tower Court) as *R. silvaticum* (KW6258), AM 1951 (Windsor) as *R. silvaticum* 'Round Wood' (KW6258), AM 1954 (Logan) as *R. silvaticum* 'Silvia'.

Epithet: Woodland.   Hardiness 3.   February–April.

This species and R. niveum *are very similar in leaf and indumentum. They may be distinguished by* R. niveum's *near-rounded apex, which contrasts with* R. lanigerum's *acute/acuminate apex. The leaves of* R. lanigerum *are usually larger than those of* R. niveum.

Branchlets moderately or densely tomentose, glandular.

Leaf upper surface convex, markedly bullate. Leaf margin recurved; underside with a thick somewhat woolly or thin-plastered, continuous, bistrate coppery indumentum.

Petiole tomentose, glandular.

Pedicel glandular.

Indumentum, though bistrate, can have a shiny plastered coppery appearance. Corolla usually crimson.

R. arboreum ssp. *zeylanicum* – the T. Hudson introduction — at Botallick, Cornwall (M.L.A. Robinson)

R. arboreum ssp. *zeylanicum* – the T. Hudson introduction — at Botallick, Cornwall (M.L.A. Robinson)

Foliage of *R. arboreum* ssp. *zeylanicum* – the T. Hudson introduction — at Botallick, Cornwall

Indumentum thick woolly, bistrate, brown or fawn.

Flower-bud ovate or nearly orbicular, large, densely tomentose.

Corolla pink, rose purple, cherry red, scarlet or crimson.

Leaf upper surface in young leaves densely tomentose with a thick creamy or creamy-white tomentum.

Inflorescence in very compact trusses of 16–35.

R. lanigerum at Inverewe

Ponticum
(Arborea)

R. niveum Hook.f.

Shrub or **tree** 2.75–6 m (9–20 ft) high, **branchlets** densely whitish tomentose, eglandular.
Leaves oblong-lanceolate, oblanceolate, lanceolate or oblong; **lamina** 8–16 cm long, 2.5–5.5 cm broad; **upper surface** dark green, sparsely tomentose or glabrous; **underside** matt, with a thick, woolly, continuous, bistrate indumentum of fawn, brown or whitish hairs.
Inflorescence very compact racemose umbel of 15–30 flowers, flower-bud oval or oblong, large densely tomentose.
Corolla tubular-campanulate, 3–3.5 cm long, lilac, mauve or purplish-lilac, with or without deeper lilac blotch, 5 nectar pouches.
Sikkim, Bhutan. 2,900–3,600 m (9,500–12,000 ft).
AM 1951 (Tower Court), FCC 1979 (Windsor) 'Crown Equerry'.
Epithet: Snow-like.   Hardiness 3.   April–May.

Subgenus *Hymenanthes*
Section *Ponticum* G.Don
Subsection *Argyrophylla* Sleumer
Argyrophylla Series

Shrubs or trees.
Branchlets floccose or tomentose.
Similar foliage to Subsection Arborea.
The leaves have a yellowish midrib which tapers gradually from the petiole to the apex. Inflorescence lax with long pedicels.
Corolla without nectar pouches, except for *R. ririei*.

For identification purposes this Subsection is divided into two groups:

Group A having a thin-plastered indumentum, appearing unistrate.

Group B having a thick woolly indumentum, appearing bistrate.

Smaller leaves than *R. lanigerum* when both are growing well, and the leaves are more rounded at the apex.

Corolla deep lilac, mauve or purplish-lilac.

Indumentum woolly, continuous, bistrate, fawn, brown or whitish.

Flower-bud oval or oblong.

R. niveum at Deer Dell

Leaves of Subsect. *Argyrophylla* at Deer Dell.
Clockwise from top right (two leaves of each): *R. hunnewellianum*, *R. farinosum*, *R. thayerianum*, *R. floribundum* and *R. insigne*.
Bar = 5 cm

R. argyrophyllum ssp. argyrophyllum Franch.

R. argyrophyllum Franch.   **Argyrophylla Gp A**
R. argyrophyllum var. *cupulare* Rehder & E.H.Wilson

Shrub or **tree** 1–12 m (3–39 ft) high, **branchlets** often with a thin tomentum.
Leaves oblong-lanceolate, lanceolate or oblong; **lamina** 5–16.3 cm long, 1.5–6 cm broad; **upper surface** olive-green or green, matt, glabrous; **underside** with a thin, plastered or somewhat plastered, continuous, unistrate indumentum of white or sometimes fawn rosulate hairs.
Inflorescence in trusses of 6–16 flowers.
Pedicel eglandular.
Corolla funnel-campanulate, base narrow, 2–4.5 cm long, white, pink, rose or white-flushed rose, with or without deeper pink spots.
W Sichuan, NW Yunnan. 1,800–3,000 m (6,300–9,800 ft).
AM 1934 (Wakehurst), flowers white-flushed rose with deeper pink spots.
Epithet: With silver leaves.   Hardiness 3.   May.

R. argyrophyllum ssp. hypoglaucum D.F.Chamb.

R. hypoglaucum Hemsl.   **Argyrophylla Gp A**

Shrub or **tree** 1–6 m (3–20 ft) high, **branchlets** floccose or glabrous.
Leaves oblong-lanceolate, oblanceolate or oblong; **lamina** 5.2–11 cm long, 1.6–2.4 cm broad, apex acuminate or acute; **upper surface** bright or dark green, shiny, glabrous; **underside** intensely glaucous white, with a thin-plastered or almost plastered, continuous unistrate indumentum of hairs.
Inflorescence in trusses of 4–8 flowers.
Pedicel rather densely or moderately glandular or sometimes eglandular.
Calyx often glandular.
Corolla funnel-campanulate, white, pink or white flushed rose, with deep rose spots, with or without reddish blotch.
Hubei, Sichuan. 1,600–2,750 m (5,300–9,000 ft).
AM 1972 (Sandling Park) 'Heane Wood', flowers white-flushed and spotted red-purple.
Epithet: Blue-grey beneath.   Hardiness 3.   May.

Leaf underside with a plastered, indumentum of usually white rosulate hairs (occasionally fawn). Inflorescence 6–16 flowers.

Pedicel eglandular.

Corolla white, white flushed rose, pink or rose.

R. argyrophyllum ssp. *argyrophyllum* at the Hillier Arboretum

Leaf underside intensely glaucous white with a thin-plastered, or almost plastered, indumentum.

Pedicel rather densely or moderately glandular.

Calyx often glandular. Corolla funnel-campanulate, white, pink or white-flushed rose.

R. argyrophyllum ssp. *hypoglaucum* at Borde Hill, Sussex

Ponticum
(Argyrophylla Group A)

R. argyrophyllum ssp. nankingense D.F.Chamb.

R. argyrophyllum var. *nankingense* Cowan.   **Argyrophylla Gp A**
(*R. argyrophyllum* var. *leiandrum* Hutch.)

Guizhou Province. 2,250 m (7,500 ft).
AM 1957 (Windsor) 'Chinese Silver', flowers rose.
Epithet: From Nanking, China.   Hardiness 3.   May.

R. argyrophyllum ssp. omeiense D.F.Chamb.

R. argyrophyllum var. *omeiense* Rehd. & Wils.   **Argyrophylla Gp A**

W Sichuan, Emei Shan. 1,800 m (6,000 ft).
Epithet: From Omei Shan.

Introduced 1980.
Differs little from ssp. argyrophyllum *would be better as a variety of it.*

R. coryanum Tagg & Forrest   Argyrophylla Gp A

Shrub or rarely a tree 2.4–6 m (8–20 ft) high, branchlets floccose, glandular.
Leaves oblanceolate, lanceolate or oblong-lanceolate; **lamina** 6.8–17 cm long,
    1.8–4 cm broad; **upper surface** dark or pale green; **underside** with a thin,
    suede-like or plastered, continuous, unistrate indumentum of fawn or whitish
    hairs, glandular with minute glands.
Petiole floccose, glandular or eglandular.
Inflorescence in lax trusses of 15–30 flowers.
Corolla funnel-campanulate, 2.5–3 cm long, creamy-white or white, with
    brownish-crimson spots.
SE Tibet. 3,600–4,250 m (12,000–14,000 ft).
Epithet: After R.R. Cory (1871–1934).   Hardiness 3.   April–May.

Broader leaves shiny and rugose on the upper surface.

Corolla usually deep pink.

R. argyrophyllum ssp. *nankingense* 'Chinese Silver' at Deer Dell

Leaves smaller and narrower than those of the species type.

Indumentum fawn, semi-plastered.

Corolla white.

R. argyrophyllum ssp. *omeiense* at Deer Dell

New foliage of *R. argyrophyllum* ssp. *omeiense* at Deer Dell

Leaves large, usually oblanceolate and V-shaped in profile.

Inflorescence 15–30 flowers.

Leaf underside usually with a whitish suede-like or plastered indumentum.

Corolla funnel-campanulate, creamy-white or white.

R. coryanum at Clyne Castle

R. formosanum Hemsl. Argyrophylla Gp A

Shrub 1.8–5.5 m (6–18 ft) high, **branchlets** sparsely hairy.
Leaves narrowly oblanceolate or lanceolate; **lamina** 7–13 cm long, 1.5–2.5 cm
 broad; **upper surface** green, glabrous, eglandular (in young leaves densely
 floccose, rather densely glandular); **underside** with a thin, plastered,
 continuous, unistrate pale buff indumentum of hairs.
Inflorescence in trusses of 7–20 flowers.
Pedicel tomentose, eglandular.
Corolla funnel-shaped, 3–4 cm long, white or pink, with purple-brown spots.
Stamens 10–12.
Ovary densely tomentose.
Taiwan (Formosa). 800–2,000 m (2,600–6,500 ft).
Epithet: From Formosa (Taiwan).   Hardiness 2–3.   April–May.

Rare in cultivation.

R. insigne Hemsl. & E.H.Wilson Argyrophylla Gp A

Shrub 1.6–6 m (5–20 ft) high, **branchlets** moderately or densely whitish
 tomentose, eglandular.
Leaves lanceolate, oblong-lanceolate or oblanceolate; **lamina** rigid, 6.8–11 cm
 long, 1.6–5 cm broad, apex acuminate; **upper surface** dark green, somewhat
 shiny, glabrous; **underside** shiny with a thin, plastered, continuous, pale
 coppery, unistrate indumentum of hairs (in young leaves woolly, white).
Inflorescence in trusses of 6–16 flowers.
Pedicel slender, eglandular.
Corolla campanulate, 2.8–4 cm long, pale or deep pink, red, or white-flushed
 rose or red down the middle of the lobes, with or without crimson spots.
W Sichuan. 2,150–3,000 m (7,000–9,800 ft).
AM 1923 (Bodnant), flowers pink.
Epithet: Remarkable.   Hardiness 3.   May–June.

Leaves narrowly oblanceolate or lanceolate, with a pale buff indumentum.

Corolla white or pink, often with purple-brown spots.

Similar to *R. thayerianum* in overall appearance, but lacks the persistent perulae.

R. formosanum at Deer Dell

Distinctive leaf underside with indumentum of 'burnished copper'. Late flowering.

Corolla white-flushed rose, pale or deep pink, rarely red.

R. insigne 'Deer Dell' at Deer Dell

R. longipes Rehder & E.H.Wilson   Argyrophylla Gp A

Shrub 1–2.5 m (3–8 ft) high, **branchlets** glabrous or sparsely puberulous, eglandular.

Leaves oblong-lanceolate or oblanceolate; **lamina** coriaceous, 8–13 cm long, 2–3.5 cm broad, apex long acuminate, base tapered; **upper surface** matt, glabrous; margin slightly recurved; **underside** with a thin, plastered, continuous indumentum of fawn rosulate hairs, sparsely glandular with minute, short-stalked glands, matt.

Petiole 1–1.5 cm long, grooved above, puberulous or glabrous, eglandular.

Inflorescence 10–15-flowered.

Pedicel slender, 3.5–4.5 cm long, sparsely hairy, moderately or sparsely glandular.

Corolla funnel-campanulate, base narrow, 3.4–3.5 cm long, pale rose, with deeper spots, without nectar pouches; **ovary** densely tomentose with a rust-coloured tomentum, glandular.

W Sichuan. 2,000–2,500 m (6,500–8,200 ft).

Epithet: With long foot-stalk.

Now in cultivation.

R. ririei Hemsl. & E.H.Wilson   Argyrophylla Gp A

Shrub or **tree** 3–12 m (10–40 ft) high, branchlets with a thin, whitish tomentum or glabrous, eglandular.

Leaves lanceolate, oblong-lanceolate, oblanceolate or oblong-elliptic; **lamina** 6.8–16 cm long, 2.3–5.1 cm broad; **upper surface** olive-green, matt, glabrous; **underside** with a thin, plastered, continuous, unistrate indumentum of silver-white hairs.

Inflorescence in trusses of 5–10 flowers.

Corolla campanulate, base broad, 4–5.3 cm long, purplish, lilac-purple or reddish-purple, 5 deep purple nectar pouches.

Sichuan (Mount Omei). 1,850 m (6,000 ft).

AM 1931 Bodnant, flowers reddish-purple.

Epithet: After Rev. B. Ririe of the Chinese Inland Missions and friend of E.H. Wilson. Hardiness 3.   February–April.

May be better in Subsection Arborea *because of the nectar pouches, large campanulate corolla and the broadly oblong capsule.*

The leaf apex has a long acuminate tip.
Similar to *R. argyrophyllum* ssp. *omeiense* and *R. chienianum* in leaf, all having the fawn indumentum.
In *R. longipes*, the pedicel is about 50% longer than the corolla, making the inflorescence very lax.

R. longipes at Glendoick, Scotland (M.L.A. Robinson)

Very early flowering and growth.
Thin shiny indumentum.
Corolla campanulate with 5 deep purple nectar pouches.

R. ririei W1808 at Deer Dell

R. simiarum Hance Argyrophylla Gp A

Shrub or **tree** 1–6 m (3–20 ft) high, branchlets floccose or glabrous, eglandular or glandular.

Leaves oblong-obovate, obovate, oblong-lanceolate or oblanceolate; **lamina** thick, rigid, 4–14.5 cm long, 1.1–4.8 cm broad; **upper surface** olive-green, glabrous, shiny; margin recurved; **underside** with a thin, somewhat plastered, continuous, unistrate indumentum of brown, fawn or white hairs, or rarely bistrate.

Inflorescence in trusses of 4–8 flowers.

Corolla campanulate, 2.5–4 cm long, pink or white, with or without a few rose spots.

Guangdong, Guangxi, Guizhou, Hong Kong to Fujian and Zhejiang. 1,000–1,600 m (3,200–5,400 ft).

Epithet: Of the monkeys.   Hardiness 2–3.   April–May.

Rarely grown. Very variable due to its wide geographical distribution.

R. thayerianum Rehd. & E.H.Wilson Argyrophylla Gp A

Shrub 1.8–4 m (6–13 ft) high, **branchlets** floccose, rather densely or moderately glandular; leaf-bud scales persistent, foliage, stiff, crowded at the ends of the branchlets.

Leaves oblanceolate or lanceolate; **lamina** thick, stiff, leathery, 6–15 cm long, 1.2–3.8 cm broad, apex acuminate; **upper surface** dark or pale green, glabrous; margin recurved; **underside** matt, with a thin-plastered or somewhat woolly, continuous unistrate indumentum of brown hairs, midrib glandular.

Petiole moderately or rather densely glandular.

Inflorescence in trusses of 10–20 flowers.

Pedicel glandular.

Calyx outside glandular: margin often glandular.

Corolla campanulate, 2.3–3 cm long, white or white suffused pink.

W Sichuan. 3,000 m (9,800 ft).

Epithet: After a well-known New England family, patrons of botany and horticulture.   Hardiness 3.   June–July.

Late flowering.

In cultivation, it has noticeably shiny leaf upper surface.

Leaves usually oblanceolate with vestiges of tomentum on the upper surface.

Leaf lamina thick, rigid. Underside with a thin, somewhat plastered, continuous, unistrate indumentum, brown, fawn or white.

R. simiarum at Deer Dell

New foliage of
R. simiarum at Deer Dell

Distinguishable out of flower by its very stiff crowded upright leaves and persistent perulae.

Branchlets floccose, rather densely or moderately glandular (sticky).

Petiole moderately or rather densely glandular. Pedicel glandular.

Calyx outside glandular.

R. thayerianum at Deer Dell

R. adenopodum Franch.   Argyrophylla Gp B

Shrub 1.2–3 m (4–10 ft) high, **branchlets** with a thin whitish or fawn tomentum.
Leaves oblanceolate or lanceolate; **lamina** 7.5–20 cm long, 2–4.6 cm broad;
  upper surface dark green, matt, glabrous; **underside** covered with a
  somewhat thick, felty or woolly, whitish, fawn or brown, continuous bistrate
  indumentum of hairs.
Petiole with a thin whitish or fawn tomentum.
Inflorescence a candelabroid umbel of 4–8 flowers.
Pedicel glandular.
Corolla funnel-campanulate, 3.6–5 cm long, pale rose, with or without deeper
  spots; **ovary** densely glandular.
Capsule glandular.
E Sichuan, W Hubei. 1,500–2,100 m (5,000–7,000 ft).
AM 1926 (Wakehurst) flowers rose-pink.
Epithet: With glandular pedicel.   Hardiness 3.   April–May.

Moved from Subsection Pontica *to* Argyrophylla *because of its geographical
distribution, stipitate glandular ovary and the form of the flowers.*

R. coeloneuron Diels   Argyrophylla Gp B

Shrub 3.6–4 m (12–13 ft) high, **branchlets** densely rufous, tomentose,
  eglandular, leaf bud scales deciduous.
Leaves oblanceolate, 8–12 cm long, 2.1–3.5 cm broad, apex acute or obtuse,
  base tapered; **upper surface** somewhat shiny, slightly rugulose, margin flat,
  glabrous, midrib grooved, moderatley or sprasely hairy; **underside** with a thick,
  woolly, rufous, continuous or discontinuous (patchy), bistrate indumentum,
  upper layer a form of stellate, under layer whitish rosulate hairs.
Petiole 1.3–2 cm long, densely tomentose, eglandular.
Inflorescence a racemose umbel of 4–8 flowers.
Pedicel 0.7–1 cm long, densely rufous tomentose, eglandular.
Corolla funnel-campanulate, 4–4.3 cm long, pink or purplish.
SE Sichuan. 2,100 m (7,000 ft).
Epithet: With impressed nerves.   Hardiness 4.   April–May.

Similar in appearance to R. floribundum: *the flowers have prominent dark
flecking. Before its introduction, this species was placed in Subsection* Taliensia
because of its similarities to R. wiltonii. *In cultivation however, its habit of growth
and general appearance indicate that it is better placed here.*

Leaves remarkably long and narrowly oblanceolate.
Inflorescence candelabroid.
Ovary densely stipitate glandular.

Leaves may sometimes be lanceolate, underside covered with a felty continuous bistrate indumentum.

R. adenopodum at Deer Dell

Leaf upper surface rugose, somewhat shiny.

Branchlets densely rufous, tomentose.

Leaves oblanceolate with a thick, woolly, rufous, continuous or patchy bistrate indumentum, upper layer a form of stellate hairs.

Pedicel densely rufous tomentose.

R. coeloneuron AC1174 at Deer Dell

R. coeloneuron SICH1198 at UBC botanic garden (M.L.A. Robinson)

R. denudatum H.Lév. Argyrophylla Gp B

Shrub 2–3.7 m (6.5–12 ft) high, **branchlets** densely floccose with stellate hairs, leaf bud scales deciduous.

Leaves lanceolate, oblong-lanceolate or oblong-elliptic, 11–20.5 cm long, 2.9–6.9 cm broad, apex acuminate, base tapered or obtuse; **upper surface** olive-green, convex, bullate, somewhat matt or shiny, glabrous, midrib grooved, margin recurved; **underside** with a somewhat thick, woolly, continuous, bistrate indumentum, upper layer brown or fawn ramiform hairs, under layer rosulate hairs widely scattered; or unistrate with ramiform hairs.

Petiole 1.5–3 cm long, moderately or rather densely tomentose with whitish stellate hairs, eglandular.

Inflorescence a lax racemose umbel of 8–12 flowers.

Pedicel 0.6–1.5 cm long, densely tomentose with fawn or whitish stellate hairs, eglandular.

Corolla campanulate, 3–4 cm long, rose with a deep crimson blotch at the base, and with spots, without nectar pouches at the base; lobes 5–6.

E Yunnan, W Sichuan. 3,000–3,350 m (9,800–11,000 ft).

Epithet: Naked.   Hardiness 4.   April–May.

Whether R. denudatum *merits specific rank is very doubtful. Variable in cultivation, some plants having a glabrous ovary, and it is likely to be a cline between* R. floribundum *and* R. coeloneuron.

R. farinosum H.Lév. Argyrophylla Gp B

Shrub 1.2–2 m (4–6 ft) high, **branchlets** densely whitish or fawn floccose.

Leaves oblong-obovate or obovate; **lamina** coriaceous, 5–8 cm long, 2–3 cm broad; **upper surface** olive-green, convex, bullate, shiny, glabrous; **underside** matt, with a thick, woolly, continuous, bistrate indumentum, upper layer fawn, large ramiform hairs with ribbon-like branches, under layer rosulate, midrib prominent, moderately or sparsely hairy.

Petiole 0.8–1 cm long, densely whitish or fawn floccose with a form of stellate hairs, eglandular.

Inflorescence 6–8 flowers.

Calyx minute, hairy with a form of stellate hairs, eglandular.

Corolla campanulate, 3–3.5 cm long, white, without nectar pouches.

Ovary densely hairy.

E Yunnan. 3,200 m (10,500 ft).

Epithet: Mealy.   Hardiness 4.   April.

This species probably comprises the clone 'Chelsea Chimes'. The 'drip tip' can be seen on the herbarium specimen of R. farinosum. *'Chelsea Chimes' has been cited as a form of* R. coryanum, *probably because of its glandular leaf blade, petiole and pedicel. However, the glands on the leaf blade of 'Chelsea Chimes' are detersile, and would not appear on an herbarium specimen as old as the Maire collection (1913). Moreover,* R. coryanum *lacks any stellate hairs, has a unistrate indumentum and larger leaves. Because of a reported mix-up of labels years ago at Borde Hill, it is doubtful that 'Chelsea Chimes' originated from KW6311.* **We suggest, therefore, that detersile glands will be found on** R. farinosum.

Leaf upper surface olive-green, convex, bullate.

Indumentum somewhat thick, woolly, continuous, fawn to yellowish.

In cultivation, some plants have a glabrous ovary.

Pedicel tomentose with fawn or whitish stellate hairs.

R. denudatum at Benmore

New foliage of *R. denudatum* at Deer Dell

Leaves rugulose with a marked turned-down cucculate 'drip tip'.

Indumentum is bistrate, the upper layer 'powdery'.

Petiole floccose with a form of stellate hairs.

Leaves oblong-obovate, olive-green.

Corolla white.

Calyx with a form of stellate hairs.

Foliage of *R. farinosum* (Chelsea Chimes) at Deer Dell (M.L.A. Robinson)

Ponticum
(Argyrophylla Group B)

R. floribundum Franch.   Argyrophylla Gp B

Shrub or **tree** 1.8–6 m (6–20 ft), **branchlets** densely floccose with whitish stellate hairs, eglandular.
Leaves oblong-lanceolate, lanceolate or oblanceolate; **lamina** 6.8–18 cm long, 2–5.5 cm broad; **upper surface** olive-green, convex, bullate, glabrous; margin recurved; **underside** with a somewhat thick, woolly, continuous, bistrate indumentum of white or fawn hairs.
Petiole densely or moderately floccose with whitish stellate hairs, eglandular.
Inflorescence in trusses of 8–12 flowers.
Pedicel densely tomentose with whitish stellate hairs, eglandular.
Calyx densely tomentose with stellate hairs, eglandular.
Corolla campanulate, 3–5 cm long, purplish-lavender, pinkish-purple, pink or rose, with a deep crimson blotch and spots, without nectar pouches.
W Sichuan. 1,300–2,600 m (4,300–8,500 ft).
AM 1963 (Exbury) 'Swinhoe', flowers rose-purple with dark crimson blotch.
Epithet: Free-flowering.   Hardiness 3.   April–May.

R. haofui Chun & W.P.Fang.   Argyrophylla Gp B

Shrub 4–6 m (13–20 ft) high, **branchlets** glabrous.
Leaves 7–10 cm long, 3–4 cm broad, apex acuminate or acute, mucronate, base obtuse or broadly cuneate; **upper surface** glabrous; **underside** covered with a thick, floccose-felted, fulvous (initially white) indumentum of woolly hairs.
Petiole 1.5–2.2 cm long, upper surface nearly flat, glabrous.
Inflorescence a racemose umbel of 5–9 flowers.
Pedicel 2.5–3.5 cm long, white hairy.
Corolla large broadly campanulate, 4–4.5 cm long, white, sometimes flushed rose.
Stamens 18–20.
N Guangxi. 1,430 m (4,700 ft).
Epithet: "Good luck" in Chinese.   Hardiness 3.   May.

Close to R. simiarum *but* R. haofui *has the distinguishing feature of 18–20 stamens.*

R. hunnewellianum Rehder & E.H.Wilson
Argyrophylla Gp B

Shrub rarely tree 1.8–5.5 m (6–18 ft) high, **branchlets** with a thin tomentum, glandular.
Leaves oblanceolate or lanceolate; **lamina** 5–17.5 cm long, 1.2–3 cm broad, apex acuminate; **upper surface** dark green, glabrous; **underside** with a somewhat thick, somewhat woolly, continuous or discontinuous bistrate, whitish or fawn indumentum of hairs.
Inflorescence in trusses of 5–7 flowers.
Pedicel hairy, often glandular.
Calyx minute, 1 mm long.
Corolla campanulate, 3.5–5 cm long, white edged with pink, with pink spots.
W Sichuan. 1,600–3,500 m (5,300–11,500 ft).
Epithet: After a well-known New England family.   Hardiness 3.   March–May.

Corolla widely campanulate, a strong colour of purplish-lavender or pinkish-purple, with prominent markings. Leaf upper surface olive green, convex, bullate.

Branchlets densely floccose with whitish stellate hairs.

Leaf underside with a somewhat thick, woolly indumentum.

Pedicel densely tomentose with whitish stellate hairs. Calyx densely tomentose with stellate hairs.

R. floribundum W4266 at Deer Dell (M.L.A. Robinson)

The foliage is long and somewhat hanging. Leaf with thick, floccose-felted, fulvous indumentum of woolly hairs. Stamens 18–20.

Corolla white.

Foliage of *R. haofui* at Deer Dell

Narrow rugulose leaves with a usually greyish felted indumentum.

Branchlets glandular. Leaves oblanceolate or lanceolate, apex acuminate.

Corolla white with pink edges.

R. hunnewellianum W1198 at Deer Dell

R. longipes var. chienianum D.F.Chamb.

R. chienianum W.P.Fang.   **Argyrophylla Gp B**

Shrub or **tree** 3–10 m (10–33 ft) high; **branchlets** moderately or sparsely tomentose with a thin, whitish or brown tomentum, glandular with short-stalked glands or eglandular.

Leaves oblanceolate or lanceolate; **lamina** coriaceous, 5–9 cm long, 1.5–2.8 cm broad, apex acuminate, base tapered or obtuse; **upper surface** dark green matt glabrous or with vestiges of hairs, midrib grooved, margin recurved; **underside** matt, with a somewhat thick, somewhat woolly, continuous, bistrate indumentum, upper layer brown ramiform hairs, under layer rosulate.

Petiole 0.8–1.2 cm long, tomentose with a thin, whitish or brown tomentum, glandular or eglandular.

Inflorescence a lax racemose umbel of 8–15 flowers.

Pedicel 2.5–3 cm long, hairy with whitish hairs, glandular.

Corolla campanulate, 2.8–3 cm long, purple or pinkish, without nectar pouches at the base; **ovary** glabrous, densely glandular.

Sichuan. 1,700–2,100 m (5,600–6,900 ft).

Epithet: After Professor S.S. Chien, Nanking, China.   Hardiness 3?

May merit full species status.

R. pingianum W.P.Fang.   **Argyrophylla Gp B**

Shrub or **tree** 4–7 m (13–23 ft) high, **branchlets** tomentose with a thin, whitish tomentum.

Leaves oblong, oblong-lanceolate, lanceolate or oblanceolate; **lamina** 8–15.3 cm long, 3–4 cm broad; **upper surface** olive-green, matt, glabrous; **underside** matt, with a somewhat thick, woolly, continuous, bistrate indumentum of white hairs.

Inflorescence in trusses of 12–20 flowers.

Pedicel with whitish hairs.

Calyx 1–2 mm long.

Corolla funnel-campanulate, 2.6–3.5 cm long, pale purple, purple, pink or pinkish-purple, without nectar pouches.

W Sichuan. 2,000–2,700 m (6,500–8,800 ft).

Epithet: After Professor C. Ping, former Director of Biological Laboratory, Science Society of China, Nanking.   Hardiness 3.   May.

Very similar to
R. longipes, but has
a thick bistrate
indumentum and is
without the long
acuminate leaf apex.

Thick, somewhat woolly,
indumentum, upper
layer **brown**.

Foliage of *R. longipes* var. *chienianum* AC4244 at Deer Dell

One of the whitest
of all leaf
indumenta.
Ovary usually rufous
tomentose.

Leaf underside matt,
with a bistrate
indumentum of white
hairs.

R. pingianum KR150 at Deer Dell

Ponticum
(Argyrophylla Group B)

Subgenus *Hymenanthes*
Section *Ponticum* G.Don
Subsection *Auriculata* Sleumer
Auriculatum Series

R. auriculatum Hemsl.

Shrub or **tree** 1.8–10 m (6–33 ft) high, **branchlets** often glandular with long-stalked glands, **foliage buds** long, conical, tapered, the outer scales acuminate with long tips.
Leaves oblong, oblong-lanceolate or oblong-oblanceolate; **lamina** 9.5–32 cm long, 2.8–12 cm broad, base auricled; **upper surface** dark green or pale green, glabrous or glabrescent; **underside** hairy and glandular with isolated, long, thread-like hairs and glands.
Petiole glandular.
Inflorescence in trusses of 6–15 flowers.
Pedicel densely glandular; **flower-buds** large, conical, tapered.
Corolla tubular-funnel shaped, 6–10 cm long, white, rose-pink or creamy-white, with a greenish blotch at the base, fragrant.
Style glandular throughout to the tip.
E Sichuan, Guizhou. 500–2,300 m (1,600–7,550 ft).
AM 1922 (Bodnant), flowers white.
Epithet: Eared or auriculate.   Hardiness 3.   July–August.

Late flowering: only likely to be confused with its hybrids. The leaves are rougher than almost all of the hybrids, and are never noticeably shiny.

R. chihsinianum Chun & W.P.Fang.

Small tree 2.5–4.5 m (8–18ft) high; **branchlets** brown or grey-brown, sparsely floccose and setose when young, partly detersile.
Leaves narrowly oblong or oblong lanceolate, 17–22 cm long, 4.5–7 cm broad, base obtuse or slightly cordate, margin slightly recurved; **upper surface** green glabrescent, **underside** glabrescent.
Petiole 1.4–2 cm long, setose glandular.
Inflorescence 7–9-flowered.
Pedicel densely floccose.
Calyx small, sparsely clad with fimbriate hairs.
Corolla funnel-campanulate, 3.8–4.1 cm long, glabrous, lobes 7.
Stamens 15, glabrous; **ovary** 6 mm long, densely yellow glandular.
Style terete, setose glandular.
Guangxi. 700–1,400 m (2,300–4,600 ft).
Epithet: From Chihsin, Hunan.   Hardiness 3.   May–June.

It is likely that this species has been introduced to cultivation recently, though until it flowers, uncertainty remains. In leaf, it is similar to R. auriculatum *but appears to lack the auricled leaf base.*

Foliage buds long, conical, tapered, the outer scales acuminate with long tips. Leaf base auricled, underside hairy and glandular. Distinctive tubular campanulate corolla.

Corolla fragrant. Style glandular throughout.

R. auriculatum at Deer Dell flowering on 1st August

Similar to *R. auriculatum* but leaf base obtuse or slightly cordate.

R. chihsinianum at Deer Dell

R. chihsinianum at Deer Dell showing setose-glandular petioles

Ponticum
(Auriculata)

Subgenus *Hymenanthes*
Section *Ponticum* G.Don
Subsection *Barbata* Sleumer
Barbatum Series

Shrubs or trees.
Bark peeling, plum-coloured or purple.
Branchlets usually bristly or bristly glandular.
Inflorescence very compact.
Corolla usually crimson or scarlet with nectar pouches.

This area has been revised radically since Davidian's descriptions. The recent inclusion of R. succothii, *but not* R. fulgens *which is very similar in appearance, in Subsection* Barbata *is odd. As this book is primarily about identification, they have been placed together here.*

Species in Subsection Barbata *usually have bristly branchlets and petioles, and were given specific status in the past if these were absent. However,* R. imberbe, *the non-bristly expression of* R. barbatum, *is now submerged into that species as there is a continuous variation from bristly to glabrous in the wild. This has led to the non-bristly* R. succothii *being included in the Subsection.*

Within the usually bristly species:
Round leaves: *R. erosum* (the former *R. argipeplum*) and *R. exasperatum.*
Long leaves: *R. argipeplum* (the former *R. macrosmithii*) and *R. barbatum.*

Within the non-bristly species:
R. succothii has cordate leaf bases, very short petioles, so almost sessile leaves.
R. fulgens has cordate leaf bases but has a distinct petiole 1–2 cm long, and indumentum.

R. argipeplum D.F.Chamb.
R. macrosmithii Davidian
(*R. smithii* Hook.f.)

Shrub or **tree** 1.2–7.6 m (4–25 ft) high; **stem** and **branches** with smooth reddish or brown flaking bark; **branchlets** setose, often setose-glandular.
Leaves oblong-lanceolate or oblanceolate; **lamina** 8–16 cm long, 2.8–6 cm broad; **upper surface** dark green, rugulose, convex, somewhat shiny, glabrous, eglandular; **underside** with a discontinuous indumentum of woolly hairs, usually in small patches and shreds.
Petiole setose, often setose-glandular.
Inflorescence a very compact racemose umbel of 10–19 flowers.
Pedicel glandular.
Corolla tubular-campanulate, 2.8–4.5 cm long, scarlet or crimson, 5 nectar pouches.
E Sikkim, Bhutan. 2,500–3,600 m (8,000–12,000 ft).
AM 1978 (Borde Hill) 'Fleurie' (shown as *R. smithii*).
Epithet: With a white covering.   Hardiness 3.   February–April.

R. argipeplum *replaces the names* R. smithii *and* R. macrosmithii.

Leaves of Subsect. *Barbata* at Deer Dell. Clockwise from top right (2 leaves each): *R. exasperatum*, *R. succothii*, *R. argipeplum*, *R. erosum*, *R. fulgens* and *R. barbatum*. Bar = 5 cm

Long oblong-lanceolate or oblanceolate rugose convex leaves.

Stem and branches with smooth reddish or brown flaking bark.

Leaf underside with a discontinuous indumentum of woolly hairs, usually in small patches and shreds.

R. argipeplum 'Fleurie' AM at Borde Hill

R. barbatum Wall & G.Don.

(*R. imberbe* Hutch.)

Shrub or **tree** 2.4–18 m (8–60 ft) high; **stem** and **branches** with smooth reddish or brown flaking bark, **branchlets** usually setose and sometimes setose-glandular.

Leaves elliptic-lanceolate or oblong-lanceolate; **lamina** 7.5–20 cm long, 2.5–7 cm broad; **upper surface** dark green, somewhat shiny; **underside** pale greenish, glabrous.

Petiole usually setose and sometimes setose-glandular.

Inflorescence very compact truss of 10–20 flowers.

Calyx 0.4–1.5 cm long.

Corolla funnular-campanulate, 2.8–4 cm long, crimson or scarlet, 5 nectar pouches.

Ovary densely glandular.

Nepal, Sikkim, Bhutan. 2,135–3,660 m (7,000–12,000 ft).

AM 1954 (Winterfold House, Cranleigh), flowers scarlet.

Epithet: Bearded.   Hardiness 3.   February–April.

Plants with glabrous petioles were previously separated as R. imberbe.

R. erosum Cowan

Shrub or **tree** 2.4–9 m (8–30 ft) high, **stem** and **branches** with smooth brown flaking bark, **branchlets** glandular with short-stalked glands or rather densely setulose-glandular.

Leaves oval, obovate or oblong-obovate; **lamina** 8–16.5 cm long, 3–9 cm broad; **upper surface** olive-green, rugulose, convex, glabrous; **underside** densely or moderately woolly with a continuous or discontinuous indumentum of hairs.

Petiole often bristly, often glandular.

Inflorescence very compact; 10–15-flowered, flower-bud scales sticky.

Pedicel glandular.

Corolla tubular-campanulate, 2.8–3.5 cm long, deep crimson, scarlet, or rose-pink, 5 nectar pouches at the base.

S Tibet. 3,000–3,800 m (10,000–12,000 ft).

Epithet: Eaten away.   Hardiness 3.   March–April.

R. erosum *includes plants with rounded leaves formerly designated* R. argipeplum.

Stem and branches with smooth reddish and brown flaking bark.

Corolla crimson or scarlet.

Petiole bristly.

Branchlets usually setose.

Leaf underside glabrous; petiole usually setose.

R. barbatum 'Leemoor' at Deer Dell

R. imberbe in the Valley Gardens

New foliage of *R. barbatum* at Deer Dell

Rugulose matt, rounded leaves – oval, obovate or oblong-obovate, convex.

Stems and branches with smooth brown flaking bark.

Branchlets glandular.

Leaf underside densely or moderately woolly with a continuous or discontinuous indumentum of hairs.

Inflorescence very compact, deep crimson, scarlet, or rose-pink.

R. erosum L&S1304 at Glenarn

New foliage of *R. erosum* at Deer Dell

Ponticum
(Barbata)

R. exasperatum Tagg

Shrub or **tree** 1.5–4.5 m (5–15 ft) high, **stem** and **branches** with smooth, brown flaking bark; **branchlets** setose, and setose-glandular; leaf-bud scales persistent, **young growths** bronzy-brown.

Leaves obovate or oval; **lamina** 9–18 cm long, 4.5–10 cm broad; **upper surface** dark green, rugulose, convex, glabrous, eglandular; **underside** pale greenish-yellow, setulose, setulose-glandular.

Petiole setose and often setose-glandular.

Inflorescence a very compact truss of 10–15 flowers, flower-bud scales persistent, sticky; pedicel often glandular.

Calyx eglandular.

Corolla tubular-campanulate, 3–4.5 cm long, brick-red, scarlet or rose-pink, 5 nectar pouches at the base.

SE Tibet, N Burma, Assam. 3,050–3,600 m (10,000–12,000 ft).

Epithet: Rough.　　Hardiness 3.　　March–May.

R. fulgens Hook.f.

Rounded **compact shrub** or sometimes a tree, 0.6–4.5 m (2–15 ft) high; **stem** and **branches** with smooth, reddish-brown flaking bark; **branchlets** glabrous.

Leaves oval, obovate or oblong-obovate; **upper surface** dark green, shiny, glabrous; **underside** covered with a thick, woolly, reddish-brown or dark brown, continuous, unistrate indumentum of hairs.

Petiole glabrous.

Inflorescence very compact rounded truss of 8–14 flowers.

Corolla tubular-campanulate, 2–3.3 cm long, deep blood-red, crimson, cherry-red, deep scarlet or scarlet, with 5 black-red or deep crimson nectar pouches.

Ovary glabrous.

Sikkim, Nepal, Bhutan, Assam, S and SE Tibet. 3,200–4,300 m (10,500–14,000 ft).

AM 1933 (Wakehurst), flowers blood-red.

Epithet: Shiny.　　Hardiness 3.　　March–April.

This species, placed into Subsection Fulgensia *(Sleumer) (Fulgens Series) by modern authors, fits best into Subsection* Barbata *for identification purposes.*

Young growths bronzy-brown to light reddish purple. Rounded smooth obovate or oval leaves with a waxy slightly glaucous upper surface.

Stems and branches with smooth, brown flaking bark.

Branchlets setose-glandular.

Leaf underside setulose, setulose-glandular.

Petiole setose and often setose-glandular.
Inflorescence compact.

R. exasperatum KW6855 at Deer Dell

New foliage of *R. exasperatum* KW8250 at Deer Dell

Shiny leaves with cordate bases and with indumentum. Bark reddish-brown flaking.

Leaf underside covered with a woolly, reddish-brown or dark brown, continuous indumentum.

Corolla deep blood-red, crimson, cherry-red, deep scarlet or scarlet, with 5 black-red or deep crimson nectar pounches.

R. fulgens at Deer Dell

New foliage of *R. fulgens* at Deer Dell

Ponticum (Barbata)

R. succothii <small>Davidian</small>

Shrub or sometimes a small tree 60 cm–6 m (2–20 ft) high; **stem** and **branches** with smooth brown flaking bark; branchlets glabrous.

Leaves subsessile, clustered in whorls on the branchlets, oblong-obovate, obovate, oblong-elliptic, elliptic or oval; **lamina** 5.3–13.5 cm long, 2.6–6.4 cm broad, base cordulate, auricled; **upper surface** dark green, glossy, glabrous; **underside** pale green, glabrous.

Petiole flat above.

Inflorescence very compact rounded truss of 10–15 flowers.

Corolla tubular-campanulate, 2.3–3.5 cm long, crimson or scarlet, 5 deep crimson nectar pouches.

Bhutan, Assam. 3,000–4,000 m (10,000–13,600 ft).

Epithet: After Sir George I. Campbell of Succoth, Bt., Crarae, Argyllshire. Hardiness 3.   March–April.

Recognisable by the arrangement of the almost sessile leaves with cordate bases. Suffers badly with powdery mildew in the South of England.

Subgenus *Hymenanthes*
Section *Ponticum* <small>G.Don</small>
Subsection *Campanulata* <small>Sleumer</small>
Campanulatum Series and Subseries

Shrubs or sometimes small trees.

Branchlets glabrous, eglandular.

Leaves with a glabrous shiny upper surface and a continuous or tufted brown to near black indumentum beneath.

Calyx usually glabrous.

Ovary eglandular, glabrous or almost so.

Flowers usually mauve or purplish.

Smooth brown flaking bark.

Leaves subsessile, clustered in whorls.

Leaf underside pale green, glabrous; petiole flat.

Inflorescence compact.

Corolla crimson or scarlet, with 5 deep crimson nectar pouches.

R. succothii L&H19850 in the Valley Gardens (supplied by Rhododendron Species Foundation)

Leaves of Subsect. *Campanulata* at Deer Dell. Left to right: *R. campanulatum* (2 leaves), *R. wallichii* (Heftii Group) (1 leaf) and *R. wallichii* (2 leaves). Bar = 5 cm

R. campanulatum ssp. campanulatum D.Don

R. campanulatum D.Don

Shrub or sometimes a tree 0.3–5.5 m (1–18 ft) or sometimes up to 9 m (30 ft) high, **branchlets** green or purple, often glabrous.

Leaves elliptic, obovate, oval or oblong-oval; **lamina** 5–15.8 cm long, 2.4–7.8 cm broad; **upper surface** dark green, somewhat shiny, glabrous; **underside** covered with a thin, suede-like, fawn, brown or rusty-brown, continuous, unistrate indumentum of capitellate (mop-like) hairs.

Petiole purple.

Inflorescence in trusses of 6–12 flowers.

Corolla broadly campanulate, 2.8–4.5 cm long, lavender-blue, pale purple, lilac, mauve, various shades of mauve, rose or white, with or without purple or red spots.

Ovary glabrous.

From Kashmir, Punjab to Nepal, Sikkim, Bhutan. 2,800–4,600 m (9,000–15,000 ft).

AM 1925 (Exbury) 'Knaphill', flowers lavender-blue.

AM 1964 (RBG, Edinburgh) 'Roland Cooper', flowers rose-purple.

AM 1965 (RBG, Edinburgh) 'Waxen Bell', flowers purple with darker spots.

Epithet: Bell-shaped.   Hardiness 3.   April–May.

R. campanulatum ssp. aeruginosum D.F.Chamb.

R. aeruginosum Hook.f.

Shrub 0.3–2.8 m (1–9 ft) high, **branchlets** glabrous.

Leaves elliptic, oblong-elliptic, obovate or oval; **lamina** 5–11 cm long, 3–5.5 cm broad; **upper surface** bluish-green, glabrous (in young foliage bluish-green with a metallic lustre, glaucous, glabrous); **underside** covered with a thick, smooth, somewhat silky bistrate indumentum of hairs.

Petiole glaucous.

Inflorescence in trusses of 8–18 flowers.

Corolla campanulate, 2.1–3.4 cm long, lilac, rose, purple, reddish-purple or pink.

Ovary glabrous.

Sikkim, Bhutan. 3,600–4,600 m (12,000–15,000 ft).

Epithet: Verdigris-coloured.   Hardiness 3.   April–May.

Evidence of plants intermediate between R. aeruginosum *and* R. campanulatum *in leaf details and in geography has been used to reduce this to subspecific status.*

New foliage of
***R. campanulatum* ssp.**
***aeruginosum* at Deer Dell**

Leaf underside covered with a thin, suede-like, brownish unistrate indumentum.
Corolla usually shades of mauve or lilac, rarely white.

R. campanulatum ssp. *campanulatum* 'Waxen Bell' AM at Deer Dell

R. campanulatum ssp. *campanulatum* Werrington, Cornwall

Leaf upper surface bluish-green, glabrous (in young foliage bluish-green with a metallic lustre). Underside with a smooth, somewhat silky bistrate indumentum.

R. campanulatum ssp. *aeruginosum* at Dawyck, Perthshire

R. wallichii Hook.f.

Shrub or small, tree 1–5 m (3–16 ft) high, **branchlets** glabrous or floccose.
Leaves elliptic, oblong-elliptic, obovate or oblong-obovate; **lamina** 5.2–12 cm long, 2.2–5.8 cm broad; **upper surface** green or pale green, glabrous; **underside** dotted with, numerous or sometimes few, powdery ferruginous or brown separate hair tufts, which do not form a continuous indumentum, rough to the touch.
Petiole often tinged with red.
Inflorescence in trusses of 5–10 flowers.
Corolla broadly campanulate, 2.5–5 cm long, mauve, pale lilac, pink, white faintly flushed purple or sometimes white with or without purple blotch, and with or without purple spots.
Nepal, Sikkim, Bhutan, S and SE Tibet. 2,750–4,300 m (9,000–14,000 ft).
Epithet: After N. Wallich, 1786–1854, a former Superintendent, Calcutta Botanic Garden.   Hardiness 3.   April–May.

The dark-tufted indumentum allows this species to be distinguished from the
R. campanulatum *subspecies.*

R. wallichii Heftii Group
R. heftii Davidian

Shrub 1.8–3 m (6–10 ft) high, **branchlets** reddish-purple.
Leaves broadly elliptic, nearly orbicular, oval or oblong-elliptic; **lamina** very thick, leathery, 7–11.8 cm long, 3.6–6.8 cm broad; **upper surface** dark green, shiny, glabrous; **underside** pale greenish, glabrous or sometimes minutely puberulous.
Petiole reddish-purple.
Inflorescence in trusses of 6–9 flowers; pedicel glabrous.
Corolla campanulate, 3.8–4.5 cm long, pure ivory-white.
Ovary slender, glabrous.
Nepal. 3,400 m (11,000 ft).
Epithet: After Dr. Lothar Heft, Bremen, Germany.   Hardiness 3. April–May.

This is now regarded as a variation of R. wallichii *with no leaf indumentum.*
Plants with pale mauve flowers have been observed in the wild. Possibly
merits varietal status.

Leaf underside dotted with separate ferruginous to very dark brown hair tufts, rough to the touch.

R. wallichii at Deer Dell

Leaf underside pale greenish, glabrous or sometimes minutely puberulous.
Flowers ivory-white.
Leaf lamina very leathery.
Petiole reddish-purple.

R. wallichii Heftii Group at Deer Dell

Ponticum
(Campanulata)

Subgenus *Hymenanthes*
Section *Ponticum* G.Don
Subsection *Campylocarpa* Sleumer
Thomsonii Series, Campylocarpum Subseries

Shrubs or rarely trees.
Leaves relatively small, roundish, glabrous on both surfaces
at maturity.
Corolla campanulate, without nectar pouches.
Ovary densely glandular.
Stamens 10.

Though related to Subsection Thomsonia, *and very similar in foliage, this
Subsection has been separated partly on the grounds of the absence of the 5
dark nectar pouches present in Subsection* Thomsonia.

R. callimorphum var. callimorphum D.F.Chamb.
R. callimorphum Balf.f. & W.W.Sm.
(*R. cyclium* Balf.f. & Forrest)

Shrub 0.6–2.75 m (2–9 ft) high, **branchlets** glandular.
Leaves orbicular, ovate or broadly elliptic; **lamina** 3–7 cm long, 2–5.3 cm broad;
 upper surface dark green, glabrous, glossy; **underside** glaucous, glabrous.
Petiole glandular.
Inflorescence 5–8-flowered.
Pedicel glandular.
Calyx 1–3 mm long, outside and margin glandular.
Corolla campanulate, 3–4.5 cm long, pink or pale to deep rose, with or without a
 crimson blotch.
W Yunnan, NE Upper Burma. 2,700–3,300 m (9,000–11,000 ft).
Epithet: With a lovely shape.   Hardiness 3.   April–May.

Similar to R. selense *but lacks the bristly branchlets. The flowers can also have
purplish tones.*

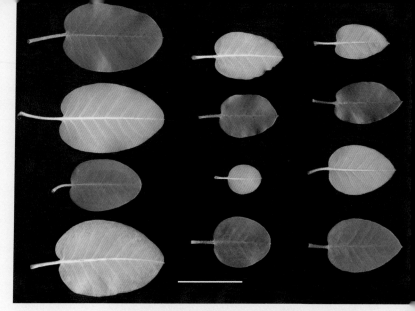

Leaves of Subsect. *Campylocarpa* at Deer Dell. Top to bottom (all 2 leaves each): left column – *R. souliei, R. wardii*; middle column – *R. campylocarpum* ssp. *caloxanthum, R. callimorphum* var. *callimorphum*; right column – two forms of *R. callimorphum* var. *myiagrum.* Bar = 5 cm

Rounded (orbicular, ovate) leaves, smaller and less cordate than the rest of this Subsection.

Corolla pink or pale to deep rose.

Branchlets glandular.
Petiole glandular.
Calyx glandular.

R. callimorphum var. *callimorphum* at Deer Dell showing the campanulate corolla and round leaf

R. callimorphum var. myiagrum D.F.Chamb.

R. myiagrum Balf.f. & Forrest

Shrub 1–1.8 m (3–6 ft) high, **branchlets** glandular.

Leaves orbicular, ovate or broadly elliptic; **lamina** 2–7 cm long, 1.5–4.8 cm broad; **upper surface** dark green, glabrous; **underside** glabrous, glaucous, papillate.

Petiole glandular or setulose-glandular.

Inflorescence in trusses of 3–5 flowers; pedicel glandular.

Calyx densely glandular.

Corolla campanulate, 2.5–3.3 cm long, white, with or without a crimson blotch, spotted crimson or unspotted.

W Yunnan, NE Upper Burma. 3,000–4,000 m (10,000–13,000 ft).

Epithet: The fly-catcher, alluding to the sticky pedicels.   Hardiness 3. April–May.

Rare in cultivation. The leaves are closest in shape and size to R. campylocarpum *ssp.* caloxanthum, *and not* R. callimorphum.

R. campylocarpum ssp. campylocarpum

R. campylocarpum Hook.f.

Shrub 1–3.7 m (3–12 ft) high, rarely a tree 4.5–6 m (15–20 ft) high, **branchlets** glandular.

Leaves elliptic, oblong-elliptic, ovate or sometimes oval; **lamina** 4–10 cm long, 2.3–5.4 cm broad; **upper surface** dark or pale green, glabrous, shiny; **underside** pale glaucous green, papillate, glabrous or minutely hairy.

Inflorescence in trusses of 6–12 flowers.

Pedicel glandular.

Calyx outside and margin glandular.

Corolla campanulate, 2.5–4 cm long, yellow, pale or bright yellow, sulphur-yellow or lemon-yellow.

Sikkim, Nepal, Bhutan, Assam, S and SE Tibet. 2,900–4,300 m (9,500–14,000 ft).

FCC 1892 (Veitch and Sons, Chelsea), flowers lemon-yellow.

Epithet: With bent fruit.   Hardiness 3.   April–May.

Pedicel glandular, very sticky.
Petiole glandular.
Corolla white.
Branchlets glandular.
Calyx densely glandular.

R. callimorphum var. *myiagrum* KW6962 at Deer Dell

Corolla campanulate, yellow. Whereas *R. wardii* has a style that is glandular along its entire length, it is never so in this species. Otherwise the two are hard to distinguish systematically.
Branchlets glandular.

R. campylocarpum ssp. *campylocarpum* at Glenarn, Rhu, W Scotland (Davidian)

Ponticum
(Campylocarpa)

R. campylocarpum ssp. caloxanthum

D.F.Chamb.

R. caloxanthum Balf.f. & Farrer

Shrub 0.6–2.5 m (2–8 ft) high, **branchlets** glandular.

Leaves orbicular, ovate or broadly elliptic; **lamina** 3.2–8 cm long, 2.5–5.6 cm broad; **upper surface** dark green, glabrous, glaucous, matt; **underside** pale glaucous green, papillate, glabrous or minutely hairy.

Petiole often glandular.

Corolla campanulate, 2.5–4 cm long, citron, yellow or deep lemon-yellow, sometimes tinged pink, bud vermilion, crimson, red or rose.

NE and E Upper Burma, NW Yunnan, SE and S Tibet. 2,800–4,000 m (9,000–13,600 ft).

AM 1934 (Exbury), flowers deep yellow, tinged red.

Epithet: Of a beautiful yellow.   Hardiness 3.   April–May.

R. campylocarpum ssp. caloxanthum D.F.Chamb.

Telopeum Group

R. telopeum Balf.f. & Forrest

Shrub 60 cm–3 m (2–10 ft) or rarely a tree 4.5 m (15 ft) high, **branchlets** glandular.

Leaves orbicular, ovate or broadly elliptic; **lamina** 2–5 cm long, 1.5–3.5 cm broad; **upper surface** dark green, glabrous, glaucous, matt; **underside** pale glaucous green, papillate, glabrous or minutely hairy.

Petiole glandular or eglandular.

Inflorescence 4–5-flowered.

Pedicel glandular.

Calyx 1–2 mm long, often glandular.

Corolla campanulate, 2.6–4 cm long, bright yellow, sulphur-yellow or creamy-white, with or without a faint crimson blotch.

SE Tibet, NW Yunnan. 3,650–4,400 m (12,000–14,500 ft).

Epithet: Conspicuous.   Hardiness 3.   April–May.

Intermediate between R. campylocarpum *ssp.* caloxanthum *and* R. campylocarpum *ssp.* campylocarpum, *the now sunk* R. telopeum *may be part of a cline between these two species or a hybrid.*

In leaf shape and size, this species is closer to *R. campylocarpum* var. *myiagrum* than to *R. campylocarpum* ssp. *campylocarpum*.

Leaves orbicular, ovate.

Corolla campanulate, citron, yellow or deep lemon-yellow.

Branchlets glandular.

R. campylocarpum ssp. *caloxanthum* in Knightshayes Court, Tiverton, Devon

Petiole glandular or eglandular.

Corolla campanulate, bright yellow, sulphur-yellow or creamy white.

Calyx not always glandular.

R. campylocarpum ssp. *caloxanthum* Telopeum Group KW6868 at Deer Dell

R. campylocarpum ssp. caloxanthum D.F.Chamb.
Panteumorphum Group
R. panteumorphum **Balf.f. & W.W.Sm.**

Shrub 0.6–3 m (2–10 ft) high, **branchlets** glandular.
Leaves oblong, oblong-elliptic or elliptic; **lamina** 3–7 cm long, 1.4–4.2 cm broad; **upper surface** dark green or pale green, shiny, glossy, glabrous; **underside** pale glaucous green, glabrous.
Petiole often glandular.
Inflorescence 4–8-flowered.
Pedicel glandular.
Calyx glandular.
Corolla campanulate, 3–3.6 cm long, yellow or pale yellow.
Ovary moderately or densely glandular.
SE Tibet, NW Yunnan. 3,400–4,500 m (11,000–14,600 ft).
Epithet: Altogether beautiful.   Hardiness 3.   April–May.

A possible hybrid of R. caloxanthum. *Rarely seen.*

R. souliei Franch.

Shrub or sometimes a **tree** 1–5 m (3–16 ft) high, **branchlets** sparsely glandular or eglandular.
Leaves ovate, ovate-elliptic or almost orbicular; **lamina** 3.5–8.2 cm long, 2.2–5 cm broad; **upper surface** dark green, shiny, glabrous; **underside** pale glaucous green, glabrous.
Inflorescence 5–9-flowered.
Pedicel glandular.
Corolla bowl- or saucer-shaped, 2.5–3.5 cm long, pink, rose, deep rose or white-tinged pink, without or rarely with a small crimson blotch.
Style glandular throughout to the tip.
W Sichuan. 3,000–4,550 m (9,900–12,000 ft).
FCC 1909 (Chelsea), flowers pale rose, deeper towards the margin.
FCC 1936 (Exbury) to a clone 'Exbury Pink', flowers of a deeper shade of pink.
FCC 1951 (Windsor) 'Windsor Park', flowers white with pink flush deepening at margins, and a small crimson blotch at the base of the three upper lobes.
Epithet: After Pere J.A. Soulie (1858–1905), French Foreign Missions, Tibet. Hardiness 3.   May–June.

A very beautiful lax truss of fairly small yellow flowers with long pedicels.

Ovary moderately or densely glandular.

Leaves often oblong.

R. campylocarpum ssp. *caloxanthium* Panteumorphum Group ('*R. panteumorphum*') at Deer Dell

Corolla bowl- or saucer-shaped, pink, rose, deep rose or white-tinged pink. The very long pedicels and open-faced flowers make this species easy to distinguish in flower.

Very similar to *R. wardii* var. *wardii* out of flower, but generally has longer petioles.

Style glandular throughout to the tip.

R. souliei 'Windsor Park' FCC at the Valley Gardens

R. wardii var. wardii

R. wardii W.W.Sm.
(*R. astrocalyx* Balf.f. & Forrest, *R. croceum* Balf.f. & W.W.Sm.)

Shrub or **tree** 0.6–7.6 m (2–25 ft) high, **branchlets** glandular or eglandular.
Leaves almost orbicular, ovate, oblong-elliptic or oblong; **lamina** 3–12 cm long, 2–6.5 cm broad; **upper surface** dark or pale green, glabrous; **underside** pale green or pale glaucous green, glabrous.
Inflorescence in trusses of 5–14 flowers.
Pedicel moderately glandular.
Calyx 0.4–1.2 cm long.
Corolla bowl- or saucer-shaped, 2.4–4 cm long, yellow, bright yellow, sulphur-yellow or lemon-yellow, with or without a small or fairly large crimson blotch.
Ovary densely glandular; **style** glandular throughout to the tip.
W and NW Yunnan, SE and E Tibet, SW Sichuan. 2,800–4,900 m (9,000–16,000 ft).
AM 1926 (Werrington) as *R. croceum*, flowers bright yellow.
AM 1926 (Werrington) as *R. astrocalyx*, flowers flat, clear lemon-yellow.
AM 1931 (Exbury), flowers bright yellow, flushed green (KW4170).
AM 1959 (Collingwood Ingram) 'Ellestee', flowers lemon-yellow with a crimson blotch (LS&T5679).
AM 1963 (Windsor) 'Meadow Pond', flowers primrose-yellow with a crimson blotch (LS&E15764).
AGM 1969.
Epithet: After F. Kingdon-Ward (1885–1958), collector and explorer.   Hardiness 3. May–June.

The late-flowering forms are usually from L&S collections.

R. wardii var. wardii Litiense Group
R. litiense Balf.f. & Forrest

Shrub or rarely tree 1–5 m (3–16 ft) high, branchlets glandular or eglandular.
Leaves oblong or oblong-oval; **lamina** 3–9.7 cm long, 1.8–4.4 cm broad; **upper surface** dark or paler green, glabrous; **underside** glabrous, markedly waxy glaucous.
Inflorescence in trusses of 5–7 flowers.
Pedicel glandular.
Calyx 5–8 mm long, glandular.
Corolla bowl- or saucer-shaped, 2–3.8 cm long, yellow or pale yellow, without blotch.
Ovary densely glandular; **style** glandular throughout to the tip.
NW and mid-W Yunnan. 2,800–4,000 m (9,000–13,000 ft).
AM 1931 (Exbury), flowers yellow.
FCC 1953 (Minterne).
Epithet: From Li-ti-ping, Yunnan.   Hardiness 3.   May–June.

Merged with R. wardii var. wardii on the grounds that intermediates exist between the leaf shapes, and between the markedly glaucous foliage of 'R. litiense' and the usually green foliage of R. wardii var. wardii. In cultivation, the extremes are so distinct that it is worth retaining the name.

Corolla bowl- or saucer-shaped, yellow.

Style glandular throughout to the tip.

Branchlets glandular or eglandular.

Pedicel moderately glandular.

R. wardii var. *wardii* KW5736 at the Valley Gardens

Opening corollas of *R. wardii* var. *wardii* at RBG Edinburgh (H.H. Davidian)

Leaves oblong or oblong-oval, underside markedly waxy glaucous.

Corolla bowl- or saucer-shaped, yellow or pale yellow, without blotch.

Style glandular throughout to the tip.

R. wardii var. *wardii* Litiense Group at Borde Hill

Ponticum
(Campylocarpa)

R. wardii var. puralbum D.F.Chamb.

R. puralbum Balf.f. & W.W.Sm.

Shrub 1.5–4.6 m (5–15 ft) high, **branchlets** sparsely glandular or eglandular.
Leaves ovate, ovate-oblong or oblong; **lamina** 5–12 cm long, 2.4–5 cm broad;
   upper surface dark green, glabrous; **underside** pale glaucous green, glabrous.
Inflorescence 5–8-flowered.
Pedicel glandular.
Corolla bowl- or saucer-shaped, 2–4 cm long, pure white or rarely ivory-white.
Ovary densely glandular; **style** glandular throughout to the tip.
NW Yunnan, SE Tibet. 3,400–4,300 m (11,000–14,000 ft).
Epithet: Pure white.   Hardiness 3.   May.

Subgenus *Hymenanthes*
Section *Ponticum* G.Don
Subsection *Falconera* Sleumer
Falconeri Series

Trees or shrubs.
Branchlets densely or moderately tomentose.
Leaves large.
Indumentum bistrate, thick, woolly or spongy.
Corolla usually ventricose, lobes 7–10, rarely 5.
Stamens usually 12–16.

For identification purposes, Subsection *Falconera* is divided into
the following groups:

Group A Species with winged petioles
Group B Species without winged petioles
   Group B1   Smaller species with smoother leaves.
                Not immediately obvious as members of
                    Subsect. *Falconera*.
                Corollas 5–7-lobed.
   Group B2   Larger-leaved species with rugulose leaves.
                Corollas usually 8-lobed.

Can only be distinguished from *R. wardii* var. *wardii* in flower.

Corolla usually saucer-shaped, pure white or rarely ivory-white.

Style glandular throughout to the tip.

R. wardii var. *puralbum* F10616 at Deer Dell

The three leaf types of Subsect. *Falconera* at Deer Dell. Left to right (2 leaves each): *R. arizelum* (Group B2), *R. coriaceum* (Group B1) and *R. semnoides* (Group A) (M.L.A. Robinson). Bar = 5 cm

R. basilicum Balf.f. & W.W.Sm.   Falconera Gp A

(*R. megaphyllum* Balf.f. & Forrest)

Shrub or **tree** 2.4–9 m (8–30 ft) high, **branchlets** densely or moderately tomentose with thin whitish or fawn tomentum.

Leaves obovate or oblong-obovate, 12–36 cm long, 6.5–18.6 cm broad; **upper surface** dark green, shiny; **underside** with thick spongy, continuous, bistrate, brown, fawn or rust-coloured broadly funnel-shaped hairs, later upper layer often falls off completely or partly, leaving an under layer of thin-plastered continuous whitish rosulate hairs.

Petiole flat above, with wings or ridges at the margins.

Inflorescence in trusses of 10–25 flowers.

Calyx minute, 1 mm long.

Corolla obliquely campanulate, ventricose, 3.5–4.8 cm long, yellow, creamy-white or pink.

W, mid-W and NW Yunnan, NE Upper Burma. 2,750–4,600 m (9,000–15,000 ft).

AM 1956 (Minterne), flowers pale whitish-cream (F24139).

Epithet: Royal.   Hardiness 3.   April–May.

R. rex *ssp.* gratum *(T.L.Ming) M.Y.Fang belongs here.*

R. preptum Balf.f. & Forrest   Falconera Gp A

Shrub or **tree** 1.8–9 m (6–30 ft) high, **branchlets** densely thin tomentose.

Leaves oblong-obovate or obovate; **lamina** 11–20 cm long, 4–9.5 cm broad, base slightly decurrent on the petiole; **upper surface** olive-green, matt, glabrous; **underside** covered with a thick or somewhat thin, woolly, continuous, bistrate indumentum of fawn, brown or dark brown hairs.

Petiole flat above, margins slightly winged or ridged.

Inflorescence in trusses of 20–22 flowers.

Corolla obliquely campanulate, ventricose, 3.5–4.6 cm long, yellowish-white, pale yellow or pale creamy-white, with crimson blotch.

Stamens 15–16.

NE Upper Burma, NW Yunnan. 3,400–3,700 m (11,000–12,000 ft).

Epithet: Distinguished.   Hardiness 3.   April–May.

This species is similar to R. arizelum. *Davidian placed all collections resembling* R. arizelum, *which have winged petioles, under* R. preptum. *However, KW5877 has winged petioles but was included under* R. arizelum. *KW5877 has an oblanceolate leaf, and closely resembles the recently introduced* **R. heatheriae**. *Other introductions of* R. preptum *are likely to be hybrids.*

Obovate or oblong-
obovate leaf
shape and matt
indumentum.

Petiole flat above,
with wings or
ridges.

Young growth with
a shiny mica-like
detersile indumentum
on the upper leaf
surface.

Leaf underside with
a thick continuous,
bistrate, brownish
indumentum of
broadly funnel-shaped
hairs; upper layer at
least partly detersile
leaving a whitish
plastered underlayer.

Corolla ventricose.

R. basilicum at RBG Edinburgh (Davidian)

New foliage of *R. basilicum* AC1853 (collected as *R. gratum*)
at Deer Dell

Leaf base slightly
decurrent on the
petiole, underside
with a woolly,
continuous, bistrate
indumentum of
brownish hairs.

Petiole flat above,
margins slightly
winged or ridged.

Indumentum thick or
somewhat thin.

Corolla obliquely
campanulate,
ventricose.

R. preptum at Benmore

R. rothschildii Davidian   Falconera Gp A

Shrub or **tree** 2.4–6 m (8–20 ft) high, **branchlets** with a thin, fawn or brown
  tomentum; **foliage-bud** large, deep crimson-purple.
Leaves oblong-obovate; **lamina** 21–36 cm long, 6–14 cm broad, base decurrent
  on the petiole; **upper surface** green, shiny, glabrous; **underside** with a thin,
  granular, discontinuous, bistrate indumentum of fawn or brown hairs.
Petiole 1.5–3 cm long, flat above, with wings at the margins, with a thin tomentum.
Inflorescence in trusses of 12–17 flowers; **flower-bud** deep crimson-purple.
Corolla campanulate, ventricose, 3.5–4.3 cm long, pale yellow or pale creamy-
  white with a crimson blotch.
NW and N Yunnan, Yunnan–Tibet border. 3,900 m (12,800 ft).
Epithet: After Lionel de Rothschild (1882–1942), Exbury.   Hardiness 3.
  April–May.

It is possible that this species is a speciating hybrid.

R. semnoides Tagg & Forrest   Falconera Gp A

Shrub or **tree** 2.4–6 m (8–20 ft) high, **branchlets** densely tomentose with a
  whitish, fawn or brown tomentum.
Leaves oblong-obovate, oblanceolate or obovate; **lamina** 10–26.5 cm long,
  4–11.5 cm broad, base decurrent on the petiole; **upper surface** olive-green,
  glabrous; **underside** covered with a thick, woolly, continuous, bistrate
  indumentum of hairs.
Petiole cylindrical or flat, margins with narrow wings, tomentose.
Inflorescence in trusses of 12–20 flowers.
Calyx minute, 1 mm long.
Corolla obliquely campanulate, ventricose, 3.5–5 cm long, white-flushed rose or
  creamy-white, with a crimson blotch.
Stamens 16.
SE Tibet. 3,700–4,000 m (12,000–13,000 ft).
Epithet: Resembling *R. semnum* (= *R. praestans*).   Hardiness 3.   April–May.

Rare in cultivation. Of uncertain status until more studies are made in the wild.

Flat winged petiole. Upper layer of the indumentum spotted or granular.

Leaf oblong or obovate.

Foliage-bud large, deep crimson-purple.

Flower-bud deep crimson purple.

Corolla campanulate, ventricose, pale yellow or pale creamy-white with a crimson blotch.

R. rothschildii F30528 at Deer Dell

New foliage of *R. rothschildii* at Deer Dell showing the granular indumentum and winged petiole

Petiole margins with narrow wings. Very similar to *R. basilicum*, but has a continuous uneven bistrate indumentum with a bobbly or tessellated appearance.

Leaf underside covered with a thick, woolly, continuous, bistrate yellowish fawn or brownish indumentum.

Corolla obliquely campanulate, ventricose, white-flushed rose or creamy-white, blotched.

R. semnoides F25639 at Deer Dell

Ponticum
(Falconera Group A)

R. coriaceum Franch. Falconera Gp B1

Shrub or **tree** 1.2–7.6 m (4–25 ft), **branchlets** densely tomentose with whitish tomentum.

Leaves oblanceolate or oblong-obovate; **lamina** 10–25 cm long, 3–8.7 cm broad; **upper surface** olive-green, glabrous; **underside** with a thick, continuous bistrate indumentum, upper layer fawn to creamy, later it falls off completely or partly, leaving an under layer of thin-plastered, continuous, fawn or silvery-white hairs.

Inflorescence in trusses of 10–20 flowers.

Pedicel densely tomentose.

Corolla campanulate, 2–4 cm long, white, white suffused with rose, or rose, with a crimson blotch, with or without crimson spots; lobes 5–7.

NW and mid-W Yunnan, SE Tibet. 3,000–4,200 m (10,000–13,600 ft).

AM 1953 (Windsor) 'Morocco', flowers white with a crimson blotch.

Epithet: Leathery.   Hardiness 3.   April–May.

R. galactinum Balf.f. & Tagg Falconera Gp B1

Shrub or **tree** 3.6–8 m (12–26 ft) high, **stem** and **branches** with rough bark, **branchlets** densely brown tomentose, eglandular.

Leaves oblong-obovate, oblong, oblong-lanceolate or oblanceolate; **lamina** 11–21.5 cm long, 3.8–8 cm broad; **upper surface** green or brownish-green, matt, glabrous; **underside** covered with a thick, soft, smooth, continuous, bistrate indumentum of brown or fawn hairs.

Inflorescence in trusses of 12–15 flowers.

Calyx minute, 1 mm long, tomentose, eglandular.

Corolla campanulate, 2.5–5 cm long, white or white suffused with pink, with deep crimson blotch and spots; lobes 7.

Stamens 14.

W Sichuan. 3,000–3,300 m (10,000–10,900 ft).

Epithet: Milky.   Hardiness 3.   April–May.

Branches thin (for this Subsection).

Leaves oblanceolate with a creamy indumentum, at least partly detersile.

Corolla campanulate-white, white flushed rose, or rose-blotched.

Ovary glabrous, glandular.

R. coriaceum at the Hiller Arboretum

Somewhat smooth, thick brown or fawn, uneven indumentum.

Corolla campanulate, white or white tinged pink, blotched, 7 lobes Ovary glabrous, eglandular.

R. galactinum at Benmore

R. arizelum Balf.f. & Forrest

(*R. rex* ssp. *arizelum* D.F.Chamb.)   **Falconera Gp B2**

Shrub or **tree** 1.8–7.6 m (6–25 ft) or rarely 12 m (40 ft) high, **branchlets** densely tomentose with a thin tomentum.

Leaves obovate, oblong-obovate, oval or sometimes oblanceolate; **lamina** 6.8–22.8 cm long, 3–11.8 cm broad; **upper surface** dark green, not rugulose, shiny; **underside** covered with a thick, velvety, continuous, bistrate indumentum; upper layer cinnamon or brown funnel-shaped hairs, under layer of thin-plastered rosulate hairs.

Inflorescence in trusses of 12–25 flowers.

Corolla obliquely campanulate, ventricose, 3–4.5 cm long, pale or deep yellow, creamy-white, white, pale or deep rose or pink, with or without a crimson blotch.

W, mid-W and NW Yunnan, NE Upper Burma, Assam, S and SE Tibet. 2,500–4,400 m (8,000–14,500 ft).

AM 1963 (Brodick) 'Brodick', flowers purple with a dark crimson blotch.

Epithet: Notable.   Hardiness 3.   April–May.

This species is closer to R. falconeri *than it is to* R. rex *by reasons of its bark being fawn (as opposed to dark brown), its funnel-shaped (as opposed to cup-shaped) indumentum hairs, its ventricose corollas and leaf buds that are similar to those of* R. falconeri. *Plants with crimson or carmine flowers have been called* **R. arizelum Rubicosum Group**.

R. falconeri ssp. falconeri Falconera Gp B2

R. falconeri Hook.

Shrub or **tree** 3–15 m (10–50 ft) high, **stem** and **branches** with smooth, brown, flaking bark; **branchlets** densely tomentose.

Leaves oblong-oval, obovate, elliptic or oval; **lamina** 12.5–32.8 cm long, 5–16.8 cm broad, **upper surface** green, rugulose, matt; **underside** with a thick, continuous, rust-coloured, dark brown or brown, bistrate indumentum of hairs, primary veins markedly raised, upper surface glabrous.

Inflorescence in trusses of 12–20 flowers.

Corolla obliquely campanulate, ventricose, 3.2–5 cm long, creamy-white, yellowish, sometimes white or pinkish, with a dark purple or pink blotch; lobes 8–10, usually 8.

Stamens 12–16.

Sikkim, Nepal, Bhutan, West Arunachal Pradesh, Assam. 2,500–3,400 m (8,000–11,000 ft).

AM 1922, flowers yellowish-white, with dark purple blotch.

Epithet: After H. Falconer (1808–1865), Supervisor, Saharanpur Gardens, India in 1832.   Hardiness 3.   April–May.

The indumentum can be somewhat detersile in older leaves.

Bark fawn.
Leaf upper surface shiny. Indumentum thick, velvety, brown or cinnamon.
Corolla ventricose, usually 8-lobed

R. arizelum at Deer Dell

R. arizelum Rubicosum Group at Deer Dell

Large ovalish leaves with prominent veins having little or no indumentum on them; upper surface rugulose, matt; underside with a thick, continuous, rust-coloured, dark brown or brown, bistrate indumentum, primary veins markedly raised.
Corolla obliquely campanulate, ventricose.

Smooth flaking bark.

R. falconeri ssp. *falconeri* at Wakehurst Place, Sussex

New foliage of *R. falconeri* ssp. *falconeri* at Deer Dell

R. falconeri ssp. eximium D.F.Chamb.

R. eximium Nutt.   **Falconera Gp B2**

Shrub or **tree** 1.5–9 m (5–30 ft); **stem** and **branches** with smooth, brown, flaking bark; **branchlets** densely tomentose with rust-coloured tomentum, eglandular.
Leaves oval or obovate-elliptic; **lamina** 12–45 cm long, 5–20 cm broad; **upper surface** dark green, rugulose, (**young leaves** up to one year old or more are densely hairy with rust-coloured woolly hairs); **underside** covered with a thick woolly continuous, bistrate indumentum of cinnamon or rust-coloured hairs; petiole densely tomentose as the branchlets.
Inflorescence in trusses of 12–20 flowers; pedicel floccose, rather densely glandular.
Calyx glandular.
Corolla obliquely campanulate, ventricose, 3–4.5 cm long, rose or pink, lobes 8–10.
Stamens 10–14.
Bhutan. 2,700–3,400 m (8,800–11,000 ft).
AM 1973 (Wakehurst), as a foliage plant.
Epithet: Excellent.   Hardiness 2–3.   April–May.

R. hodgsonii Hook.f.   **Falconera Gp B2**

Shrub or **tree** 3–12 m (10–40 ft) high; **stem** and **branches** with smooth, brown, flaking bark; **branchlets** with densely thin tomentum, eglandular.
Leaves oblong-obovate, obovate or oblong-elliptic; **lamina** 11–40 cm long, 4.8–15.6 cm broad; **upper surface** dark green, glabrous or sparsely tomentose appearing as spots (in **young leaves** up to one year old or more, tomentose with a thin white tomentum appearing as mica-like metallic spots); **underside** with a thick, continuous bistrate indumentum, later the upper layer falls off leaving an under layer of a thin-plastered, continuous indumentum of hairs.
Inflorescence compact rounded trusses of 12–20 flowers.
Calyx eglandular.
Corolla tubular-campanulate, 3–5.8 cm long, crimson, crimson-purple, purple or rose, lobes 7–8.
Stamens 14–18.
Nepal, Sikkim, Bhutan, Assam, S Tibet. 2,900–4,300 m (9,500–14,000 ft).
Epithet: After B.H. Hodgson, former East India Company Resident in Nepal.
   Hardiness 3.   April.

A *R. falconeri* with
persistent upper leaf
indumentum.

This and the rose
or pink flower
colour are the
distinguishing
features.

R. falconeri ssp. eximium at
Borde Hill

New foliage of *R. falconeri*
ssp. *eximium* at Deer Dell

Stems and branches
with smooth, brown,
flaking bark.

Foliage bud pointed
like the roof of a
pagoda.

Indumentum pale
fawn, detersile.

Corolla tubular-
campanulate,
crimson, crimson-
purple, purple
or rose.

Leaf upper surface
glabrous or sparsely
tomentose (in young
leaves up to one year
old, with mica-like
silvery spots).

New foliage of *R. hodgsonii* at
Deer Dell

R. hodgsonii at UBC
botanic garden,
Vancouver
(M.L.A. Robinson)

R. rex ssp. rex D.F.Chamb.   Falconera Gp B2
R. rex Levl.

Shrub or **tree** 3–12 m (10–40 ft) high, **stem** and **branches** with rough bark; **branchlets** densely tomentose with a thin, fawn, whitish or brown tomentum.

Leaves oblong-obovate or oblanceolate; **lamina** 20.8–37 cm long, 7.1–11.5 cm broad; **upper surface** dark green or pale green, shiny, glabrous; **underside** with a thick, rough, continuous, bistrate indumentum of hairs.

Inflorescence in trusses of 15–24 flowers.

Pedicel densely tomentose.

Corolla tubular-campanulate, 3.8–5 cm long, rose, pink, white or creamy-white, with or without a crimson blotch, and with or without crimson spots.

Stamens 14–16.

NE and NW Yunnan, SW Sichuan. 3,200–4,300 m (10,400–14,000 ft).

AM 1946 (Bodnant), flowers pale rose, with a small blotch (KW4509).

AM 1955 (Windsor) 'Quartz', flowers pale pink with a crimson blotch and spots (R 03800 USDA)

FCC 1935 (Embley Park), as *R. fictolacteum*, flowers white with a crimson blotch (KW4509).

Epithet: King.   Hardiness 3.   April–May.

R. rex ssp. fictolacteum D.F.Chamb.
R. fictolacteum Balf.   Falconera Gp B2

Shrub or **tree** 1.8–12 m (6–40 ft) high; **stem** and **branches** with rough bark; **branchlets** densely tomentose with a thin dark brown, fawn or whitish tomentum, eglandular.

Leaves oblong-obovate or oblanceolate; **lamina** 10–24.5 cm long, 3.5–8.5 cm broad; **upper surface** dark green, shiny, glabrous; **underside** covered with a thick, rough continuous, bistrate, indumentum of rust-coloured, dark brown or brown hairs.

Petiole densely tomentose as the branchlets, eglandular.

Inflorescence in trusses of 12–25 flowers.

Pedicel rather densely tomentose.

Calyx 7–8-lobed.

Corolla obliquely campanulate, 3–4 cm long, white, white-flushed rose, pink or creamy, often with a deep crimson blotch or spots; lobes 7–8.

Stamens 14–16.

Mid-W and NW Yunnan, NE Upper Burma, E and SE Tibet. 3,000–4,500 m (10,000–14,500 ft).

AM 1923 (Reuthe), flowers white with a crimson blotch.

AM 1953 (Minterne) 'Cherry Tip', flowers white with pink margins and a deep crimson blotch and numerous spots (R59255).

Epithet: False *R. lacteum*.   Hardiness 3.   April–May.

The variety R. fictolacteum *var.* miniforme *has smaller leaves and flowers. In cultivation, the subspecies* rex *and* fictolacteum *seem to merge in leaf and flower characteristics.*

Leaf upper surface dark green or pale green, shiny, often larger than those of *R. rex* ssp. *fictolacteum*.

Upper surface of the bistrate indumentum partially detersile.

Corolla tubular-campanulate, rose, pink, white or creamy-white, often blotched.

R. rex ssp. *rex* 'Quartz' AGM at Deer Dell

Leaf upper surface dark green, shiny. The leaves are often smaller and narrower than in *R. rex* ssp. *rex*, the indumentum is usually darker, and the upper layer is persistent.

Corolla obliquely campanulate, white, white-flushed rose, pink or creamy, often blotched.

R. rex ssp. *fictolacteum* at Deer Dell

Ponticum
(Falconera Group B2)

R. sinofalconeri Balf.f.   Falconera Gp B2

Tree 6 m (20 ft) high; **stem** and **branches** with smooth, brown flaking bark; **branchlets** densely tomentose with a thin, fawn tomentum, eglandular.

Leaves elliptic, obovate or oval; **lamina** coriaceous, 14.5–28 cm long, 8.6–17.2 cm broad, broadest at middle, apex rounded, base obtuse or rounded; **upper surface** pale green, slightly rugulose, matt, glabrous, **underside** with a thick, continuous, bistrate indumentum, upper layer light brown, funnel-shaped hairs, under layer of thin, plastered, whitish rosulate hairs.

Petiole 3–4.5 cm long, cylindrical, slightly, not winged or slightly winged at the margins, tomentose with a thin, fawn or whitish tomentum, eglandular.

Inflorescence about 12 flowers.

Pedicel 2.8–4 cm long, densely tomentose with a fawn tomentum, eglandular.

Corolla obliquely campanulate, ventricose, 4.6–5.5 cm long, yellow or pale yellow; lobes 8.

SE Yunnan and N Vietnam. 2,800 m (9,000 ft).

Epithet: Chinese *R. falconeri.*   Hardiness 2–3.   April–May.

R. sinofalconeri **(S. Hootman)**

Smooth, brown flaking bark. Distinguished from the brownish *R. falconeri* by the general silvery ambience, oval leaf shape, paler indumentum, the glandular calyx, and the usually glandular pedicel and ovary.

Corolla obliquely campanulate, ventricose, yellow or pale yellow.

New foliage of *R. sinofalconeri* at Deer Dell

Subgenus *Hymenanthes*
Section *Ponticum* G.Don
Subsection *Fortunea* Sleumer
Fortunei Series

Shrubs or trees.
Leaf underside almost always glabrous.
Corolla lobes 5–8.
Stamens usually 12–25.

The Edinburgh revision does not maintain the old Subseries in Subsection Fortunea. However, as identification out of flower is one of the aims here, it is useful to divide the Subsection by leaf shape.

	Group A	Group B	Group C
Leaf morphology	**Leaves lanceolate or oblanceolate; usually more than 3 times as long as broad. Leaves often large. Style glandular to the tip.*	*Leaves oblong, oblong-obovate or obovate.*	*Leaves orbicular or rounded, usually relatively small. Style usually not glandular to the tip.*
Taxa	R. asterochnoum, R. calophytum, R. praevernum, R. sutchuenense, R. × geraldii, R. davidii, R. decorum ssp. diaprepes, R. fortunei ssp. fortunei, R. fortunei ssp. discolor, R. fortunei ssp. discolor (Houlstonii Group), R. glanduliferum, R. griffithianum, R. huianum, R. maoerense, R. vernicosum.	R. decorum, R. decorum ssp. cordatum, R. oreodoxa, R. oreodoxa var. fargesii, R. praeteritum, R. serotinum.	R. hemsleyanum, R. orbiculare ssp. orbiculare, R. orbiculare ssp. cardiobasis, R. platypodum, R. yuefengense.

*Within Group A, R. calophytum, R. praevernum, R. sutchuenense and R. × geraldlii have comparatively large hanging leaves: R. asterochnoum is closely related to R. calophytum. The remainder of Group A is mainly the former Fortunei Subseries in which the corollas are almost always 7-lobed and usually fragrant.

Leaves of Subsect. *Fortunea* Group A at Deer Dell. Top to bottom (one leaf each): *R. davidii*, *R. praevernum*, *R. griffithianum*, *R. maoerense* and *R. asterochnoum*. Bar = 5 cm

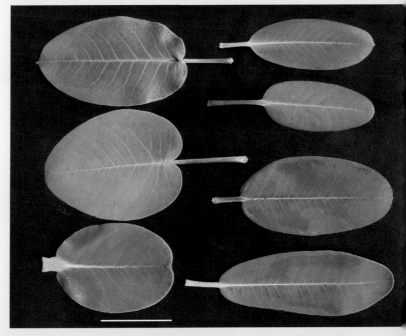

Leaves of Subsect. *Fortunea* Groups B and C at Deer Dell. Left column (top to bottom): *R. orbiculare* ssp. *cardiobasis*, *R. orbiculare* ssp. *orbiculare* and *R. platypodum*. Right column (top to bottom: *R. oreodoxa* ssp. *oreodoxa*, *R. oreodoxa* ssp. *fargesii*, *R. decorum* ssp. *cordatum* and *R. decorum* ssp. *decorum*. Bar = 5 cm

R. asterochnoum Diels.   Fortunea Gp A

Shrub or small **tree**.
Leaves oblanceolate; **lamina** leathery, 18–25 cm long, 5–6 cm broad, apex obtuse or rounded, base cuneate; **upper surface** glabrous; **underside** with closely or widely scattered stellate hairs, midrib prominent, hairy with stellate hairs.
Petiole 1.5–2.5 cm long, stout, flattened above, margins with narrow ridges, thinly floccose.
Inflorescence many-flowered, a somewhat lax racemose umbel of 15–20 flowers.
Pedicel somewhat slender, elongate, 3.5–5 cm long, not hairy or sparsely hairy with short hairs, eglandular.
Corolla funnel campanulate with narrow base, set oblique to pedicel 4–4.5 cm long, white-suffused rose.
Ovary glabrous.
Style glabrous.
C and S Sichuan, 3,000–3,600 m (10,000–12,000 ft).
Epithet: With star-like down.   Hardiness 4?   Flowering time unknown.

R. calophytum Franch.   Fortunea Gp A

Shrub or **tree** 4.5–15 m (15–49 ft) high, **branchlets** with a thin white tomentum.
Leaves oblanceolate; **lamina** 17–36 cm long, 4–8.5 cm broad, base decurrent on petiole; **upper surface** bright green; **underside** pale green, glabrous, midrib glabrous or tomentose.
Petiole flat above, margins with narrow wings.
Inflorescence in trusses of 15–30 flowers.
Pedicel somewhat stout, elongate, 2.6–8 cm long, often scarlet.
Calyx minute, 1 mm long.
Corolla widely campanulate, ventricose, 4.3–5.5 cm long, white, pink, rose or purple, with a dark blotch and with dark spots; lobes 5–7.
Stamens 15–22.
Style somewhat stout, stigma large, 5–8 mm across.
W Sichuan. 2,400–4,000 m (8,000–13,000 ft).
AM 1920 (Reuthe), flowers white, heavily suffused with pink.
FCC 1933 (South Lodge), flowers pale pink.
Epithet: Beautiful plant.   Hardiness 3.   March–April.

R. calophytum var. openshawianum *is smaller in all its parts, and said to be from a distinct population in South Sichuan.*

Differs from *R. calophytum* in that:

the underside of the leaves is hairy with stellate hairs, some of which persist;

the petiole is always winged;

the pedicels are longer.

Corolla funnel campanulate and set oblique to pedicel.

New foliage of *R. asterochnoum* at Deer Dell

Leaves often V-shaped in profile, oblanceolate, base decurrent on petiole.

Petiole flat above, margins with narrow wings.

Corolla widely campanulate, ventricose, with a dark blotch.

Stigma large.

R. calophytum at Deer Dell

Ponticum
(Fortunea Group A)

R. praevernum Hutch.   Fortunea Gp A

Shrub or **tree** 1.8–4.6 m (6–15 ft) high, branchlets tomentose with a thin, white
   tomentum or glabrous.
Leaves oblanceolate, oblong-obovate or oblong; **lamina** 10–21.5 cm long,
   2.5–7.5 cm broad, base slightly decurrent on petiole; **upper surface** olive-
   green or dark green, matt, glabrous; **underside** pale green, glabrous,
   midrib glabrous.
Petiole flat above, margins with very narrow wings or ridges.
Inflorescence in trusses of 8–10 flowers.
Pedicel stout.
Calyx glabrous, eglandular.
Corolla tubular-campanulate, 4–5 cm long, white or white suffused with pink or
   rose with a large dark purple or crimson blotch, lobes 5.

Hubei. 1,600–2,500 m (5,000–8,200 ft).

AM 1954 (Minterne), flowers white-flushed pink, with crimson blotch.
Epithet: Before the spring.   Hardiness 3.   February–April.

Note that the leaf size is smaller than that of R. calophytum *or* R. sutchuenense,
and that the corolla is not ventricose.

R. sutchuenense Franch.   Fortunea Gp A

Shrub or **tree** 2–6 m (6–20 ft) high, **branchlets** with a thin, white
   tomentum, eglandular.
Leaves oblong-obovate, oblanceolate or oblong-oval; **lamina** 9–28 cm long,
   3–7.5 cm broad; **upper surface** dark green or pale green, matt, glabrous;
   underside pale green, glabrous, midrib moderately or densely woolly or hairy in
   its entire length.
Petiole flat above, margins with very narrow wings or ridges.
Inflorescence in trusses of 8–20 flowers.
Pedicel stout, 1–3.6 cm long.
Calyx eglandular.
Corolla widely campanulate, 5–7.5 cm long, white suffused with pink, rose-pink
   or deep rose, with crimson spots, without a blotch at the base.
Stamens 13–17 or rarely 22.

W Hubei, E Sichuan. 1,400–2,500 m (4,500–8,000 ft).

Epithet: From Sichuan.

AM 1978 (Borde Hill) 'Seventh Heaven'.   Hardiness 3.   February–April.

A smaller version of
R. sutchuenense, but
with a glabrous
lower leaf midrib.
Corolla tubular-
campanulate with a
large dark purple or
crimson blotch.

R. praevernum at Deer Dell

Can be hard to
distinguish from
R. calophytum out
of flower:
R. sutchuenense
(and usually
R. praevernum)
have more
convex leaves
with a rounded
apiculate tip.
Leaf underside
glabrous, midrib
woolly or hairy
along its entire
length.
Corolla widely
campanulate
without a blotch at
the base.

R. sutchuenense W1232 'Seventh Heaven' AM at Deer Dell

R. × geraldii

Fortunea Gp A

R. sutchuenense var. *geraldii* Hutch.

W Hubei.
AM 1945 (South Lodge), flowers rose with a deep purple blotch.
AM 1971 (Sunte House) 'Sunte Rose', flowers rose with crimson blotch.
Epithet: After Gerald Loder, Wakehurst Place.   Hardiness 3.   February–April.

Generally considered to be a natural hybrid of R. sutchuenense × R. praevernum.
Plants identified as this often seem to be open-pollinated seedlings of one or other
of the related species, and the name is something of a 'catch all'!

R. davidii Franch.

Fortunea Gp A

Shrub or **tree** 1–6 m (3–20 ft) high, **branchlets** glabrous, eglandular.
Leaves oblanceolate; **lamina** coriaceous, 9–17 cm long, 2.3–4 cm broad, apex
   abruptly acuminate, base tapered, slightly decurrent on petiole; **upper surface**
   green, matt, glabrous; **underside** paler, glabrous.
Petiole 1.6–2.5 cm long, flat above, grooved, rounded below, margins with
   narrow wings, tomentose with a thin, white tomentum or glabrous, eglandular.
Inflorescence racemose, 8–10-flowered.
Pedicel somewhat stout, 2.5 cm long, glabrous, moderately or rather densely
   glandular.
Calyx densely glandular.
Corolla widely campanulate or tubular-campanulate, not ventricose, 4–5 cm long,
   rose, light purple, bright rosy-red or lilac, with or without purple spots, outside
   glabrous, sparsely glandular at the base of the tube or eglandular, inside
   glabrous; lobes 7–8.
Stamens 14–16.
Ovary glabrous, densely glandular.
Style glabrous or sparsely glandular.
W Sichuan. 1,300–4,000 m (4,300–13,200 ft).
Epithet: After L'Abbe Armand David (1826–1900), early collector in W China.
   Hardiness 3?   April–May.

Very rare in cultivation. Not introduced until about 1990: older plants are
incorrectly named.

Corolla with a large crimson or reddish blotch at the base.

R. × geraldii at Deer Dell

Leaves oblanceolate, long and narrow curving down to the acuminate tip. Long rhachis – usually 3–10 cm. Ovary densely glandular.

Corolla rose, light purple, bright rosy-red or lilac.

Foliage of *R. davidii* EN4213 at Deer Dell

R. davidii at Tregrehan (M.L.A. Robinson)

R. decorum ssp. diaprepes T.L.Ming

R. diaprepes Balf.f. & W.W.Sm. **Fortunea Gp A**

Shrub or **tree** 1.2–8 m (4–25 ft) high, **branchlets** pale green, often slightly glaucous, glabrous, eglandular.

Leaves oblong-elliptic, oblong-obovate, oblong or oblong-lanceolate; **lamina** 10–30 cm long, 4–12.5 cm broad; **upper surface** olive-green or dark green, wax-coated, becoming glossy when rubbed or heated; **underside** pale glaucous green, glabrous.

Petiole glabrous, eglandular.

Inflorescence in trusses of 7–10 flowers.

Pedicel glabrous, glandular.

Corolla widely funnel-campanulate, 6.5–10.5 cm long, fragrant, white or white-flushed rose; lobes 7–8.

Stamens 18–20, puberulous at the base.

Ovary densely glandular.

Style glandular throughout to the tip with white or yellow glands.

W and mid-W Yunnan, NE Upper Burma. 1,800–3,000 m (6,000–11,000 ft).

AM 1926 (Exbury), flowers white-tinged pink.

AM 1953 (Tower Court).

FCC 1974 (Windsor) 'Gargantua', very large white flowers with a green basal flush (F11958).

Epithet: Distinguished.   Hardiness 3.   June–July.

Larger than R. decorum *ssp.* decorum *in all its parts, and later flowering. Now reduced to a subspecies as intermediate forms between this and* R. decorum *ssp.* decorum *have been found in the wild.*

R. fortunei ssp. fortunei D.F.Chamb.

R. fortunei Lind. **Fortunea Gp A**

Shrub or **tree** 1.8–9 m (6–30 ft) high, **branchlets** glabrous.

Leaves obovate, oblong-obovate, oblong-elliptic or oblong; **lamina** 7–17 cm long, 2.8–8 cm broad; **upper surface** olive-green or dark green, matt, glabrous; **underside** pale glaucous green, glabrous, under magnification with minute glands and hairs.

Petiole purple, glabrous, eglandular.

Inflorescence in trusses of 6–12 flowers.

Pedicel moderately or densely glandular.

Calyx 6–7-lobed, glandular.

Corolla widely funnel-campanulate, 4–6 cm long, fragrant, pale rose, lilac or pink; lobes 7.

Stamens 14.

Ovary densely glandular.

Style glandular throughout to the tip with white or yellowish glands.

Chekiang, Kiangsi, Anhwei, Hunan, E China. 600–1,200 m (2,000 –4,000 ft).

Epithet: After Robert Fortune (1812–1880).   Hardiness 3.   May–June.

Widely distributed throughout China.

Larger, longer leaves and a larger corolla than *R. decorum* ssp. *decorum*.

Stamens 18–20.

Calyx sparsely or moderately glandular or eglandular.

Corolla widely funnel-campanulate, to 10.5 cm long, fragrant; lobes 7–8.

Leaves oblong-elliptic, oblong-obovate, oblong or oblong-lanceolate.

Style glandular throughout to the tip with white or yellow glands.

R. decorum ssp. *diaprepes* AC730 at Hindleap Lodge (M.L.A. Robinson)

Pure *R. fortunei* is rare in cultivation: check the leaf shape, stamens and corolla.

Leaves obovate, oblong-obovate, oblong-elliptic or oblong, underside pale glaucous green.

Corolla widely funnel-campanulate.

Stamens 14, glabrous.

Petiole purple.

Corolla fragrant, usually pale lilac.

Style glandular throughout to the tip with white or yellowish glands.

Foliage of *R. fortunei* ssp. *fortunei* at Deer Dell showing emergent foliage with flowers

R. fortunei ssp. discolor D.F.Chamb.

R. discolor Franch.
(*R. kwangfuense* Cun & Fang)

Fortunea Gp A

Shrub or **tree** 1.2–8 m (4–26 ft) high, branchlets pale green, glabrous.

Leaves oblanceolate, oblong or oblong-obovate; **lamina** 10–20 cm long, 2.4–7 cm broad; **upper surface** dark or pale green; **underside** pale glaucous green, glabrous.

Petiole glabrous.

Inflorescence in trusses of 6–10 flowers.

Pedicel glandular or eglandular.

Calyx margin glandular.

Corolla funnel-campanulate, 4.6–8 cm long, white, pink or rosy-pink, with or without a yellowish-green blotch, lobes 7.

Stamens 12–14.

Style glandular throughout to the tip with white or yellowish glands.

E Sichuan, W Hubei. 1,000–2,300 m (3,500–7,500 ft).

AM 1921 flowers white, tinged pink.

AM 1922 (Bodnant), flowers pale pink, with crimson blotch.

FCC 1922 (RBG Kew), flowers white, tinged pink.

AGM (1969) 2002.

Epithet: Of various colours.   Hardiness 3.   June–July.

Some doubt exists as to the exact taxonomic placing of this species: in cultivation, many hybrids shelter under this name.

R. fortunei ssp. discolor Houlstonii Group

R. houlstonii Hemsl. & Wils.

Fortunea Gp A

Shrub 1.5–4.6 m (5–15 ft) high, **branchlets** glabrous, eglandular.

Leaves oblanceolate, oblong-obovate or oblong; **lamina** 5.6–15 cm long, 2–5 cm broad; **upper surface** olive-green or dark green, matt, glabrous; **underside** pale glaucous green, glabrous.

Petiole 1.5–3.4 cm long, purple, glabrous, eglandular.

Inflorescence in trusses of 6–10 flowers.

Pedicel glandular.

Calyx 1–2 mm long.

Corolla, widely funnel-campanulate, 4–6 cm long, pink; lobes 7.

Stamens 14.

Style glandular throughout to the tip with white or yellowish glands.

Hubei, E Sichuan. 1,400–2,200 m (4,600–7,000 ft).

AM 1977 (Windsor) 'John R. Elcock'.

AM 1981 (Borde Hill) 'Random Harvest'.

Epithet: After G. Houlston, Chinese Imperial Maritime Customs, friend of E.H. Wilson.   Hardiness 3.   April–June.

The clone 'John R. Elcock' has oblong leaves, whereas the clone 'Random Harvest' has narrow oblanceolate leaves.

Similar to *R. fortunei*, but with narrower often recurved leaves.

In cultivation, it is late flowering. Stamens 12–14.

Corolla white, pink or rosy-pink.

Pedicel glandular or eglandular.

Style glandular throughout to the tip with white or yellowish glands.

Foliage of *R. fortunei* ssp. *discolor* PW38 at the Millais Nursery

Most similar to *R. fortunei* ssp. *discolor*. Flowers earlier than ssp. *discolor*, the flowers white or pink, not always fragrant.

R. fortunei ssp. *discolor* Houlstonii Group (Davidian)

R. fortunei ssp. *discolor* Houlstonii Group W648A 'Random Harvest' AM at Deer Dell

Ponticum
(Fortunea Group A)

R. glanduliferum Franch.   Fortunea Gp A

Shrub to 2 m or more; **branchlets** glabrous, sparsely glandular with short-stalked glands.

Leaves oblong-lanceolate or oblanceolate, 12–20 cm long, 2–6 cm broad, apex shortly acuminate, base tapered; **upper surface** dark green; glabrous; petiole **underside** pale glaucous green, glabrous.

Petiole 2–2.5 cm long, stout, rounded above, round, margins without wings, glabrous.

Inflorescence 5–18-flowered.

Pedicel 2–4 cm long, glabrous, densely setulose-glandular.

Corolla funnel-campanulate, 5–9 cm long, about 4.8–8.0 cm across, white shading to pale yellow in the throat with or without some orange markings; outside glabrous; densely setulose-glandular; inside puberulous at the base of the tube; lobes 7–8.

Stamens 14–16; glabrous.

Ovary conoid, 4 mm long, glabrous, densely glandular.

Style slender, glabrous, rather densely glandular throughout to the tip.

NE Yunnan, 2,200 m (7,250 ft).

Epithet: Gland-bearing.   Hardiness unknown.   Flowering time unknown.

It is difficult to be certain of its identity out of flower.

R. griffithianum Wight.   Fortunea Gp A
(*R. aucklandii* Hook.f.)

Shrub or **tree** 1–15 m (3–50 ft) high; **stem** and **branches** with smooth, brown, flaking bark; **branchlets** glabrous, eglandular.

Leaves oblong, oblong-elliptic or oblong-oval; **lamina** 10–30 cm long, 3.5–10 cm broad; **upper surface** bright green or pale green, matt, glabrous; **underside** pale glaucous green, glabrous.

Petiole 2–5 cm long, glabrous.

Inflorescence a lax racemose umbel of 3–6 flowers, **rhachis** 3.2–8 cm long.

Corolla widely campanulate, 5–8.3 cm long, 6.5–15 cm across, white or white suffused with pink, 5-lobed, fragrant.

Stamens 12–18.

Style glandular throughout to the tip.

E Nepal, Sikkim, Bhutan, Assam. 1,800–2,900 m (6,000–9,500 ft).

FCC 1866 (Standish, Ascot) as *R. griffithia*.

Epithet: After W. Griffith (1810–1845), a former superintendent of Calcutta Botanic Garden.   Hardiness 2–3.   April–May.

Formerly in its own Subseries. Note the variation in height. In flower, only likely to be confused with its hybrids. The former name R. aucklandii *still occurs occasionally.*

Corolla outside densely setulose-glandular.

Glandular pedicels.

Less oblong and longer leaves than *R. fortunei*, and without purple petioles on the specimen examined.

Densely glandular pedicels and calyx

R. glanduliferum, collected by E. Millais, at Loder Plants, Sussex (M.L.A. Robinson)

R. glanduliferum detail showing the glandular corolla (M.L.A. Robinson)

Corolla 5-lobed.

Rhachis long, inflorescence lax: (3–6) flowers in the truss.

The bark is distinct – a shiny mahogany not usually transmitted to its hybrids.

Leaf undersides with a pale yellow 'wash'.

R. griffithianum at Meerkerk Gardens, Oregon (M.L.A. Robinson)

Ponticum
(Fortunea Group A)

R. huianum Fang.   Fortunea Gp A

Shrub or **tree** 2–9 m (6–30 ft) high, **branchlets** glabrous, eglandular.

Leaves oblanceolate; **lamina** coriaceous, 8.5–14.5 cm long, 1.8–3.5 cm broad, apex acuminate, base tapered, slightly decurrent on the petiole; **upper surface** pale green, matt, glabrous; **underside** paler, glabrous.

Petiole 1.5–3 cm long, flat above, grooved, rounded below, margins with very narrow wings, glabrous, eglandular.

Inflorescence 6–12-flowered.

Pedicel 2–3.4 cm long, glabrous, sparsely glandular with short-stalked glands or eglandular.

Calyx 7-lobed, large, coloured, 4–10 mm long, outside glabrous, margin glandular.

Corolla widely campanulate, not ventricose, 3.5–5.9 cm long, 4–6.8 cm across, lilac, deep lilac, light violet, pale red, purplish or pale rose, outside glabrous, eglandular, inside glabrous; lobes 7.

Stamens 12–14.

Ovary conoid, densely glandular.

Style slender, glabrous, glandular throughout to the tip.

S and SE Sichuan, NE Yunnan, NE Guizhou. 1,000–2,700 m (3,250–9,000 ft).
Epithet: After Professor Hu, China.   Hardiness 3?   April

R. maoerense W.P.Fang & G.Z.Li   Fortunea Gp A

Tree 8–12 m (25–40 ft); **branchlets** purplish at first, maturing to grey, glabrous.

Leaves oblanceolate, rarely obovate, apex nearly obtuse and cuspidate, base cuneate, 14–16 cm long, 4–5 cm broad, with 19–21 pairs of veins; **underside** pale green, glabrous.

Petiole pale purple, glabrous.

Inflorescence 9–12-flowered.

Pedicel 3–4.5 cm long; purplish, markedly glandular.

Corolla campanulate, 6–6.5 cm long, pale pink fading to white, lobes 7, reflexed.

Stamens 14–17, filaments glabrous.

Calyx persistent, glandular.

Ovary densely glandular.

Style sparsely glandular.

NE Guangxi. 1,800–1,900 m (6,000–6,300 ft).
Epithet: From Maoershan, Guangxi.   Hardiness 3?   May.

Recently introduced from Maoershan by Alan Clark. On plants in cultivation, the young growth is maroon or deep red, but this may well be a juvenile characteristic.

Resembles
R. discolor in
general appearance,
but differs in that:
the leaves are
narrower with an
acuminate apex;
the leaf-base is
slightly decurrent on
the petiole with very
narrow wings;
the calyx is often
larger, with
persistent lobes; and
the corolla is more
deeply coloured.

R. huianum showing the calyces
(S. Hootman)

R. huianum at the RSF
showing the characteristic
hanging foliage
(M.L.A. Robinson)

Leaves similar
to those of
R. glanduliferum,
but more noticeably
oblanceolate.
Petiole pale purple.
Pedicel noticeably
glandular.
Corolla pale pink.
Style sparsely
glandular.

R. maoerense AC4207 at Deer Dell

R. vernicosum Franch. Fortunea Gp A

Shrub or **tree** 1–7.6 m (3–25 ft) high, **branchlets** glabrous.

Leaves oblong-elliptic, elliptic, oblong-oval or oval; **lamina** 4–13.5 cm long, 2–5.6 cm broad; **upper surface** olive-green or dark green, matt, wax-covered becoming glossy when rubbed or heated, glabrous; **underside** pale glaucous green, glabrous.

Inflorescence in trusses of 5–10 flowers.

Pedicel moderately or rather densely glandular.

Calyx densely glandular.

Corolla widely funnel-campanulate, 3.8–5.4 cm long, white, deep or pale rose, rose-lavender or purplish-red, with or without crimson spots; lobes 7.

Stamens 12–14.

Style glandular throughout to the tip with dark red or crimson glands.

W and SW Sichuan, NW and mid-W Yunnan, SE Tibet. 2,750–4,500 m (9,000–15,000 ft).

AM 1964 (Benmore) 'Loch Eck', flowers pure white.

AM 1976 (Bodnant) 'Spring Sonnet', flowers white, flushed red-purple, with red spots (R11408).

Epithet: Varnished.   Hardiness 3.   April–May.

Davidian distinguishes the following four forms as distinct, but these probably fit within the spectrum of variation of the species: identification of them can be a source of amusement among experts, but they have little botanical significance.

R. vernicosum Franch. f. araliiforme (Balf.f. & Forrest) Tagg.

Small, narrow, oblong **leaves**.
NW Yunnan.
Epithet: Aralia-like.   Hardiness 3.   April–May.

R. vernicosum Franch. f. euanthum (Balf.f. & W.W.Sm.) Tagg.

Oblong-oval or oval **leaves**. **Flowers** large, inside white in the lower half, pink above with crimson spots.
NW Yunnan.
Epithet: With beautiful flowers.   Hardiness 3.   April–May.

R. vernicosum Franch. f. rhantum (Balf.f. & W.W.Sm.) Tagg.

Smaller **leaves**, usually of thinner texture. **Corolla** small, 4–4.5 cm long, usually pale rose with crimson spots.
NW Yunnan.
Epithet: Sprinkled.   Hardiness 3.   April–May.

R. vernicosum Franch. f. sheltonae (Hemsl. & Wils.) Tagg.

Smaller **leaves**. Small **corolla** 3–3.5 cm long. **Capsule** small.
NW Yunnan.
Epithet: After Mrs. Shelton, wife of Dr. Shelton, of the Chinese Missions.
Hardiness 3.   April–May.

The dark red glands on the style are unusual in Subsection *Fortunea*. Easily *confused* with *R. decorum* out of flower: the average leaf size of *R. vernicosum* is somewhat smaller.

R. vernicosum at Deer Dell

Small, narrow, oblong leaves.

R. vernicosum f. *araliiforme* in the Valley Gardens

Ponticum
(Fortunea Group A)

R. decorum ssp. decorum Franch.   Fortunea Gp B

R. decorum Franch.

Shrub or **tree** 1–9 m (3–30 ft) high, **branchlets** glabrous, eglandular.

Leaves oblong, oblong-obovate, obovate or oblong-oval; **lamina** 5–16.5 cm long, 2–7 cm broad; **upper surface** olive-green or bright green, wax-coated, becoming glossy when rubbed or heated; **underside** pale glaucous green, glabrous.

Petiole often glaucous, glabrous, eglandular.

Inflorescence in trusses of 8–12 flowers.

Pedicel moderately or rather densely glandular.

Calyx densely moderately or densely glandular.

Corolla widely funnel-campanulate, 3.5–6 cm long, fragrant, white, white suffused with rose, pink or rose, lobes 6–8.

Stamens 12–16.

Style glandular throughout to the tip with white or yellowish glands.

W and SW Sichuan, SE Tibet, NW and S Yunnan, NE Upper Burma, N Vietnam, 1,800–4,500 m (6,000–15,000 ft).

AM 1923 (Nymans) 'Mrs. Messel', flowers pure white.

Epithet: Ornamental.   Hardiness 3.   May–June.

As would be expected from its wide geographical distibution, this species is very varied in height and growth. Note the distinctive fragrance – different from that of R. fortunei *and its hybrids, and the extended flowering season.*

R. decorum ssp. cordatum W.K.Hu   Fortunea Gp B

Similar to *R. decorum* ssp. *decorum* but with shorter, broader **leaves** and with a shorter **petiole** and cordate bases.

Leaves smaller but broader in comparison to their length than those of *R. decorum* ssp. *diaprepes*.

Calyx densely moderately or densely glandular. Stamens 12–16.

Corolla widely funnel-campanulate, to 6 cm long, fragrant, white, lobes 6–8.

Leaves oblong, oblong-obovate, obovate or oblong-oval.

Style glandular throughout to the tip with white or yellowish glands.

R. decorum ssp. *decorum* 'Greeneye' at Deer Dell

R. decorum ssp. *cordatum* AC1105 at Deer Dell

Ponticum
(Fortunea Group B)

R. oreodoxa var. oreodoxa Franch.

R. oreodoxa Franch.   **Fortunea Gp B**

Shrub or small **tree** 1.5–5 m (5–23 ft) high, **branchlets** often with a thin whitish tomentum.

Leaves oblong or oblong-oval; **lamina** 5–10.5 cm long, 1.8–4 cm broad; **upper surface** bright green or olive-green, matt, glabrous; **underside** pale glaucous papillate, glabrous.

Inflorescence in trusses of 5–12 flowers.

Pedicel densely or sparsely glandular.

Calyx minute, 1 mm long, glandular.

Corolla tubular-campanulate, 2.6–4.8 cm long, pale rose or pink or rarely carmine, with or without purple spots; lobes 7–8.

Stamens 14; glabrous or puberulous at the base.

Ovary glabrous, eglandular.

W Sichuan, Hubei, Gansu. 2,300–3,800 m (7,500–12,500 ft).

AM 1937 (Exbury), flowers pale rose.

Epithet: Glory of the mountains.   Hardiness 3.   February–April.

Similar to R. fargesii, *but has narrower usually oblong leaves about twice as long as broad: both have eglandular styles. As the leaf shapes occasionally intergrade,* **the ovary must be examined to make a proper distinction from** R. fargesii.

R. oreodoxa var. fargesii D.F.Chamb.

R. fargesii Franch.   **Fortunea Gp B**
(*R. erubescens* Hutch.)

Shrub or small **tree** 1–6 m (3–20 ft) high, **branchlets** glabrous.

Leaves oblong, oblong-oval, oblong-elliptic, ovate or rounded; **lamina** 5–8.2 cm, long, 2–4.8 cm broad; **upper surface** dark green or olive-green, matt, glabrous; **underside** pale glaucous papillate, glabrous.

Petiole glabrous, eglandular.

Inflorescence in trusses of 5–10 flowers.

Pedicel densely or moderately glandular.

Calyx glandular.

Corolla widely campanulate or tubular-campanulate, 2.5–4.5 cm long, white, pink, rose, deep rose or deep rosy-red, lobes 7, rarely 6.

Stamens 13–14, rarely 12; glabrous or puberulous at the base.

Ovary densely glandular.

Style glabrous, eglandular.

E and SW Sichuan, W Hubei, NW Yunnan. 2,000–4,000 m (6,500–13,000 ft).

AM 1926 (Wakehurst), flowers rose-pink with crimson spots.

AM 1969 (Bodnant) 'Budget Farthing', flowers white suffused with purple.

Epithet: After Pere Farges (1844–1912), French foreign missions in NW Sichuan. Hardiness 3.   February–April.

The ovary must be examined to make a proper identification.

R. oreodoxa *var.* shensiense *D.F.Chamb. has rufous-tomentose pedicels.*

Ovary glabrous, eglandular.

Leaves usually oblong, about twice as long as broad.

Corolla lobes 7–8. Style glabrous, eglandular.

R. oreodoxa var. *oreodoxa* at Nymans Sussex

Ovary densely glandular.

Usually broader leaves than ssp. *oreodoxa*.

Corolla lobes 6–7. Style glabrous, eglandular.

R. oreodoxa var. *fargesii* at RBG Edinburgh

Ponticum
(Fortunea Group B)

R. praeteritum Hutch.   Fortunea Gp B

Shrub 3 m (10 ft) high or more, **branchlets** rather densely floccose.
Leaves oblong, oblong-oval or oblong-elliptic; **lamina** 4.5–9.5 cm long, 2–4.1 cm broad; **upper surface** dark green, matt, glabrous; **underside** pale green, glabrous.
Calyx minute, 1 mm long, eglandular.
Corolla widely campanulate, 3–4 cm long, pink or white suffused with pink, with a few carmine spots; lobes 5.
Stamens 10.
Ovary glabrous, eglandular.

W Sichuan.

Epithet: Passed over.   Hardiness 3.   February–April.

Very rare in cultivation. Similar to R. oreodoxa *but has a 5-lobed corolla. Ignored by recent authors until 'The Flora of China' 2005.*

R. serotinum Hutch.   Fortunea Gp B

Shrub 2.5–3 m (8–10 ft), branchlets glabrous.
Leaves oblong-obovate, oblong, oblong-elliptic or oblong-oval; **lamina** 8–16 cm long, 4–7 cm broad; base rounded, obtuse or slightly cordulate, not decurrent; **upper surface** olive-green, matt, wax-coated, becoming glossy when rubbed or heated, glabrous; **underside** pale glaucous green, glabrous, under magnification punctulate with minute hairs.
Inflorescence in trusses of 7–8 flowers.
Pedicel glandular with short-stalked or almost sessile glands.
Calyx minute, 1 mm long, glandular.
Corolla widely funnel-campanulate, 4.8–6 cm long, 7-lobed, fragrant, white slightly flushed rose, with a red blotch, and with red spots, outside glandular.
Stamens 14–16, puberulous at the base.
Ovary densely glandular with short or almost sessile glands.
Style glandular throughout to the tip with white or yellowish glands.

W Yunnan.

AM (RBG Kew), flowers white, flushed rose, with a blotch.

Epithet: Late, i.e. autumnal.   Hardiness 3.   July–September, sometimes June.

This species appears to be a late-flowering form of R. decorum *with red throat markings. It has never been recollected in the wild, the original being received from Paris as a seedling from a collection by Delavay. It has been stated that the recently introduced C&H7189 is also this species.*
See R. hemsleyanum *and the subsequent entry.*

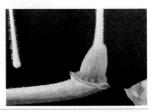

Stamen, style,
ovary and calyx of
R. serotinum from
Hawkwood West
(M.L.A. Robinson)

Branchlets sparsely floccose.
Leaves very similar to *R. oreodoxa*.
Corolla 5-lobed.
Ovary glabrous, eglandular.

R. praeteritum in the Valley Gardens (M.L.A. Robinson)

In cultivation, and out of flower, indistinguishable from a somewhat glaucous-leaved *R. decorum*.
Corolla with a red blotch.
Very late flowering: July–September.

R. serotinum at Hawkwood West

R. hemsleyanum Wils.   Fortunea Gp C

(*R. chengianum* Fang)

Shrub or **tree** 2–8 m (7–26 ft) high, **branchlets** glabrous, often glandular.

Leaves oblong-oval, broadly oblong or oval; **lamina** 10–20 cm long, 4–10.2 cm broad, base auricled, deeply cordate; **upper surface** green, glabrous; **underside** pale glaucous green, glabrous, minutely papillate.

Petiole often glandular.

Inflorescence in trusses of 8–10 flowers.

Pedicel rather densely setulose-glandular.

Calyx rather densely setulose-glandular.

Corolla widely funnel-campanulate, 5–6 cm long, 7-lobed, fragrant, white or white suffused with pink, lobes 7.

Stamens 10–14.

Ovary densely glandular.

Style glandular throughout to the tip with white glands.

W Sichuan. 1,100–2,000 m (3,600–6,600 ft).

Epithet: After W.B. Hemsley (1843–1924), English botanist.   Hardiness 3. May–July.

There are recent related introductions with narrower undulate leaves from N Vietnam and Yunnan. C&H1789 is similar.

C&H7189, a probable new subspecies

The leaf has an auricled and deeply cordate base with a corrugated lamina. Late flowering. Stamens 10–14, glabrous.

R. hemsleyanum from Mt Emei at Hindleap Lodge (M.L.A. Robinson)

Recent introductions related to *R. hemsleyanum*

C&H7189, AC1239, and a taxon from N Vietnam (herbarium material AC368, AC436, AC477 and KR2015).

Three taxa with similar leaf shapes to *R. hemsleyanum* were introduced in the 1990s – two from SE Yunnan and one from N Vietnam. They are of superficially very similar appearance, **all having oblong leaves with undulate margins, a markedly cordate leaf base and late, white, scented flowers**. However, they differ in botanical detail. For example, the Vietnam collection is glandular on the leaf underside and the corolla has 9 lobes and 20 puberulous stamens (*R. hemsleyanum* has 10–14 glabrous stamens and a 7-lobed corolla). C&H7189 is very fast growing and has 7 lobes and 18 puberulous stamens. It may be found in gardens labelled as *R. serotinum*.

Both of the Yunnan introductions and Vietnam taxon are significant introductions by Alan Clark and by P.A. and K.N.E. Cox.

They differ from both R. hemsleyanum *and the original introduction of* R. serotinum *(see page 120), but their status awaits clarification that can only be provided in a formal description by a botanist.*

AC436 a taxon with 9 lobes and 20 stamens, related to *R. hemsleyanum*, at Deer Dell

R. serotinum, R. decorum and *R. hemsleyanum* foliage.
Top row (left to right): all *R. hemsleyanum*: aff. AC436 from
N Vietnam; aff. AC1239 from Wensham, Yunnan; and from Mt
Emei, Sichuan. Bottom row (left to right): CH7189 listed as
R. serotinum; *R. serotinum* from the original introduction; and
R. decorum. Bar = 5 cm

R. orbiculare ssp. orbiculare Decne.

R. orbiculare Decne.   **Fortunea Gp C**

Compact rounded **shrub** or **tree** 1.5–3 m (5–10 ft) high or more, **branchlets**
   bright green, glaucous, glabrous, eglandular.
Leaves orbicular or ovate-orbicular; **lamina** 4–10 cm long, 4–8 cm broad, base
   deeply cordate, auricled, auricles usually overlapping; **upper surface** bright
   green, matt, glabrous; **underside** glaucous, papillate, glabrous, under
   magnification punctulate with minute glands.
Inflorescence in trusses of 7–16 flowers.
Pedicel often glandular.
Corolla widely campanulate with a broad base, 3.2–4 cm long, rose, rosy-red,
   deep red or reddish-purple, lobes 7.
Stamens 14.

W Sichuan. Border between Sichuan and Yunnan. 2,300–4,000 m (7,500–13,000 ft).
AM 1922 (Bodnant), flowers rose-pink.
Epithet: Circular, alluding to the leaves.   Hardiness 3.   April–May.

R. orbiculare ssp. cardiobasis D.F.Chamb.

R. cardiobasis Sleumer   **Fortunea Gp C**

Shrub 1.2–3 m (4–10 ft) high, **branchlets** glabrous.
Leaves ovate-elliptic or ovate-orbicular; **lamina** 8–12 cm long, 5–9 cm broad;
   upper surface bright green, matt, glabrous, base cordulate or deeply
   cordate, slightly or moderately auricled, auricles not overlapping; **underside**
   papillate, glabrous.
Petiole glabrous.
Inflorescence in trusses of 7–13 flowers.
Corolla funnel-campanulate, base narrow, 3.8–5.5 cm long, white or red, lobes 7.
Stamens 12–14.
Style glandular throughout to the tip.
Guangxi, S China. 1,650 m (5,400 ft).
Epithet: With a heart-shaped base.   Hardiness 3.   April–May.

Recent introductions seem to confirm this as a distinct subspecies.

Orbicular or ovate-orbicular leaves.
Leaf base usually with auricles overlapping.
Corolla lobes 7.
Style eglandular.

R. orbiculare ssp. *orbiculare* at Deer Dell

Growth habit of *R. orbiculare* ssp. *orbiculare* at Deer Dell

Leaves ovate-elliptic or ovate-orbicular. Base cordulate or deeply cordate, slightly or moderately auricled, auricles not overlapping.
Corolla lobes 7.
Style glandular throughout to the tip.

R. orbiculare ssp. *cardiobasis* at Deer Dell

Ponticum
(Fortunea Group C)

R. platypodum Diels. Fortunea Gp C

Shrub or **tree** 2–8 m (6–26 ft) high, **branchlets** glabrous, eglandular.
Leaves broadly elliptic, rounded, or oval; **lamina** coriaceous, 7.6–12.5 cm long,
5–7.4 cm broad, apex rounded, base obtuse or rounded, broadly decurrent on
petiole; **upper surface** dark green, minutely rugulose, matt glabrous; **underside**
pale green, glabrous, under magnification punctulate with minute glands and
sometimes with vestiges of juvenile hairs, midrib prominent, glabrous.
Petiole 1–2 cm long, 0.6–2 cm broad, green, flat above, grooved or not grooved;
margins with broad wings, glabrous, eglandular.
Inflorescence 10–15-flowered.
Pedicel 2–3.8 cm long, glabrous, eglandular or glandular.
Corolla funnel-campanulate, 3.6–5.6 cm long, 5.3–7 cm across, pinkish-red or
pink, without spots; outside glabrous, eglandular; inside glabrous or rather
densely puberulous at the base of the tube; lobes 7.
Style slender, glabrous, glandular throughout to the tip.
SE Sichuan. 1,900–2,200 m (6,000–7,000 ft).   Hardiness 3.   April–May.
Epithet: Broad-stalked.

Recently introduced and still rare in cultivation.

R. yuefengense G.Z.Li Fortunea Gp C

Shrub 0.5–1.5 m (1.5–5 ft) high, branches red-brown.
Leaves glabous, oval or broadly elliptic, 4.5–9.5 cm long, 4–9 cm broad; **upper
surface** shiny, somewhat convex at the margins, apex rounded, mucronate,
base obtuse or cordate; **underside** glabrous.
Petiole flattened above, rounded below, 1–3 cm long, slightly winged near the
lamina.
Inflorescence 5–10-flowered, **rhachis** long, glabrous.
Pedicel 2–3.8 cm long, glabrous at maturity.
Calyx small, glandular.
Corolla funnel-campanulate, 4–4.8 cm long, very pale purple, white at base
within, lobes 7, initially glandular.
Stamens 13–15.
Ovary densely glandular.
Style at least sparsely glandular throughout to the tip.
Guangxi. 1,800–2,150 m (5,950–7,100 ft).   Hardiness 3.   May–June.
Epithet: From Yuefeng, Guangxi.

A newly introduced species with a habit of growth that is wider than tall.

Wide oval leaves decurrent onto the petiole.
Petiole flat above and with very broad wings.

R. platypodum foliage at Deer Dell

Distinctively round leaves, usually with a cordate base.
Petiole flattened above but significantly longer than in *R. platypodum*.

R. yuefengense at Deer Dell

Ponticum
(Fortunea Group C)

Subgenus *Hymenanthes*
Section *Ponticum* G. Don
Subsection *Fulva* Sleumer
Fulvum Series

Branchlets densely or moderately fulvous or greyish tomentose.
Leaves oblanceolate, oblong obovate or sometimes obovate.
Indumentum coarsely or finely granular.
Inflorescence usually compact and globose.
Calyx small.
Stamens 10.

R. fulvum ssp. fulvoides D.F.Chamb.
R. fulvoides Balf.f. & Forrest

Shrub or **tree** 1.2–7.6 m (4–25 ft) high, **branchlets** densely or moderately
tomentose with a thin tomentum, eglandular.
Leaves oblanceolate or oblong-obovate; **lamina** 6–26.5 cm long, 2–8.6 cm
broad; **upper surface** pale green or olive-green, matt, glabrous; **underside**
covered with a coarsely granular, yellowish, fawn or brown, continuous, bistrate
indumentum of hairs, upper layer capitellate (mop-like).
Inflorescence in compact, globose trusses of 8–20 flowers.
Calyx glabrous, eglandular.
Corolla campanulate, 2.6–4.4 cm long, white, white flushed with rose, pink or
deep rose, with or without crimson blotch and spots.
Ovary slender, glabrous, eglandular.
Mid-W, W and NW Yunnan, SE and E Tibet, Assam. 3,200–4,400 m
(10,500–14,500 ft).
Epithet: Resembling *R. fulvum*.   Hardiness 3.   March–April.

Intermediate in the wild between R. fulvum *ssp.* fulvum *and* R. uvariifolium.

Leaves of Subsect. *Fulva* at Deer Dell. Left to right (two leaves each): *R. fulvum* ssp. *fulvoides*, *R. uvariifolium* var. *uvariifolium* and *R. fulvum* ssp. *fulvum*. Bar = 5 cm

Very similar to *R. fulvum* ssp. *fulvum* but with granular (spotted) yellowish, fawn or brown indumentum. Leaf upper surface matt, somewhat regulose. Leaves often narrower than *R. fulvum* ssp. *fulvum*. Corolla white, pink, rose, usually with a crimson blotch. Inflorescence compact, globose.

R. fulvum ssp. *fulvoides* R10931 in the Valley Gardens

R. fulvum ssp. *fulvoides* F8989 at Deer Dell

New growth of *R. fulvum* ssp. *fulvoides* in the Valley Gardens

Ponticum
(Fulva)

R. fulvum ssp. fulvum Balf.f. & W.W.Sm.

R. fulvum Balf.f. & W.W.Sm.

Shrub or **tree** 1–12 m (3–40 ft) high, **branchlets** densely or moderately
tomentose with a thin tomentum, eglandular.

Leaves oblanceolate or oblong-obovate; **lamina** 8–21 cm long, 2.8–8.5 cm
broad; **upper surface** dark green, shiny, glabrous; **underside** covered with a
soft, finely granular, suede-like, cinnamon or cinnamon-brown, continuous,
bistrate indumentum of hairs, upper layer capitellate (mop-like).

Inflorescence compact or fairly compact, globose trusses of 8–15 flowers.
Calyx glabrous, eglandular.

Corolla campanulate, 2.5–4.5 cm long, white, white flushed rose, pink, rose or
deep rose, with or without crimson blotch and spots.

Ovary slender, glabrous, eglandular.

W, mid-W and NW Yunnan, E Tibet, SW Sichuan. 2,400–4,000 m (8,000–13,000 ft).
AM 1933 (Bodnant), flowers pink with a crimson blotch.

Epithet: Tawny.   Hardiness 3.   March–April.

R. uvariifolium var. uvariifolium Diels.

R. uvariifolium Diels.
(*R. niphargum* Balf.f. & Kingdon Ward)

Shrub or **tree** 1–11 m (4–35 ft) high, **branchlets** with a thin white or grey
tomentum; **young growth** creamy-white.

Leaves oblanceolate or oblong-obovate; **lamina** 7–24.5 cm long, 2.2–7.2 cm broad;
upper surface pale green or dark green, glabrous; **underside** covered with a thick,
woolly, white or ash-grey or rarely fawn, continuous, bistrate indumentum of hairs.

Inflorescence compact, globose, in trusses of 6–18 flowers.

Corolla campanulate or funnel-campanulate, 2.7–4 cm long, white, white flushed
rose, pale rose or pink, with or without a crimson blotch, with or without
crimson spots.

NW and mid-W Yunnan, SW Sichuan, SE Tibet. 2,200–4,300 m (7,000–13,000 ft).
AM 1965 (RBG Edinburgh) 'Yangtze Bend', flowers rose-pink with purple blotch
and spots.

Epithet: With leaves like a *uvaria*.   Hardiness 3.   March–April.

R. uvariifolium var. griseum Cowan

The variety differs from the species type in that the indumentum on the **underside**
of the leaves is plastered, silky to the touch, unistrate and thin, and in that the
hairs are rosulate.

SE Tibet. 2,300–4,000 m (7,500–13,000 ft).

Epithet: Grey.   Hardiness 3.   March–April.

*Descriptions vary, and more work is needed to determine the status of this taxon.
Very likely a natural hybrid of* R. uvariifolium *and perhaps* R. vellereum. *The
epithet and name are unhelpful.*

Indumentum suede-like, cinnamon or cinnamon-brown. Leaf upper smooth dark and shiny.

Corolla white, pink, rose, usually with a crimson blotch.

Inflorescence compact or fairly compact, globose.

R. fulvum ssp. *fulvum* F18310 at Deer Dell

New foliage of *R. fulvum* ssp. *fulvum* at Deer Dell

Leaves long, obovate and semi-rugulose. Indumentum grey, sometimes thin.

Young growths creamy white.

Inflorescence compact, globose.

R. uvariifolium var. *uvariifolium* F10639 'Yangtze Bend' AM at Deer Dell

New foliage of *R. uvariifolium* var. *uvariifolium* F10639 'Yangtze Bend' AM at Deer Dell

Lanceolate leaves with rounded bases.

Indumentum thin, plastered and silky.

R. uvariifolium var. *griseum* LS&E15817 in the Valley Gardens

Ponticum
(Fulva)

Subgenus *Hymenanthes*
Section *Ponticum* G.Don
Subsection *Glischra* Tagg & Chamberlain
Barbatum Series, Glischrum Subseries

Shrubs or trees.
Glandular bristles on branches, petioles, leaves etc.
**Foliage noticeably rugose or at least rugulose, except for
R. spilotum.**
Corolla usually pink.

R. adenosum Davidian
(*R. kuluense* D.F.Chamb.)

Shrub or **tree** 2–5 m (6–16 ft) high, **branchlets** rather densely setulose-glandular.
Leaves ovate-lanceolate or lanceolate; **lamina** 7–12.8 cm long; 2–5 cm broad;
   upper surface olive-green, matt, often setulose-glandular; **underside** rather
   densely hairy, rather densely or moderately glandular.
Petiole setulose-glandular.
Inflorescence in trusses of 4–8 flowers.
Pedicel setulose-glandular.
Corolla campanulate, 3–5.3 cm long, white-tinged pink or white, with or without
   a crimson blotch, and with or without crimson spots.
SW Sichuan. 3,400–3,500 m (11,000–11,600 ft).
Epithet: Glandular.   Hardiness 3.   April–May.

Rare in cultivation.

Leaves of Subsect. *Glischra* at Deer Dell. Left to right (two leaves each): *R. rude*, *R. vesiculiferum*, *R. recurvoides* (the small leaves at the top), *R. glischroides* and *R. crinigerum*. Bar = 5 cm

Leaves relatively small for this Subsection, recurved, with an upper surface which is roughish to the touch, underside rather densely or moderately glandular.

Corolla pink or white.

R. adenosum R18228 at Deer Dell

R. crinigerum var. crinigerum Franch.

R. crinigerum Franch.

Shrub or **tree** 1–6 m (3–20 ft) high, **branchlets** rather densely setulose-glandular and often setose-glandular.

Leaf-bud scales persistent, leaf-buds very sticky.

Leaves lanceolate or oblanceolate; **lamina** 6.8–19 cm long, 1.8–5.2 cm broad, apex acuminate; **upper surface** dark green, shiny; **underside** densely hairy with a somewhat thick continuous indumentum of brown, fawn, yellowish, cinnamon or whitish hairs, midrib glandular.

Petiole rather densely setulose-glandular and often setose-glandular.

Inflorescence in trusses of 8–15 flowers.

Calyx glandular.

Corolla campanulate, 2.8–4 cm long, white, white suffused with rose, pink, reddish-purple or red, with or without a crimson blotch, often with crimson spots.

W Sichuan, SE Tibet, NE Upper Burma, NW Yunnan. 3,000–4,400 m (10,000–14,500 ft).

AM 1935 (Exbury), flowers pale pink, with red blotch and spots.

Epithet: Bearing hairs.   Hardiness 3.   April–May.

R. crinigerum var. euadenium (Franch.)
Tagg & Forrest

NW Yunnan, SE Tibet, NE Upper Burma. 3,300–4,300 m (11,000–14,100 ft).

Epithet: Well-developed glands.   Hardiness 3.   April–May.

This appears to be a speciating hybrid of R. glischrum *and* R. crinigerum *var.* crinigerum, *with some plants tending to resemble the glabrous* R. glischrum *whereas others are closer to the heavily indumented* R. crinigerum. *Some plants cultivated as* R. glischrum *belong here.*

R. diphrocalyx Balf.

Shrub 1–4.6 m (3–15 ft) high, **branchlets** hairy, often setose and often setose-glandular.

Leaves oblong-obovate, oblong or obovate; **lamina** 5–15 cm long, 1.9–5.8 cm broad; **upper surface** bright green or dark green, glabrous, eglandular, midrib often floccose; **underside** moderately or sparsely hairy or glabrous.

Petiole hairy, often setose and setose-glandular.

Inflorescence in trusses of 6–20 flowers.

Pedicel densely hairy.

Calyx cup-shaped divided to about the middle, 1.8–2.7 cm long.

Corolla broadly campanulate, 3–4.1 cm long, deep wine-crimson, deep crimson-rose, rose or bright red, often with a deep crimson blotch and spots, 5 nectar pouches.

NW and W Yunnan. 3,000–3,400 m (10,000–11,000 ft).

Epithet: Calyx broadly cup-shaped.   Hardiness 3.   April–May.

Very rare in cultivation. Derived from F15665: likely to be a natural hybrid between R. glischrum *(bristles) and a Subsection* Neriiflora *species (calyx).*

Leaf upper surface bullate and shiny.
Leaf underside densely hairy with a somewhat thick continuous yellowish or brownish indumentum.

Branchlets rather densely glandular.
Leaf-bud scales persistent.
Leaf-buds very sticky.
Petiole rather densely glandular.

R. crinigerum var. *crinigerum* R59065 at Deer Dell

Indumentum shows as a thin veil of hairs: the primary veins on the leaf underside are not concealed.

R. crinigerum var. *euadenium* at Deer Dell

R. crinigerum var. *euadenium* F25818 at Borde Hill, Sussex

Calyx cup-shaped divided to about the middle.
Corolla deep wine-crimson, deep crimson-rose, rose or bright red, often blotched.

Branchlets hairy, often glandular.
Petiole hairy, often glandular.

R. diphrocalyx F15665 at Deer Dell

Ponticum
(Glischra)

R. glischroides Tagg & Forrest

Shrub 1.8–4.6 m (6–15 ft) high; **stem** and **branches** with rough bark; **branchlets** rather densely brown or crimson setose-glandular and setulose-glandular.

Leaves oblong-lanceolate or oblanceolate; **lamina** 7–15 cm long, 2.5–4.8 cm broad, apex long acuminate or acuminate; **upper surface** bright green, rugose or markedly bullate, convex, matt, glabrous, not setose; not setose-glandular; margin setulose; **underside** pale green, rather densely setulose, white hairy on the veins, not vesicular, primary veins markedly raised.

Petiole rather densely brown or crimson setose-glandular and setulose-glandular.

Inflorescence in trusses of 7–10 flowers.

Corolla campanulate 3.5–4.5 cm long, white suffused with deep rose, pale rose or creamy-white, often with a crimson blotch.

NE Upper Burma, W Yunnan. 2,800–3,400 m (9,000–11,000 ft).

Epithet: Resembling *R. glischrum*.   Hardiness 3.   March–April.

R. glischroides var. arachnoideum Tagg & Forrest

NE Upper Burma. 3,050 m (10,000 ft).

Epithet: Cobwebby.   Hardiness 3.   April–May.

Extremely rare in cultivation.

R. glischrum Balf.f. & W.W.Sm.

Shrub or **tree** 1.2–9 m (4–30 ft) high; **stem** and **branches** with rough bark; **branchlets** rather densely green or pale green, setose-glandular and setulose-glandular; **leaf-buds** sticky.

Leaves oblanceolate, oblong-lanceolate, lanceolate or oblong-obovate; **lamina** somewhat chartaceous, 9–27 cm long, 2.5–7.8 cm broad, apex long acuminate or acuminate; **upper surface** olive-green, slightly rugulose, flat, matt, glabrous, not setose, not setose-glandular; margin setulose; **underside** pale green, rather densely setulose, setulose-glandular or not setulose-glandular.

Petiole rather densely green or pale green, setose-glandular and setulose-glandular.

Inflorescence in trusses of 8–12 flowers.

Corolla campanulate, 3–4.6 cm long, plum-rose, rose, pink, white, or white flushed deep rose, with crimson blotch.

NW Yunnan, NE Upper Burma, SE Tibet. 2,500–4,300 m (8,000–14,000 ft).

Epithet: Sticky.   Hardiness 3.   April–May.

Rare in cultivation, with many so-called examples of R. glischrum *having some indumentum and being referable to* R. crinigerum *var.* euadenium.

Leaves extremely rugose and recurved in two dimensions. Upper surface markedly bullate.

Branchlets rather densely brown or crimson glandular.

Petiole rather densely brown or crimson glandular.

R. glischroides at Deer Dell

Underside of the leaves is rather densely hairy, forming a thin veil or cobweb, with long and short, branched white or fawn hairs.

R. glischroides var. *arachnoideum* at Leith Hill, Surrey (M.L.A. Robinson)

Leaf upper surface olive-green, slightly rugulose, **flat, glabrous and not setulose-glandular**. *R. glischrum* has **no indumentum** beneath but is setulose-glandular on the midrib and veins and sometimes on the lamina.

Branchlets rather densely green or pale green glandular.

Leaf-buds sticky.

Leaf upper surface not setose, not setose-glandular.

R. glischrum F12901 at Werrington, Cornwall

Leaf underside detail of *R. glischrum* DGEY15 at Glendoick (M.L.A. Robinson)

R. habrotrichum Balf.f. & W.W.Sm.

Shrub 1–3.7 m (3–12 ft) high; **stem** and **branches** with rough bark, **branchlets** rather densely crimson-purple setose-glandular with reddish purple bristles.

Leaves elliptic, oblong-elliptic or ovate; **lamina** 6.5–16 cm long, 3–7.3 cm broad; **upper surface** dark green, slightly rugulose, matt, glabrous; margin fringed with setae; **underside** glabrous, midrib prominent, setose-glandular in the lower half or entire length.

Petiole rather densely crimson-purple setose-glandular.

Inflorescence in trusses of 8–15 flowers.

Pedicel rather densely crimson-purple setulose-glandular.

Calyx crimson-purple setulose-glandular.

Corolla campanulate, 3.5–6 cm long, pale or deep rose, white or white suffused with crimson or rose.

W and NW Yunnan, NE Upper Burma, SE Tibet. 2,500–3,700 m (8,000–12,000 ft).

AM 1933 (Sunningdale Nurseries), flowers pink.

Epithet: With soft hairs.   Hardiness 3.   April–May.

R. recurvoides Tagg & Ward.

Shrub 0.6–1.5 m (2–5 ft) high, **branchlets** densely bristly, densely bristly glandular; **leaf-bud scales** persistent.

Leaves lanceolate, oblanceolate, oblong or sometimes oval; **lamina** 3–7 cm long, 1–2.2 cm broad; **upper surface** dark green, shiny, rugulose, glabrous; margin recurved; **underside** covered with a thick, woolly, yellowish-brown or dark brown, continuous, bistrate indumentum of hairs.

Petiole densely bristly, densely bristly glandular.

Inflorescence in trusses of 4–7 flowers.

Pedicel densely bristly glandular or glandular.

Calyx large, 5–8 mm long, rather densely glandular with long-stalked glands.

Corolla funnel-campanulate, 2.6–3 cm long, rose or white or white suffused with pink, with reddish spots.

Upper Burma. 3,400 m (11,000 ft).

AM 1941 (Trengwainton), flowers pale rose, flushed deep pink.

Epithet: Resembling *R. recurvum* (*R. roxiaenum* var. *oreanastes*).   Hardiness 3. April–May.

Branchlets rather densely crimson-purple with reddish purple bristles.
Recurved oblong/ovate leaves.
Petiole rather densely crimson-purple setose-glandular.

Leaf underside glabrous, midrib setose-glandular.
Pedicel rather densely crimson-purple setulose-glandular.
Calyx crimson-purple setulose glandular.

R. habrotrichum F27343 at Deer Dell

New foliage of *R. habrotrichum* F27343 at Deer Dell

Much more densely growing and with smaller leaves than the other species in this Subsection.
Leaf-bud scales persistent.
Leaf rugulose, shiny.

Branchlets densely bristly.
Leaf underside with yellowish or dark brown indumentum.
Petiole densely bristly.

R. recurvoides KW7184 at Deer Dell

R. rude Tagg & Forrest

Shrub 1.8–2.75 m (6–9 ft) high, **branchlets** rather densely green setose-glandular and moderately green setulose-glandular; **leaf-buds** sticky.

Leaves oblong-obovate, oblanceolate or oblong-lanceolate; **lamina** 9–19 cm long, 2.5–7.5 cm broad, apex abruptly acuminate or long acuminate; **upper surface** olive-green or dark green, matt, setose and setose-glandular; margin setose and setose-glandular; **underside** pale green, rather densely setulose and rather densely setulose-glandular.

Petiole setose-glandular and setulose-glandular.

Inflorescence in trusses of 8–10 flowers.

Pedicel setose-glandular; **flower buds** sticky.

Calyx setulose-glandular.

Corolla campanulate, 2.5–3.8 cm long, purplish-crimson or pinkish-purple, with darker bands outside, with crimson spots.

NW Yunnan. 2,250–3,700 m (7,400–12,000 ft).

AM 1968 (Windsor) 'High Flier', flowers white flushed red-purple.

AM 1969 (Glenarn) 'Frank Kingdon Ward', flowers pinkish-purple with crimson spots.

Epithet: Rough.   Hardiness 3.   April–May.

R. spilotum Balf.f. & Farrer

Shrub 1–1.5 m (3–5 ft) high or a small **tree**; **branchlets** setose-glandular and setulose-glandular.

Leaves oblong-lanceolate, lanceolate or oblong-elliptic; **lamina** 5–12.8 cm long, 2–4.5 cm broad; **upper surface** dark green, glabrous, eglandular; **underside** glabrous or with a thin veil of hairs, punctulate with minute sessile glands or not punctulate, midrib glandular at the base or in the lower half.

Petiole setose-glandular and setulose-glandular.

Inflorescence in trusses of 5–12 flowers.

Pedicel setulose-glandular or glandular.

Calyx glandular.

Corolla campanulate, 2.8–4 cm long, pink or white suffused with pink, with a crimson blotch.

NE Upper Burma.

Epithet: Stained.   Hardiness 3.   April–May.

Very rare in cultivation. Likely to be a natural hybrid.

Leaf upper surface very bristly.

Branchlets rather densely green setose-glandular.

Leaf underside rather densely glandular. Flower-buds sticky.

R. rude LS&T6569 at Deer Dell

New foliage of *R. rude* LS&T6569 at Deer Dell

Similar to a small *R. glischrum*.

Leaf small, flat, underside glabrous or with a thin veil of hairs.

Corolla pink or white-tinged pink, with a crimson blotch.

R. spilotum at RBG Edinburgh (Davidian)

Ponticum
(Glischra)

R. vesiculiferum Tagg

Shrub or **small tree** 1.5–3 m (5–10 ft) high, **branchlets** rather densely setose-glandular and setulose-glandular.

Leaves oblong-lanceolate or oblanceolate; **lamina** 8–16 cm long, 2–5.5 cm broad, apex acuminate or long acuminate; **upper surface** green or olive-green, rugose or markedly bullate, matt, glabrous; margin setulose; **underside** pale green, rather densely setulose, hairy with vesicular white hairs on the veins.

Petiole with vesicular white hairs or glabrous; rather densely setose-glandular and setulose-glandular.

Inflorescence in trusses of 10–15 flowers.

Pedicel with vesicular white hairs, setulose-glandular and setose-glandular.

Corolla campanulate, 3–3.5 cm long, purplish-rose, pinkish-purple or almost white, with a deep crimson or deep purple blotch.

N Burma, Burma–Tibet Frontier; NW Yunnan. 2,500–3,400 m (8,000–11,000 ft).
Epithet: Bearing vesicles.   Hardiness 3.   April–May.

Extremely rare in cultivation. Similar to R. glischroides *in appearance; the white hairs on the veins of the leaf underside are a determining factor, but it is only under magnification that their vesicular nature can be determined, and it is this that is the essential diagnostic feature.*

Subgenus *Hymenanthes*
Section *Ponticum* G.Don
Subsection *Grandia* Sleumer
Grande Series

Trees or shrubs.
Large leaves.
Indumentum usually shiny, plastered, unistrate, but sometimes woolly, bistrate.
Corolla ventricose, lobes usually 8, but occasionally 6–7 or 10.
Stamens usually 16–20.

For ease of identification Subsection Grandia *can be split into two groups:*

Group A Species with a winged petiole.
Group B Species without winged petioles.

Similar to a wider leaved form of *R. glischroides*.

Leaf underside hairy with white hairs on the vein.

Petiole with white hairs, glandular.

Pedicel with white hairs, glandular.

Leaf apex acuminate or long acuminate; upper surface rugose or markedly bullate.

R. vesiculiferum KW9485 at Deer Dell

Leaves of Subsect. *Grandia* at Deer Dell. Top to bottom (one leaf each): *R. montroseanum* (Gp B), *R. sidereum* (Gp B), *R. grande* (Gp B), and *R. praestans* (syn. *R. coryphaeum*) (Gp A). Bar = 5 cm

Ponticum
(Glischra/Grandia)

R. praestans Balf.f. & W.W.Sm. Grandia Gp A

(*R. coryphaeum* Balf.f. & Forrest, *R. semnum* Balf.f. & Forrest)

Shrub or **tree** 1.2–9 m (4–30 ft) high, trunk 5–2 ft in diameter; **branchlets** with a thin, whitish or fawn tomentum.

Leaves oblong-obovate, oblanceolate or obovate; **lamina** 14–38 cm long, 4.5–14 cm broad, base tapered, decurrent on the petiole; **upper surface** dark green or green, glabrous; **underside** with a thin, plastered, shiny, silvery-white or fawn, continuous, unistrate indumentum of hairs.

Petiole short, 1.5–2.5 cm long, flat, margins with wings.

Inflorescence in trusses of 10–25 flowers.

Corolla obliquely campanulate, ventricose 3–5 cm long, white suffused with rose, creamy-white, white, yellow or deep rose, with or without a crimson blotch, and with or without crimson spots.

Yunnan, SE Tibet, Burma–Tibet frontier. 3,400–4,300 m (11,000–14,000 ft).

AM 1963 (Exbury) as *R. coryphaeum* 'Exbury', flowers white-tinged pale yellow, with a crimson blotch.

Epithet: Excellent.   Hardiness 3.   April–May.

The former R. coryphaeum *is identical to this species except that it has a silvery indumentum and creamy white flowers.*

R. watsonii Hemsl. & Wils. Grandia Gp A

Shrub or **tree** 1.5–10 m (5–33 ft), **branchlets** with a thin, whitish or fawn tomentum or glabrous.

Leaves oblong-obovate, obovate, oblong-elliptic or oblanceolate; **lamina** 10.5–23 cm long, 4–10 cm broad, base decurrent on the petiole; **upper surface** olive-green or green, glabrous; **underside** with a thin, plastered, shiny, silvery-white or fawn, continuous, unistrate indumentum of hairs.

Petiole short, flat, margins with broad or narrow wings.

Inflorescence in trusses of 10–16 flowers.

Corolla ventricose, 3–5 cm long, white or white suffused with pink, with or without a crimson blotch, and with or without crimson spots.

W Sichuan. 2,600–4,000 m (8,500–13,000 ft).

Epithet: After W.C. Haines-Watson, Chinese Customs.   Hardiness 3.   March–April.

Leaf base tapered, decurrent on the very broadly winged petiole.

Leaf underside with a shiny copper indumentum.

Corolla obliquely campanulate, ventricose.

R. praestans (syn. *R. coryphaeum*) at Deer Dell

Tapering yellow midrib on a flat upper leaf surface. Leaf base decurrent on the short winged petiole.

Leaf underside with a plastered, shiny, silvery-white or fawn, indumentum.

Corolla ventricose, white or white-tinged pink.

R. watsonii W1872 at Deer Dell

R. grande Wight   Grandia Gp B

(*R. argenteum* Hook.f.)

Shrub or **tree** 3–15 m (10–50 ft) high; **stem** and **branches** with rough bark; **branchlets** with a thin whitish tomentum, eglandular.

Leaves oblong-obovate, oblong, oblong-lanceolate or oblong-elliptic; **lamina** 14–30 cm, long, 5–13 cm broad; **upper surface** dark or pale green, shiny, glabrous; **underside** with a thin, plastered, shiny, silvery-white or fawn, continuous, unistrate indumentum of hairs.

Inflorescence in trusses of 15–25 flowers.

Calyx minute, 1 mm long.

Corolla obliquely campanulate, ventricose, 4–7cm long, pale yellow, lemon-yellow, creamy-white or white, with dark purple blotch, 8 nectar pouches.

Nepal, Sikkim, Bhutan, Assam. 1,700–3,600 m (5,600–12,000 ft).

FCC 1901 (South Lodge), flowers creamy-white with a purple blotch.

Epithet: Large.   Hardiness 1–3.   February–April.

Do not expect the leaves to be as large as those of R. sinogrande, *the same shape, or with the same indumentum. In drier areas, the leaves are usually less than 20 cm long.*

R. kesangiae D.G.Long & Rushforth   Grandia Gp B

Upright **shrub** or **small tree** 3–12 m (10–40 ft), bark mid-brown, not peeling.

Branchlets usually sparsely whitish floccose; **terminal buds** rounded, green, reddish or maroon.

Leaves 19–30.5 cm by 9.5–16 cm (larger on juvenile plants), broadly elliptic to obovate, apex rounded or almost truncate with a short mucro; **upper surface** with a detersile discontinuous thin white tomentum appearing as mica-like patches, glabrous at maturity; **underside** with a densely matted white indumentum, sometimes with a floccose layer of whitish or brownish-fawn dendroid hairs, often brownish on old leaves.

Petiole stout not flattened, glabrous or white tomentose.

Inflorescence usually compact, 15–25-flowered.

Pedicel glandular.

Corolla campanulate, 3–4.7 cm long, rose or pink, fading to pale pink or almost white, with nectar pouches.

Ovary densely glandular.

Bhutan. 2,900–3,500 m (9,500–11,500 ft).

Epithet: After the queen of Bhutan.   Hardiness 4.   April–May.

Leaf 14–30 cm long, with a **plastered, shiny, silvery-white indumentum.**

Corolla obliquely campanulate, ventricose, pale yellow, lemon-yellow, creamy white or white, with a dark purple blotch.

R. grande at Arduaine

Close to *R. hodgsonii,* but without the peeling bark or pointed growth buds.

Buds are very rounded, wider than tall and sometimes coloured red to maroon.

Indumentum greyish white, often brownish on old leaves.

Corolla campanulate, rose pink, or almost white.

A pale *R. kesangiae* at Meerkerk Gardens, Oregon (M.L.A. Robinson)

R. kesangiae at Meerkerk Gardens, Oregon (M.L.A. Robinson)

R. macabeanum Watt ex Balf.f. Grandia Gp B

Shrub or **tree** 3–15 m (10–50 ft) high, **branchlets** densely tomentose with a thin
fawn or whitish tomentum, **foliage-bud** large, oblong-oval, reddish-green.
Young growth densely white woolly.
Leaves oval, obovate, elliptic or oblong-elliptic; **lamina** 13.2–38.5 cm long,
8–20 cm broad; **upper surface** dark green, glabrous (in young leaves, with a
white tomentum); **underside** with a thick, woolly, whitish or fawn,
continuous bistrate indumentum of hairs.
Petiole tomentose.
Inflorescence in trusses of 12–20 flowers, flower-bud scales often persistent red
during early flowering period.
Corolla obliquely tubular-campanulate, ventricose, 5–7.3 cm long, yellow,
yellowish-white or pale greenish-yellow, with or without purple blotch, 8 nectar
pouches, lobes 8.
Stamens 16–19.
Manipur, Assam. 2,450–3,800 m (8,000–12,500 ft).
AM 1937, FCC 1938 (Trengwainton), flowers yellowish-white.
Epithet: After Mr. McCabe, a former Deputy Commissioner. Naga Hills, NE India.
Hardiness 3.   March–May.

*Recently, this species has been found at higher elevations (on Mt Saramati,
Nagaland) than those recorded by Davidian. In the wild, the highest altitude
forms are low growing.*

R. magnificum Ward. Grandia Gp B

Shrub or **tree** 3–18 m (10–60 ft) high, **branchlets** densely tomentose with a thin
fawn tomentum.
Leaves oblanceolate; **lamina** 20–46 cm long, 6–14.3 cm broad; **upper surface**
green, matt, glabrous; **underside** covered with a thin, somewhat plastered,
matt, brown or fawn, continuous, unistrate indumentum of hairs.
Petiole with a thin tomentum.
Inflorescence in trusses of 12–30 flowers.
Calyx minute, 1 mm long, tomentose.
Corolla obliquely tubular-campanulate, ventricose, 5.3–7.5 cm long, crimson-
purple or rosy-purple, without blotch, 8 nectar pouches; lobes 8.
Stamens 16.
Burma–Tibet frontier. 1,500–2,000 m (5,000–8,000 ft).
AM 1950 (Corsewell, Stranraer), flowers pink.
FCC 1966 (Brodick) 'Kildonan', pink (KW9200).
Epithet: Magnificent.   Hardiness 1–3.   February–April.

*Tender, and extremely rare in cultivation. There are many hybrids labelled with
this name.*

Shiny upper leaf surface

Thick, whitish matt indumentum.

Young growth densely white woolly.

Corolla obliquely tubular-campanulate, ventricose, yellow, yellow-white or pale greenish-yellow.

R. macabeanum at Deer Dell

Very large oblanceolate leaves with a thin but continuous indumentum.

Corolla ventricose, 5.3–7.5 cm long, crimson purple or rosy-purple.

Leaves oblanceolate.

Indumentum continuous, thin, somewhat plastered, **matt,** brown or fawn.

R. magnificum KW9200 at Brodick Castle, Isle of Arran

Ponticum
(Grandia Group B)

R. montroseanum Davidian   Grandia Gp B

Shrub or **tree** 3–15 m (10–50 ft) high, **branchlets** densely thin, whitish tomentose.

Leaves oblong or oblanceolate; **lamina** thick, coriaceous, rugulose, 12.5–61 cm long, 4.6–23 cm broad; **upper surface** dark green, shiny, glabrous; **underside** with a thin, plastered, shiny, silvery-white, continuous, unistrate indumentum of hairs.

Inflorescence in trusses of 15–20 flowers.

Corolla ventricose, 3.5–5 cm long, deep pink, with a crimson blotch; lobes 8.

Stamens 15–16.

SE Tibet. 2,500–2,800 m (8,000–9,000 ft).

FCC 1957 (Benmore) 'Benmore', flowers deep pink (KW6261A).

Epithet: After the Duchess of Montrose.   Hardiness 2–3.   March–May.

Formerly known as R. mollyanum.

R. protistum var. giganteum D.F.Chamb.

R. giganteum Forrest ex Tagg   Grandia Gp B
(R. giganteum var. seminudum Tagg & Forrest)

Shrub or **tree** 6–30 m (20–100 ft) high; **stem** and **branches** with rough bark; **branchlets** densely tomentose with a thin fawn or whitish tomentum, eglandular.

Leaves oblong-obovate, elliptic or oblanceolate; **lamina** 20–37 cm long, 6.5–16.5 cm broad; **upper surface** green, matt, glabrous; **underside** covered with a thin woolly, brown or fawn, continuous, bistrate indumentum of hairs.

Inflorescence in trusses of 20–25 flowers.

Pedicel densely tomentose.

Calyx minute, 1 mm long.

Corolla obliquely tubular-campanulate, 5–6.8 cm long, deep rose-crimson or deep crimson, with or without crimson blotch, 8 nectar pouches at the base.

Stamens 16–20.

SW, W, and NW Yunnan, NE and N Burma. 2,750–4,300 m (9,000–14,000 ft).

FCC 1953 (Brodick).

Epithet: Gigantic.   Hardiness 1–3.   February–April.

Tender and rare in cultivation.

Long rugulose shiny leaves with a shiny thin silvery white indumentum.

May have **very** narrow wings on the petiole. Corolla ventricose deep pink.

R. montroseanum at KW6262A at Deer Dell

R. montroseanum KW6262A FCC at RBG Edinburgh

New foliage of *R. montroseanum* KW6262A at Deer Dell

Very large leaves. Thin woolly indumentum covering the entire underside of leaf.

Corolla obliquely tubular-campanulate, deep rose-crimson or deep crimson, **not ventricose**.

R. protistum var. *giganteum* at Brodick Castle, Isle of Arran

Ponticum
(Grandia Group B)

R. protistum var. protistum   Grandia Gp B

R. protistum Balf.f. & Forrest

Shrub or **tree** 6–30 m (20–100 ft) high, **branchlets** densely tomentose with a
   thin, fawn or whitish tomentum.
Leaves oblong-obovate, obovate or elliptic; **lamina** 20–4(–5) cm long, 7–21 cm
   broad; **upper surface** dark green, matt, glabrous; **underside** glabrous, or with
   a very thin veil of hairs or with closely scattered short hairs, or with a marginal
   border having a thin continuous indumentum of hairs.
Inflorescence in trusses of 20–30 flowers.
Calyx minute, 1 mm long, densely tomentose.
Corolla obliquely tubular-campanulate, 5–7.6 cm long, crimson-purple, rosy-
   crimson or creamy-white suffused with rose, with or without a small crimson
   blotch, 8 nectar pouches.
NW, mid-W and SW Yunnan, SE Tibet, NE Upper Burma, Assam. 2,600–4,300 m
   (8,500–14,000 ft).
Epithet: First of the first.   Hardiness 1–2.   February–March.

R. protistum *var.* **giganteum** *is likely to be just the adult form of* R. protistum
var. protistum*, which may take up to 60 years to develop its full indumentum.*

R. pudorosum Cowan   Grandia Gp B

Shrub or **tree** 1.8–12 m (6–40 ft) high, **branchlets** densely tomentose with a thin
   tomentum, **leaf-bud scales** long, persistent.
Leaves oblanceolate, oblong or oblong-obovate; **lamina** 8–35 cm long,
   3–13.2 cm broad; **upper surface** dark green, shiny, glabrous; **underside**
   with a thin, plastered, shiny silvery-white or fawn, continuous unistrate
   indumentum of hairs. On young plants, glabrous.
Inflorescence in fairly compact trusses of 10–24 flowers.
Calyx minute, 1 mm long.
Corolla obliquely campanulate, ventricose, 3–4 cm long, fragrant, bright pink,
   pink, rose or mauve-pink, with or without a purple blotch.
Stamens 16.
S and SE Tibet. 3,400–3,800 m (11,000–12,500 ft).
Epithet: Very bashful.   Hardiness 3.   March–April.

Leaf underside glabrous or with a marginal border of thin indumentum.

Corolla obliquely tubular-campanulate crimson purple, rosy crimson or creamy-white-flushed rose, **not ventricose**.

The young growth stands out because of the bright red leaf bud scales.

The flowers may fade to almost cream.

R. protistum var. *protistum* KW21498 at Pukeiti, New Zealand (M.L.A. Robinson)

Very persistent leaf bud scales (perulae). Corolla obliquely campanulate, ventricose, pink, rose or mauve pink.

Indumentum plastered, shiny silvery-white or fawn.

R. pudorosum L&S2752 at Deer Dell

R. sidereum Balf.f.　　　　　　　　Grandia Gp B

Shrub or **tree** 1.8–12 m (6–40 ft) high, **branchlets** with a thin fawn or whitish tomentum.

Leaves lanceolate, oblong-lanceolate or oblanceolate; **lamina** very coriaceous, 9–25 cm long, 2.5–7.2 cm broad; **upper surface** olive-green, matt, glabrous; **underside** with a thin, plastered, shiny, silvery-white, fawn or coppery, continuous, unistrate indumentum of hairs.

Inflorescence in trusses of 12–20 flowers.

Calyx minute, 1 mm long.

Corolla campanulate, ventricose, 3–5 cm long, creamy-white, yellow or lemon-yellow, with a crimson blotch, 8 nectar pouches.

Stamens 16.

NE and E Upper Burma, mid-W and NW Yunnan, Assam. 2,450–4,000 m (8,000–13,000 ft).

AM 1964 (Brodick) 'Glen Rosa', flowers primrose yellow, with a deep crimson blotch.

Epithet: Excellent.　　　Hardiness 2–3.　　　April–May.

Because of its relatively small leaves, this species may sometimes be confused with R. arboreum *when not in flower.*

R. sinogrande Balf.f. & W.W.Sm.　　　　Grandia Gp B

Shrub or **tree** 3–15 m (10–50 ft) high, **branchlets** with a thin, whitish or fawn tomentum.

Leaves obovate, oblong-obovate, elliptic or sometimes oval; **lamina** 20–70 cm long, 8–30 cm broad; **upper surface** dark green, shiny, glabrous; **underside** with a thin, plastered, shiny silvery-white or fawn, continuous, unistrate indumentum of rosulate hairs.

Inflorescence in trusses of 15–30 flowers.

Corolla campanulate, ventricose, 4.5–6 cm long, creamy-white, creamy-yellow, or white, with a deep crimson blotch, 8–10 nectar pouches.

W, mid-W and NW Yunnan, E Tibet, NE Upper Burma, Assam. 2,100–4,000 m (7,000–13,000 ft).

AM 1922 (South Lodge), flowers creamy-white, with a crimson blotch.

FCC 1926 (Trewithen), flowers ivory-white, with a crimson blotch.

Epithet: Chinese *R. grande*.　　　Hardiness 1–3.　　　April–May.

R. sinogrande var. *boreale*

Differs from the species type in that the leaves are usually shorter and narrower.

Leaves often small for this Subsection, lamina very coriaceous; underside with a thin, plastered, shiny, silvery-white or coppery, indumentum.

Corolla campanulate, ventricose, creamy-white, yellow or lemon-yellow.

R. sidereum F24563 at Deer Dell

Very large leaves with a rounded apex, upper leaf midrib yellow.
Indumentum shiny silvery-white or fawn.

Corolla campanulate, ventricose, creamy-white, creamy-yellow, or white.

R. sinogrande at Lochinch, Scotland

R. suoilenhense D.F.Chamb. & K.Rushforth Grandia Gp B

Large shrub or **tree** to 12 m (40 ft), **branchlets** tomentose.

Mature leaves elliptic to slightly oblanceolate: 20–35 cm long by 9–16 cm broad: apex rounded; **upper surface** dark green, somewhat shiny, almost flat, not rugose, sparsely pubescent with dendroid hairs; **underside** with a continuous persistent suede-like brown unistrate indumentum of dendroid hairs. **Juvenile foliage:** leaves widely oblong, chartaceous, margin curved, markedly rugose; the same size or larger than the mature leaves; **underside** initially glabrous, a dendroid indumentum appearing in later years, starting at the rim of the lamina.

Petiole round, not grooved.

Inflorescence 15-flowered.

Pedicel 4.2–5.3 cm long, densely white floccose.

Corolla broadly tubular campanulate, ventricose, fleshy, glabrous, 8-lobed, almost white or pale yellow, with maroon basal flashes.

Calyx an undulate rim, sparsely floccose, eglandular.

Ovary densely floccose, eglandular.

Style curved, glabrous.

Lao Cai province, Vietnam. 2,200–3,140 m (7,200–10,300 ft).

*Introduced 1992 by A. Clark and K. Rushforth, and quite widely seen in collections. The phenomenon of juvenile foliage differing in characteristics from the foliage of the mature flowering plant, other than the indumentum covering, is most unusual. **The juvenile foliage has a very distinctive wavy and chartaceous appearance**. As the leaves have no rosulate or cup-shaped hairs, this species does not fit comfortably into either Subsection Falconera or Subsection Grandia, but we have placed it in Subsection Grandia where a few species have a dendroid indumentum.*

Mature (left) and juvenile (right) foliage of *R. suoilenhense* AC432 at Hindleap Lodge (M.L.A. Robinson)

Juvenile foliage of large widely oblong **rugose** and chartaceous leaves, underside initially glabrous.

Mature leaves elliptic to slightly oblanceolate; upper surface dark green, somewhat shiny, almost **flat**, underside with a continuous suede-like brown unistrate indumentum.

Corolla tubular campanulate, ventricose, almost white.

R. suoilenhense **AC432 at Hindleap Lodge (M.L.A. Robinson)**

R. suoilenhense **AC432 at Tregrehan (M.L.A. Robinson)**

Ponticum
(Grandia Group B)

Subgenus *Hymenanthes*
Section *Ponticum* G. Don
Subsection *Griersoniana* Sleumer
Griersonianum Series

R. griersonianum Balf.f. & Forrest

Lax, broadly upright **shrub** 1.2–3 m (4–10 ft) high, **branchlets** densely or
   moderately bristly-glandular, densely floccose; **foliage-buds** long, conical,
   tapered, the outer scales with long tapering tips.
Leaves lanceolate; **lamina** 6.4–20 cm long, 1.2–5.3 cm broad; **upper surface**
   pale green, matt, glabrous; **underside** covered with a thick woolly, fawn or
   brown, continuous or discontinuous unistrate indumentum of hairs.
Inflorescence in trusses of 5–12 flowers; **flower-buds** large, long, conical, tapered.
Corolla funnel-shaped, 5–8 cm long, bright geranium-scarlet or rich carmine
   almost vermilion or bright rose, outside hairy.

W and mid-W Yunnan, N and E Upper Burma. 2,100–2,750 m (7,000–9,000 ft).

FCC 1924 (Exbury and Sunninghill), flowers salmon-coloured.

Epithet: After R.G. Grierson, Chinese Maritime Customs at Tengyuh, friend of
   George Forrest.   Hardiness 2–3.   June–July.

*Can only be confused with its hybrids. The species has long matt leaves and the
whole plant appears hairy, especially on the new growth. The growth buds are
similar to those of R. auriculatum.*

R. griersonianum at Deer Dell

Foliage-buds long, conical, tapered.
Leaf upper surface matt.
Indumentum easily rubbed off.

Lax shrub.

Leaf underside with a woolly, brownish, continuous or discontinuous indumentum.

Flower-buds conical, tapered. Corolla funnel-shaped in tones of brick red, outside hairy.

R. griersonianum at Deer Dell

Ponticum
(Griersoniana)

Subgenus *Hymenanthes*
Section *Ponticum* G. Don
Subsection *Irrorata* Sleumer
Irroratum Series

Shrubs or small trees.
Leaf upper surface glabrous at maturity, shiny or with a thin indumentum underneath.
Leaf apex acuminate.
Corolla almost always 5-lobed, with or without nectar pouches.
Style usually glandular.

For identification purposes, this subsection is best divided into six principal groups depending on the leaf underside and the presence or absence of nectar pouches:

Group A1. *Leaf underside whitish and matt or matt, without hairs. Corolla* **without** *nectar pouches:*

R. aberconwayi, R. annae ssp. annae, R. annae ssp. laxiflorum, R. brevinerve, R. hardingii and R. ziyuanense.

Group A2. *Leaf underside whitish and matt or matt. Corolla* **with** *nectar pouches:*

R. anthosphaerum, R. irroratum ssp. irroratum, R. irroratum ssp. pogonostylum, R. irroratum ssp. yiliangense and R. ningyuenense.

Group B1. *Leaf underside shiny green. Corolla* **without** *nectar pouches:*

R. araiophyllum ssp. araiophyllum and R. lukiangense.

Group B2. *Leaf underside shiny green. Corolla* **with** *nectar pouches:*

R. gongshanense, R. kendrickii, R. ramsdenianum and R. tanastylum var. tanastylum.

Group C1. *Leaf with indumentum.* **Corolla without** *nectar pouches:*

R. papillatum and R. wrayi.

Group C2. *Leaf with indumentum. Corolla* **with** *nectar pouches:*

R. agastum and R. tanastylum var. pennivenium.

Leaves of Subsect. *Irrorata* at Deer Dell. From top to bottom (all 2 leaves each): *R. aberconwayi* (Gp A1), *R. anthosphaerum* (Gp A2), *R. irroratum* (Gp A2), *R. lukiangense* (Gp B1), *R. ramsdenianum* (Gp B2), *R. kendrickii* (Gp B2) and (1 leaf) *R. tanastylum* var. *pennivenium* (Gp C2). Bar = 5 cm

R. aberconwayi Cowan

Irrorata Gp A1

Shrub 0.3–2.5 m (1–8 ft) high, **branchlets** glandular.
Leaves lanceolate, oblong-elliptic or elliptic; **lamina** markedly rigid, 2.8–7 cm
  long, 0.8–3.3 cm broad; **upper surface** dark green, shiny, glabrous; margin
  markedly recurved; **underside** whitish, matt, glabrous, punctulate with minute
  red glands, papillate.
Inflorescence in trusses of 5–12 flowers.
Corolla saucer-shaped or flatly campanulate, 2–3 cm long, white or white
  suffused with pink, with a few crimson spots.
Style glandular throughout to the tip.
NE Yunnan.
AM 1945 (Windsor) 'His Lordship', flowers white with crimson spots.
Epithet: After the first Lord Aberconway (1879–1953), former President of the
  RHS.   Hardiness 3.   April–May.

R. annae ssp. annae Franch.

Irrorata Gp A1

R. annae Franch.

Shrub 1–2 m (3–6 ft) high, **branchlets** glandular.
Leaves lanceolate; **lamina** rigid, 7–12.5 cm long, 1.5–3.2 cm broad; **upper
  surface** olive-green or green, matt, glabrous; **underside** paler, matt, glabrous,
  punctulate with minute red glands, papillate.
Petiole glandular or eglandular.
Inflorescence in trusses of 8–15 flowers.
Pedicel rather densely or moderately glandular, 1.2–2.5 cm long.
Corolla broadly cup-shaped, 2–2.5 cm long, rose or white suffused with rose,
  with or without purple spots.
Ovary densely glandular.
Style glandular throughout to the tip.
Guizhou, NW Yunnan. 1,400–1,500 m (4,600–5,000 ft).
Epithet: After a French Lady.   Hardiness 1–3.   April–June.

Closer to R. aberconwayi *than to either* R. annae *ssp.* laxiflorum *or the former*
R. hardingii *(which are closely related to each other).*

Markedly rigid leaves that have tips which are almost stiff enough to prick the hand. Leaf margin markedly recurved.

Corolla saucer-shaped or flat-campanulate, white or white-tinged pink.

Style glandular throughout to the tip.

R. aberconwayi McL.T41 'His Lordship' AM at Deer Dell

Long narrow leaf, recurved in both directions; underside glabrous, punctulate with minute red glands, papillate.

Style glandular throughout to the tip.

Leaves lanceolate, lamina rigid.

Petiole glandular or eglandular.

Pedicel rather densely or moderately glandular.

Corolla broadly cup-shaped, white, small.

R. annae ssp. *annae* at Deer Dell

New foliage of *R. annae* ssp. *annae* at Deer Dell

R. annae ssp. *annae* at Deer Dell

R. annae ssp. laxiflorum Ming.   Irrorata Gp A1
R. laxiflorum Balf.f. & Forrest

Shrub 1.2–6 m (4–20 ft) high, **branchlets** glandular.

Leaves lanceolate or oblong-lanceolate; **lamina** 9–15.3 cm long, 2–4.5 cm broad; **upper surface** green or olive-green, matt, glabrous; **underside** paler, matt, glabrous; punctulate with minute red glands, papillate.

Petiole glandular.

Inflorescence in trusses of 8–12 flowers.

Pedicel moderately or rather densely glandular, 1.8–3.5 cm long.

Calyx minute, 1–2 mm long, margin fringed with glands.

Corolla broadly cup-shaped, 3.5–4.5 cm long, white or creamy-white, flushed or not flushed rose, with or without crimson spots.

Ovary densely glandular.

Style glandular throughout to the tip.

W and NW Yunnan. 2,400–3,300 m (8,000–11,000 ft).

AM 1977 (Sandling Park) 'Folks Wood', flowers white.

Epithet: Loose-flowered.   Hardiness 3.   April–May.

Rare in cultivation. Lacks the elegance in leaf of R. annae.

R. annae ssp. laxiflorum Ming.   Irrorata Gp A1
R. hardingii Forrest

Shrub 1.5–2.6 m (5–8 ft) high; **branchlets** glandular.

Leaves lanceolate; **lamina** 6–12.8 cm long, 1.5–3.5 cm broad; **upper surface** green or olive-green, matt, glabrous; **underside** paler, matt, glabrous, papillate.

Petiole glandular.

Inflorescence in trusses of 8–14 flowers; rhachis glandular.

Pedicel 2–4.5 cm long, glandular.

Corolla broadly cup-shaped, 3–4 cm long, white or white-flushed pink, with or without crimson spots.

Ovary densely glandular.

Style glandular throughout to the tip.

Mid-W Yunnan. 1,800–2,200 m (6,000–7,000 ft).

Epithet: After H.I. Harding, HM Consul at Tengyueh, who supplied the type flowering material to Forrest.   Hardiness 1–3.   April–May.

Rare in cultivation.

Larger leaves and flowers than ssp. *annae*, and with the longer pedicels which give rise to the name.

Style glandular throughout to the tip.

Branchlets glandular.

Leaves lanceolate or oblong-lanceolate, underside punctulate with minute red glands, papillate.

Petiole glandular.

Corolla broadly cup-shaped, white or creamy white, larger.

R. annae ssp. *laxiflorum* at Nymans, Sussex

Differs from the former *R. laxiflorum* Balf.f. & Forrest by having rough leaf margins caused by glands and gland bases.

R. annae ssp. *laxiflorum* (*R. hardingii*) at Deer Dell

Ponticum
(Irrorata Group A1)

R. brevinerve Chun & Fang   Irrorata Gp A1

A small **tree** up to 5 m (16 ft) high; **branchlets** setose glandular or eglandular.
Leaves oblanceolate or lanceolate; **lamina** coriaceous, 10–15 cm long, 2.5–4 cm broad, apex acuminate, base tapered or cuneate; **upper surface** green, matt, glabrous, eglandular; **underside** pale green, matt, glabrous, punctulate with minute red glands.
Corolla broadly campanulate, 3–3.5 cm long, purplish or pink, without spots, outside glabrous, eglandular.
Ovary glabrous, densely setulose-glandular.
Guangxi, SE Guizhou and SW Hunan. 500 m (1,650 ft) or more.
Epithet: With short nerves.

It is by no means certain that this species is in cultivation.

R. ziyuanense Tam   Irrorata Gp A1

Tree or **shrub** 2.5–5 m (4–30 ft) high.
Leaves 5–10 cm long, 1.8–2.8 cm wide; **upper surface** somewhat shining.
Inflorescence 3–5 flowers.
Ovary glandular, sparsely setose, pubescent.
Corolla white or white flushed pink.
Ovary glandular, sparsely setose, pubescent.
Guangxi. 1700 m (5600 ft).
Epithet: from Ziyuan, Guangxi.   Hardiness 3–4?   May.

Recently introduced by Alan Clark (as R. stamineum) *from Maoersham, Guilin, Guangxi.*

R. anthosphaerum Diels.   Irrorata Gp A2
(*R. gymnogynum* Balf.f. & Forrest, *R. heptamerum* Balf.f. and *R. persicinum* Hand.-Mazz)

Shrub or **tree** 1.5–9 m (5–30 ft) high; **branchlets** floccose or glabrous, glandular or eglandular.
Leaves oblong-lanceolate, oblanceolate or lanceolate; **lamina** 5.2–18.5 cm long, 1.5–5 cm broad; **upper surface** green or olive-green, matt, glabrous; **underside** paler, matt, glabrous or sometimes with a thin veil of hairs, papillate.
Inflorescence in trusses of 8–15 flowers.
Calyx 5–7-lobed, minute, 1–2 mm long.
Corolla tubular-campanulate, 3–5.9 cm long, purple, lilac, pale lavender-blue, pink or deep rose-magenta, lobes 7, rarely 5.
Stamens 14 or sometimes 10.
NW and mid-W Yunnan, NE Upper Burma, SE Tibet. 2,750–4,000 m (9,000–13,000 ft).
Epithet: With round flowers.   Hardiness 3.   March–May.

Note the variation in flower colour.

Narrow, pointed, leathery leaves, apex long, acuminate, underside matt.

Leaves elliptic to oblanceolate, often convex.

Inflorescence lax 2–5-flowered.

Corolla broadly campanulate, white, without nectar pouches.

R. ziyuanense AC4211 at Hindleap Lodge (M.L.A. Robinson)

Seven (rarely 5)-lobed corolla and an irrorata leaf type, rarely with a very thin veil of hairs.

Leaf underside glabrous, papillate.

Corolla purple, lilac, pale lavender-blue, pink or deep rose-magenta.

R. anthosphaerum F25984A in the Valley Gardens

Ponticum
(Irrorata Group A1/A2)

R. anthosphaerum Eritimum Group   Irrorata Gp A2

R. anthosphaerum var. eritimum Davidian

Shrub or **tree** 1.5–12 m (5–40 ft) high.
The variety also differs from the species in the often larger leaves.
NW Yunnan, NE Upper Burma. 2,600–3,950 m (8,500–13,000 ft).
Epithet: Highly prized.   Hardiness 3.   March–May.

The flower colour is a useful distinction in cultivation and for identification purposes, but apparently of no botanical significance. Worthy of Group status.

R. irroratum ssp. irroratum Franch.   Irrorata Gp A2

R. irroratum Franch.

Shrub or **tree** 1–9 m (3–30 ft) high; **branchlets** often glandular.
Leaves lanceolate, oblanceolate, oblong-lanceolate or narrowly elliptic; **lamina**
    rigid, 5–15.5 cm long, 1.5–5 cm broad; **upper surface** pale green, matt,
    glabrous; **underside** paler, matt, glabrous, punctulate with minute red glands,
    papillate.
Petiole often glandular.
Inflorescence in trusses of 8–15 flowers.
Pedicel glabrous, rather densely or moderately glandular.
Calyx outside rather densely glandular, margin fringed with glands.
Corolla tubular-campanulate, 3–5.5 cm, long, yellowish-white, creamy-yellow,
    deep rose or pink, with or without a blotch, with or without deep purple,
    greenish or crimson spots, 5 nectar pouches.
Ovary glandular, glabrous.
Style glandular throughout to the tip.
NW and W Yunnan, Yunnan–Tibet border. 1,800–3,700 m (6,000–12,000 ft).
AM 1957 (Minterne), flowers white faintly tinged pink.
AM 1957 (Exbury) 'Polka Dot', flowers white suffused with pink, heavily spotted
    deep purple.
Epithet: Covered with dew, i.e. minutely spotted.   Hardiness 3.   March–May.

The commonest species in cultivation from this Subsection, and this should be considered first when trying to identify an Irrorata species.

Striking deep plum-crimson or crimson flowers.

R. anthosphaerum Eritimum Group at Deer Dell

Leaf underside matt, sometimes whitish (as with most of Gp A2).

Calyx margin fringed with glands.

Ovary glandular, glabrous.

Leaf lamina rigid, glabrous, punctulate with minute red glands, papillate.

Petiole often glandular.

Calyx outside rather densely glandular.

Style glandular throughout to the tip.

Corolla yellowy-white, pink or rose.

R. irroratum ssp. *irroratum* at Deer Dell

New foliage of *R. irroratum* ssp. *irroratum* AC1225 at Deer Dell

Ponticum
(Irrorata Group A2)

R. irroratum ssp. pogonostylum D.F.Chamb.

R. pogonostylum Balf.f. & W.W.Sm.   Irrorata Gp A2

Shrub or **tree** 1–6 m (4–20 ft) high; **branchlets** glandular.
Leaves lanceolate, oblong-lanceolate or rarely oblong-oval; **lamina** rigid,
    6–12.3 cm long, 1.8–4.5 cm broad; **upper surface** olive green, matt,
    glabrous; **underside** paler, matt, glabrous, punctulate with minute red
    glands, papillate.
Petiole 1.2–2.5 cm, floccose.
Inflorescence in trusses of 8–14 flowers.
Pedicel rather densely or moderately glandular.
Calyx outside floccose, densely or moderately glandular, margin glandular.
Corolla tubular-campanulate, 3.5–4.5 cm, long, white, light yellow, white suffused
    with pink, pink, or red, with or without crimson spots; 5 nectar pouches.
Ovary densely or moderately tomentose, glandular.
Style glandular throughout to the tip.
SE and W Yunnan. 2,150–3,000 m (7,000–10,000 ft).
Epithet: With bearded style.   Hardiness 3.   March–May.

R. irroratum ssp. yiliangense N.Lancaster

Irrorata Gp A2

Shrub (in cultivation); **branchlets** sparsely glandular, detersile.
Leaves lanceolate; **lamina** rigid, 7–12 cm long, 1.9–3.9 cm broad; **upper
    surface** pale green, matt, glabrous; **underside** paler, glabrous, epapillate.
Petiole 1.8–2.5 cm, glabrous when mature.
Inflorescence in trusses of 10–12 flowers.
Pedicel glandular.
Calyx glabrous, margin moderately to densely glandular.
Corolla tubular-campanulate, 5 cm long, pale yellow or white shading to pale pink
    at the lobes, sometimes striped pink but all have a pale yellow throat, with or
    without maroon spotting.
Ovary densely puberulous with adpressed setose hairs and a few scattered
    clumps of glands.
Style glandular throughout with red-tipped glands.
Xiaocoaba, Yiliang, NE Yunnan. 1,800–2,000 m (6,000–6,600 ft).
Epithet: From Yiliang.   Hardiness 3?   April–May.

*Collected by Alan Clark under numbers AC1142, AC1152, AC1175 and AC1211,
this subspecies differs from* R. irroratum *ssp.* pogonostylum *and* R. irroratum *ssp.*
ningyuenense *as the calyx is glandular, from* R. irroratum *ssp.* irroratum *as the
ovary is floccose, and from* R. araiophyllum *as the corolla is tubular campanulate.*

Calyx floccose, margin glandular.

Ovary tomentose and glandular.

Leaf lamina rigid, underside matt, glabrous, punctulate with minute red glands, papillate.

Petiole floccose.

Style glandular throughout to the tip.

Corolla white, yellowish, pink or red.

R. irroratum ssp. *pogonostylum* at Deer Dell

Leaf underside somewhat shiny.

Calyx margin densely to moderately glandular.

Ovary densely puberulous, with a very few scattered clumps of glands.

Leaf underside epapillate.

Petiole glabrous when mature.

Style glandular throughout to the tip.

Corolla yellow shades or yellow flushed pink.

R. irroratum ssp. *yiliangense* AC1175 at Deer Dell

Ponticum
(Irrorata Group A2)

R. ningyuenense Hand.-Mazz.   Irrorata Gp A2

Shrub medium-sized; **branchlets** thinly tomentose, eglandular.

Leaves lanceolate or oblanceolate; **lamina** leathery, 4.5–9.5 cm long, 1.2–3 cm broad; **upper surface** matt, glabrous; **underside** paler, matt, glabrous, punctulate with minute red glands, papillate.

Petiole 0.8–1.2 cm, glabrous.

Inflorescence 3–6-flowered.

Pedicel moderately or sparsely floccose, glandular.

Calyx rather densely glandular, margin glandular.

Corolla tubular-campanulate, 4–4.2 cm, long, whitish rose, unspotted, 5 nectar pouches.

Ovary tomentose at the base, densely glandular.

Style glabrous, glandular throughout to the tip.

Ningyuen, Sichuan. 2,900–3,300 m (9,500–11,000 ft).

Epithet: From Ningyuen.

Alan Clark's recent introductions collected as R. ningyuenense *have a yellow corolla and a puberulous ovary with few glands: they are referable to* R. irroratum *ssp.* yiliangense.

Probably not in cultivation.

R. araiophyllum ssp. araiophyllum
Balf.f. & W.W.Sm.   Irrorata Gp B1

R. araiophyllum Balf.f. & W.W.Sm.

Shrub or **tree** 1.2–9 m (4–30 ft) high; **branchlets** glabrous or sometimes floccose, eglandular.

Leaves lanceolate; **lamina** 5–12.5 cm long, 1.6–3.2 cm broad, apex long acuminate or acuminate; **upper surface** green or olive-green, matt, glabrous; **underside** somewhat shiny, glabrous.

Inflorescence in trusses of 6–10 flowers.

Corolla broadly cup-shaped, 2.7–3.5 cm long, white, creamy-white or white suffused with rose, with or without a crimson blotch, and with or without crimson spots.

W, NW, and mid-W Yunnan, and NE Upper Burma. 2,500–3,500 m (8,000–11,000 ft).

AM 1971 (Wakehurst) 'George Taylor', flowers white with red blotch and spots.

Epithet: With slender leaves.   Hardiness 2–3.   April–May.

Rare in cultivation. Reminiscent of R. annae. Another subspecies (R. araiophyllum ssp. lapidosum) is described in the Flora of China (2005) but is probably not in cultivation.

Narrow, pointed leaves, apex long acuminate, shiny underneath.

Leaves lanceolate.

Corolla broadly cup-shaped, creamy-white or white-flushed rose.

R. araiophyllum ssp. *araiophyllum* AC589 at Deer Dell

R. lukiangense Franch.   Irrorata Gp B1

R. admirabile Balf.f. & Forrest, *R. adroserum* Balf.f. & Forrest, *R. ceracium* Balf.f. & W.W.Sm.,
R. gymnanthum Diels.

Shrub or **tree** 1–6 m (3–20 ft) high; **branchlets** glabrous or floccose, eglandular
or glandular.
Leaves oblong-lanceolate, lanceolate or oblanceolate; **lamina** 8–18.5 cm long,
1.8–5 cm broad, apex acute or acuminate; **upper surface** olive green or green,
somewhat glossy, glabrous; **underside** paler, shiny, glabrous, not papillate.
Inflorescence in trusses of 6–14 flowers.
Pedicel glabrous, eglandular.
Calyx 1–2 mm long, eglandular.
Corolla tubular-campanulate, 3.2–4.8 cm long, rose, deep rose, pale rose-pink or
pinkish-purple, magenta-rose, dark magenta-crimson, crimson-purple or red
with or without crimson blotch, and with or without crimson spots.

NW and mid-W Yunnan, SW Sichuan, SE Tibet. 1,800–4,000 m (5,900–13,000 ft).
Epithet: From Lukiang, Yunnan.   Hardiness 3.   March–April.

Rare in cultivation.

R. lukiangense Franch.   Irrorata Gp B1

R. adroserum Balf.f. & Forrest

Shrub 1–4.6 m (3–15 ft) high; **branchlets** glabrous or floccose.
Leaves oblong-elliptic, oblong-obovate or oblanceolate; **lamina** 6.8–12.8 cm
long, 2–4.6 cm broad; **upper surface** olive-green or green, somewhat
glossy, waxy, glabrous; **underside** shiny, glabrous, not papillate.
Inflorescence in trusses of 6–12 flowers.
Calyx minute, 1–2 mm long.
Corolla tubular-campanulate, 3.5–4 cm long, white flushed with magenta-
rose towards margins or white with pink tinges, with crimson spots.

NW Yunnan. 3,000–3,700 m (10,000–12,000 ft).
Epithet: Eglandular.   Hardiness 3.   April–May.

Identical to R. lukiangense *except for the usually rounded or obtuse
leaf apex. This has been submerged into the* R. lukiangense *without
explanation, and may be a form of that species at the extreme end of
the geographical range.*

Waxy shiny lower leaf surface, bronze shiny young growth.

Leaves oblong-lanceolate, lanceolate or oblanceolate, somewhat glossy above; underside paler, shiny, glabrous, not papillate.

Pedicel eglandular.

Calyx eglandular.

Corolla rose to pinkish-purple to magenta, or red.

R. lukiangense R11275 in the Valley Gardens

R. adroserum F16353, now submerged in *R. lukiangense*

R. gongshanense T.L.Ming   Irrorata Gp B2

Shrub 2.5–4 m high; young **branchlets** setose glandular.

Leaves leathery, narrowly lanceolate or oblanceolate, 12–21 cm long, 2.8–4.2 cm broad, apex acuminate, base cuneate; **upper surface** glabrous, rugose; **underside** with veins prominently raised, sparsely glandular with setose glands; detersile; papillose.

Petiole sparsely setose glandular or glabrous.

Inflorescence 18–21-flowered; **rhachis** sparsely glandular.

Pedicel sparsely glandular and pubescent.

Calyx small, glabrous.

Corolla 5-lobed, funnel-campanulate, 3–3.5 cm long, deep red with nectar pouches.

Stamens 10.

Ovary puberulous, style glabrous.

Gaoligongshan, Yunnan. 2,100–2,500 m (6,950–7,250 ft).

Epithet: From the Gongshan.   Hardiness 2–3.   March.

Recently introduced by P.A. and K.N.E. Cox.

R. kendrickii Nutt.   Irrorata Gp B2
R. pankimense Cowan.

Shrub or **tree** 1.8–7.6 m (6–25 ft) high; **branchlets** glabrous or floccose, eglandular.

Leaves lanceolate or oblong-lanceolate; **lamina** 7–17 cm long, 1.7–3.2 cm broad, apex long acuminate; **upper surface** olive-green or dark green, matt, glabrous; margin undulate; **underside** green, shiny, glabrous, not papillate.

Inflorescence in trusses of 8–16 flowers.

Pedicel moderately or rather densely floccose, eglandular.

Calyx minute, 1 mm long, eglandular.

Corolla tubular-campanulate, 2.6–3.6 cm long, pink, rose, scarlet or crimson, with or without darker spots, 5 nectar pouches.

Bhutan, S Tibet, Assam. 2,100–2,800 m (7,000–9,200 ft).

Epithet: After Dr G. Kendrick (1771–1847), a friend of the botanist Dr Nuttall.   Hardiness 1–3.   April–May.

Rare in cultivation. Closely related to R. ramsdenianum, *but has a narrower leaf, some with a crinkly margin. One could well be a variety of the other.*

Long narrow leaves with pointed tips and deeply impressed veins.

Leaf underside with veins prominently raised, sparsely glandular with setose glands which are detersile.

Corolla deep red with nectar pouches.

New foliage of *R. gongshanense* BASE9636 at Deer Dell

Stiff, narrow leaves with wavy edges: apex long, acuminate.

Pedicels moderately or rather densely floccose, eglandular.

Leaves lanceolate or oblong-lanceolate; underside green, noticeably shiny, glabrous, not papillate.

Corolla usually scarlet or crimson.

R. kendrickii at Brodick

New foliage of *R. kendrickii* KW11378 at Deer Dell

R. ramsdenianum Cowan   Irrorata Gp B2

Shrub or **tree** 1.8–12 m (6–40 ft) high; **branchlets** glabrous or floccose, eglandular.
Leaves oblong-lanceolate, lanceolate or oblanceolate; **lamina** 8–12.5 cm long,
    2.5–4 cm broad, apex acute or acuminate; **upper surface** dark green or olive-
    green, matt, glabrous; **underside** paler, shiny, glabrous, not papillate.
Inflorescence in trusses of 10–15 flowers.
Calyx 1 mm long, glabrous, eglandular.
Corolla tubular-campanulate, 2.8–4 cm long, crimson, scarlet-crimson or rose,
    with or without a dark crimson blotch, without spots, 5 nectar pouches.
SE and S Tibet. 2,100–2,600 m (7,000–8,500 ft).
Epithet: After Sir John Ramsden, who cultivated rhododendrons at Muncaster.
    Hardiness 3.   April–May.

Broader leaves than *R. kendrickii*.

R. tanastylum var. tanastylum Balf.f. & Kingdon Ward
R. tanastylum Balf.f. & Kingdon Ward   **Irrorata Gp B2**
(*R. cerochitum* Balf.f. & Forrest, *R. ombrochares* Balf.f. & Kingdon Ward)

Shrub or **tree** 1–12 m (3–40 ft) high; **branchlets** glabrous, eglandular or glandular.
Leaves oblanceolate, lanceolate, oblong-lanceolate or ovate-lanceolate; **lamina**
    7–19 cm long, 2–6.5 cm broad, apex acuminate; **upper surface** olive-green or
    green, matt, glabrous; **underside** paler, shiny, glabrous, punctulate with minute
    red glands, not papillate.
Inflorescence in trusses of 5–8 flowers.
Calyx minute, 1–2 mm long.
Corolla tubular-campanulate, 4–5.3 cm long, crimson, deep crimson, cherry-
    crimson or black-crimson, with or without deeper spots, 5 nectar pouches.
E and NE Upper Burma, W and mid-W Yunnan, Assam. 1,800–3,500 m
    (6,000–11,500 ft).
Epithet: With a long style.   Hardiness 2–3.   April–May.

R. papillatum Balf.f. & Cooper   Irrorata Gp C1

Recently said to be in cultivation. Strangely, R. epapillatum *(probably not in
cultivation) is said to be closely related.*

R. wrayi King & Gamble   Irrorata Gp C1

*Not hardy and not in cultivation in the UK. Introduced to warmer climates such as
the USA and Australia. Grows among vireyas in the wild.*

Pedicel eglandular, usually glabrous. Corolla usually red in cultivation.

Leaves oblong-lanceolate, lanceolate or oblanceolate, underside, noticeably shiny, glabrous.

R. ramsdenianum KW6284 at Deer Dell

The fewer flowers in the truss separate this species from the closely related *R. ramsdenianum*. Leaf underside noticeably shiny and comparable to *R. ramsdenianum*.

Corolla cherry-crimson, crimson to black crimson.

R. tanastylum var. *tanastylum* TH505 at Botallick (M.L.A. Robinson)

Foliage of
R. tanastylum
var. *tanastylum*
TH505 at Botallick,
Cornwall
(M.L.A. Robinson)

Ponticum
(Irrorata Group B2)

R. × agastum Balf.f. & W.W.Sm. Irrorata Gp C2

Shrub or **tree** 1.2–6 m (4–20 ft) high; **branchlets** sparsely floccose or glabrous.

Leaves oblong, oblong-obovate or oblong-elliptic; **lamina** 6–13.5 cm long, 2–5 cm broad; **upper surface** olive-green or green, matt, glabrous; **underside** matt, with a thin veil of hairs.

Inflorescence in trusses of 8–20 flowers.

Pedicel rather densely glandular.

Calyx glandular.

Corolla tubular-campanulate, 4–5 cm long, pale or deep rose, or white suffused with pink, with or without a crimson blotch, and with or without a few crimson spots, 5 nectar pouches.

Ovary densely glandular; style glandular throughout to the tip or in lower half or rarely just at the base.

W Yunnan. 2,100–3,700 m (7,000–12,000 ft).

Epithet: Charming.   Hardiness 3.   March–May.

Extremely rare in cultivation. This species has disappeared from most modern texts, and is very likely to be a hybrid, although it appears to have been collected by the Chinese. See Rhododendrons of China, Vol. 1. *According to the* Rhododendron Handbook (1998), *some plants labelled* R. agastum *may be referable to* R. papillatum.

R. tanastylum var. pennivenium D.F.Chamb.

R. pennivenium Balf.f. & Forrest   Irrorata Gp C2

Shrub 2.4–6 m (8–20 ft) high; **branchlets** glandular.

Leaves oblong-lanceolate, lanceolate or oblong-elliptic; **lamina** 8–15.5 cm long, 2.1–5.3 cm broad, apex acuminate or acute; **upper surface** olive-green or green, matt, glabrous, eglandular; **underside** matt, with a thin veil of fawn or whitish hairs.

Petiole floccose, glandular.

Inflorescence in trusses of 8–10 flowers.

Pedicel floccose, glandular.

Calyx 1–2 mm long.

Corolla tubular-campanulate, 4–4.5 cm long, rose-crimson or deep crimson, with or without crimson spots, 5 nectar pouches.

W and mid-W Yunnan, NE Upper Burma. 2,750–3,400 m (9,000–11,000 ft).

Epithet: Pinnately veined.   Hardiness 2–3.   April–May.

Leaf oblong-obovate or oblong-elliptic, underside matt with a thin veil of hairs.

Pedicel rather densely glandular.

Calyx glandular.

Ovary densely glandular.

Corolla pale or deep rose, or white tinged pink.

R. × *agastum* at Nymans

The indumentum – a thin veil of fawn or whitish hairs – is more marked than in any other species in the Subsection.

Petiole floccose, glandular.

Pedicel floccose, glandular.

Ovary tomentose.

Corolla rose-crimson or deep crimson.

R. tanastylum var. *pennivenium* at Deer Dell

Ponticum
(Irrorata Group C2)

Subgenus *Hymenanthes*
Section *Ponticum* G.Don
Subsection *Lanata* D.F.Chamb.
Campanulatum Series, Lanatum Subseries

Shrubs.
Branches and buds densely tomentose.
Leaves relatively small, elliptic to obovate with a dense indumentum almost obscuring the veins below, and often with persistent tomentum above.
Indumentum of dendroid or ramiform hairs.

Since Davidian's writings, R. lanatum *and its relatives have rightly been placed in a separate Subsection.*

R. circinnatum Cowan & Kingdon Ward

Often a bushy **shrub** or **tree** 6–7.6 m (20–25 ft) high; **branchlets** with a thick, buff or whitish-buff tomentum.
Leaves oblong or oblong-lanceolate; **lamina** coriaceous, 10–14 cm long, 2.8–4 cm broad, apex acute or shortly acuminate, base rounded or cordulate; **upper surface** olive-green, matt, glabrous, midrib grooved, densely woolly with whitish wool at the base or on the lower one-third of its length; **underside** covered with a thick, woolly, brownish-yellow, continuous, bistrate indumentum, upper layer ramiform with very short stem and narrow curly branches, lower layer with widely scattered rosulate hairs, midrib prominent, densely woolly, eglandular.
Petiole densely woolly with whitish wool, eglandular.
Inflorescence 12-flowered.
Pedicel densely floccose, glandular with short-stalked glands.
Corolla funnel-campanulate, 2.5–3 cm long, pale lemon yellow with reddish spots.
SE Tibet. 4,000–4,270 m (13,000–14,000 ft).
Epithet: Made round, i.e. coiled, alluding to the curled hairs.   Hardiness 4?
April–May.

Recently introduced. The indumentum is bistrate and brownish yellow on young leaves. Plants in cultivation as R. luciferum *are probably* R. circinnatum. R. luciferum *has a unistrate rust-coloured indumentum and is not glandular in any of its parts. It is because* R. circinnatum *has a glandular ovary and pedicel that it is best placed in subsection* Taliensia, *as Davidian suggested, rather than here, as placed by more modern authorities. It would be included in Group B of subsection* Taliensia.

Leaves of Subsect. Lanata at Deer Dell. From left to right
(two leaves each): *R. flinckii*, *R. lanatum*, *R. lanatoides*, *R. tsariense*
var. *tsariense*, *R. tsariense* var. *trimoense* and *R. poluninii*, and
(one leaf of) *R. tsariense* var. *tsariense* from KW8288. Bar = 5 cm

Note the epithet: the
bistrate indumentum
of hairs has narrow
curly branches and is
thick, continuous,
woolly and
brownish-yellow.

Corolla yellow on one
recent introduction.

Petiole densely woolly
with whitish wool,
eglandular.

R. circinnatum (as *R. luciferum*) (J.C. Birck)

Foliage of
R. circinnatum
(as *R. luciferum*)
(J.C. Birck).
Bar = 5 cm

R. flinckii Davidian

Shrub 1.5–2.5 m (5–8 ft) high; **branchlets** densely tomentose, eglandular.

Leaves oblong-oval, oblong-lanceolate, oblong-elliptic or elliptic; **lamina** somewhat thin, 4–9.8 cm long, 2–4.4 cm broad; **upper surface** pale green, matt, hairy or with vestiges of hairs (in young leaves, densely rusty-brown tomentose); **underside** covered with a thin, felty, bright rusty-brown, continuous, unistrate indumentum of hairs (in young leaves, white or creamy-white).

Petiole densely tomentose.

Inflorescence in trusses of 3–8 flowers.

Pedicel densely tomentose.

Corolla campanulate, 3.5–5 cm, yellow, with crimson spots.

Bhutan. 3,000 m (10,000 ft).

Epithet: After Mr. K.E. Flinck, Sweden.   Hardiness 3.   April–May.

Not distinguished as a separate species from R. lanatum *by Chamberlain, but in cultivation,* **the thinner leaves with their thin indumentum and the young foliage are distinct***. The type form (Cooper 3990) has pink flowers, and, in cultivation, seems to be a variable species in flower colour, leaf size and indumentum. These differences lend weight to the suggestion that this species may be part of a cline between* R. lanatum *and* R. tsariense.

R. lanatoides D.F.Chamb.

Dense **shrub** or a small **tree** 2–4 m high; **branchlets** densely tomentose.

Leaves lanceolate 9–11 cm long, 2.1–7 cm broad; **upper surface** dark green with some persistent indumentum, otherwise glabrous and shiny; **underside** covered with a thick brown to fawn indumentum of dendroid hairs.

Petiole 1–1.5 cm long, densely tomentose.

Inflorescence 10–15 flowers.

Corolla campanulate, 3.5–4 cm long, white flushed with pink, with a few faint markings.

SE Tibet. 3,200–3,700 m (10,500–12,000 ft).

Epithet: Resembling *R. lanatum*.   Hardiness 4.   February–April.

Early flowering.

Leaf lamina somewhat thin. Indumentum thin, felty, continuous. Corolla yellow or pink.

Young leaves densely rusty-brown tomentose. Underside of young leaves white or creamy-white.

R. flinckii Cooper 3990 at Deer Dell

R. flinckii L&S2858 in the Valley Gardens

R. flinckii (Glenarn form) at Deer Dell

Corolla white-flushed pink. Easily distinguished by its lanceolate leaf shape and thick mustard-coloured indumentum

R. lanatoides KW5971 at Deer Dell

New foliage of *R. lanatoides* KW5971 at Deer Dell

Ponticum

(Lanata)

R. lanatum Hook.

Shrub or sometimes a small **tree** 0.3–3 m (1–10 ft) high; **branchlets** densely tomentose with white, tawny or brown tomentum.

Leaves obovate, oblong-obovate or elliptic; **lamina** 6–12 cm long, 1.8–5 cm broad; **upper surface** dark green, shiny, glabrous; **underside** covered with a thick coffee- or deep rust-coloured, woolly, continuous, unistrate indumentum of ramiform hairs.

Petiole densely tomentose.

Corolla broadly campanulate, 3.2–4.8 cm long, yellow, pale yellow, pale sulphur-yellow or lemon-yellow, with red spots.

Sikkim, Bhutan, East Himalayas. 3,000–4,500 m (10,000–15,000 ft).

Epithet: Woolly.   Hardiness 2–3.   April–May.

R. luciferum (Cowan) Cowan

Shrub or **tree** 1.2–8 m (4–25 ft) high; **branchlets** densely tomentose.

Leaves oblong, elliptic or oblong-elliptic; **lamina** 8–14 cm long, 2.5–5.4 cm broad, apex abruptly acute or abruptly acuminate; **upper surface** dark green, shiny, glabrous; **underside** covered with a thick, dark brown or rust-coloured, continuous, unistrate indumentum of hairs.

Petiole densely tomentose.

Inflorescence in trusses of 8–10 flowers.

Pedicel densely tomentose.

Calyx densely tomentose.

Corolla funnel-campanulate, 3–4.3 cm long, pale lemon-yellow, lemon-yellow or pale yellow, with or without red spots.

S and SE Tibet. 3,400–4,000 m (11,000–13,000 ft).

Epithet: Light-bringing.   Hardiness 3.   April–May.

Almost certainly not in cultivation (see R. circinnatum).

R. poluninii Davidian

Shrub; **branchlets** densely rusty-brown tomentose, eglandular.

Leaves lanceolate, oblong-lanceolate or oblong; **lamina** 6.2–8 cm long, 2.6–2.8 cm broad; **upper surface** dark green, shiny, glabrous, midrib densely tomentose; **underside** covered with a thick woolly, pale rusty-brown, continuous, unistrate indumentum of hairs (in young leaves, yellowish-brown).

Petiole densely rusty-brown tomentose.

Inflorescence in trusses of 5–11 flowers.

Corolla tubular-campanulate, 3.5–3.8 cm long, creamy-white or creamy-white lightly suffused pink, with numerous crimson spots.

Central Bhutan. 4,000 m (13,000 ft).

Epithet: After Oleg Polunin, notable plant collector.   Hardiness 3.   April–May.

Rare in cultivation.

Distinct thick usually coffee-coloured indumentum. Corolla yellow.

R. lanatum at Deer Dell

Distinguished from *R. tsariense* by its larger and usually lanceolate or oblong lanceolate leaves and the mustard-coloured indumentum; these characteristics make *R. poluninii* a close relative of *R. lanatum*. Corolla creamy-white or creamy-white slightly tinged pink.

R. poluninii (J.C. Birck)

Ponticum

(Lanata)

Forms of *R. tsariense*

*Field studies by P.A. and K.N.E. Cox have shown that Davidian's varieties of
R. tsariense are the extremes of a continuous variation in size and leaf shape
within the wild population. R. flinckii and R. tsariense, however, may be be
distinguished from each other by the thickness of the indumentum and size
of leaf.*

R. tsariense var. tsariense Cowan

R. tsariense Cowan

Shrub 0.6–3.7 m (2–12 ft) high; **branchlets** densely tomentose with brown or
cinnamon tomentum.

Leaves obovate, elliptic or oblong-elliptic; **lamina** leathery, 1.8–5.5 cm long,
1–2.8 cm broad; **upper surface** dark green, shiny, glabrous (in young leaves,
brown tomentose); **underside** covered with a thick, woolly, rust-coloured or
cinnamon, continuous, unistrate indumentum of hairs.

Petiole densely tomentose.

Inflorescence in trusses of 2–5 flowers.

Pedicel densely tomentose.

Corolla campanulate, 2.6–4 cm long, pink, white, or white suffused with pink,
with or without red spots.

Bhutan, S and SE Tibet, Assam. 3,000–4,500 m (10,000–14,500 ft).

AM 1964 (Tremeer) 'Yum-Yum', flowers white, flushed pink.

Epithet: From Tsari, SE Tibet.   Hardiness 3.   March–May.

*A form in cultivation under the collector's number KW8288, having a very thick
indumentum on a tsariense-type leaf, may be a new species.*

R. tsariense Cowan **var. *magnum*** Davidian

This variety differs from the species type in that the leaves are large,
laminae 5.6–8.5 cm long, 2.4–4 cm broad, and in that it is a large plant
2.4–3 m (8–10 ft) high.

Bhutan. 3,800–4,100 m (12,500–13,500 ft).

Epithet: Large.   Hardiness 3.   April–May.

Status uncertain.

R. tsariense var. trimoense Davidian

S Tibet. 3,400 m (11,000 ft).

Epithet: From Trimo, S Tibet.   Hardiness 3.   April–May.

Small roundish leaves and more dwarf habit distinguish this species from *R. flinckii*. Indumentum woolly, rust-coloured or cinnamon. Corolla pink, white, or white-tinged pink.

R. tsariense var. *tsariense* (J.C. Birck)

Larger growing and with larger leaves than usual for *R. tsariense*. Could be confused with *R. flinckii*, but var. *magnum* variety has thick indumentum.

R. tsariense var. *magnum* Cooper 2148 at RBG Edinburgh

Leaf indumentum, branchlets, petioles, pedicels, calyx and ovaries are pale cream, pale fawn coloured, or almost white.

R. tsariense var. *trimoense* at Deer Dell

Ponticum
(Lanata)

Subgenus *Hymenanthes*
Section *Ponticum* G.Don
Subsection *Maculifera* Sleumer
Barbatum Series, Maculiferum Subseries

Shrubs or trees with rough bark.
Branchlets usually at least moderately floccose.
Leaf lamina usually glabrous.
Midrib on the leaf underside floccose.
Petiole usually at least moderately floccose

R. anwheiense E.H.Wilson
(*R. maculiferm* ssp. *anwheiense* D.F.Chamb.)

Rounded **shrub** 1–3.6 m (3–12 ft) high; **branchlets** floccose.
Leaves ovate-lanceolate, elliptic or oblong; **lamina** 3–8.5 cm long, 1.5–3.6 cm broad; **upper surface** green, glabrous; margin recurved; **underside** paler, matt, glabrous.
Petiole floccose.
Inflorescence in trusses of 6–12 flowers.
Calyx minute 1 mm long.
Corolla funnel-campanulate, 2.4–3.5 cm long, white, white suffused with pink, or pink, with crimson spots.
Anhuei, China. 1,200–1,800 m (4,000–6,000 ft).
AM 1976 (Windsor), flowers white flushed pink, with red spots.
Epithet: From Anhwei, China.   Hardiness 3.   April–May.

Shows affinity to Subsection Irrorata *where Davidian placed it, but morphologically it looks more like some Subsection* Pontica *species.*

Leaves of Subsect. *Maculifera* at Deer Dell. From left to right (two leaves each): *R. strigillosum, R. pachytrichum, R. maculiferum* and *R. longesquamatum.* Bar = 5 cm

This species purports to have hairs and glands on the midrib: these are certainly detersile with age.

Calyx and ovary glabrous.

Petiole floccose.

Leaf upper surface green, glabrous; margin recurved; underside paler, matt, glabrous. Corolla usually white, tinged pink.

R. anwheiense AM form in the Valley Gardens

The following two species are newly introduced and closely related; they may not be distinct.

R. leishanicum Fang & S.S.Chang

Shrub to approx. 3 m; **young shoots** setulose.

Leaves broadly oblong, 3.5–6 cm by 2.5–3.5 cm, usually less than twice as long as broad, apex rounded, apiculate, base rounded to sub-cordate, **underside** with glabrous, **midrib** setulose up to half its length with a dark brown matted indumentum of a form of dendroid or of fagellate hairs.

Petioles 5–10 mm, with a dense setulose sticky tomentum.

Inflorescence 2–3-flowered.

Pedicel densely setulose.

Calyx tomentose.

Corolla open-campanulate, purple with a dark blotch and 5 nectar pouches.

Ovary densely setulose-tomentose.

Style glabrous.

E Guizhou, China. 1,850 m (6,000 ft).

Epithet: From Leishan, Guizhou.   Hardiness 4?   April–May.

As yet very rare in cultivation. This description is based on that of Chamberlain and on specimens of a plant in cultivation.

R. oligocarpum Fang ex X.Y.Zhang

Shrub or small **tree** 4–6 m (13–20 ft) high, **branchlets** tomentose when young.

Leaves leathery, oblong, oblong-elliptic or obovate-elliptic, apex rounded, apiculate, base rounded, 4–6.5 cm long, 2–3 cm broad, **upper surface** dark green, glabrous, **underside** pale green, the midrib brownish setulose near the base otherwise glabrous.

Petiole tomentose, grooved above.

Inflorescence 3–4-flowered.

Pedicel 1–5 cm long, densely pubescent.

Calyx densely pubescent.

Corolla campanulate, purple with darker purple spots.

Stamens 10, puberulous at base; **ovary** densely tomentumose; **style** glabrous.

Guizhou, Guangxi, China. 1,800–2,500 m (5,900–8,250 ft).

Epithet: Few fruited.   Hardiness 4?   April–May.

This species is newly introduced and closely related to R. leishanicum. *It is not clear whether the corolla should have nectar pouches or not. If it has, then* R. oligocarpum *would seem to be identical to* R. leishanicum.

Leaves broadly oblong, oblong, oblong-elliptic or obovate-elliptic; underside midrib setulose up to half its length with a dark brown matted indumentum.

Corolla campanulate, purple with a dark blotch.

A probable *R. leishanicum* at Hindleap Lodge (M.L.A. Robinson)

A purported *R. oligocarpum* at Glendoick (M.L.A. Robinson)

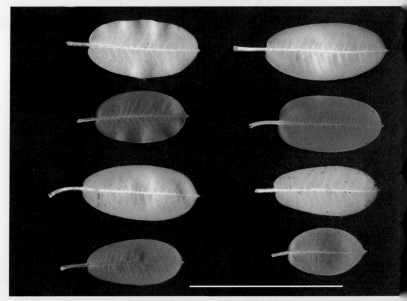

Leaves of *R. leishanicum* (left column) and *R. oligocarpum* (right column): leaves of these two species do not differ discernably.
Bar = 5 cm

Ponticum
(Maculifera)

R. longesquamatum Schneider

(*R. brettii* Hemsl. & E.H.Wilson)

Shrub 2–6 m (6–20 ft) high; **branchlets** densely hairy with very long, coarse, shaggy, branched, rusty hairs, eglandular; leaf-bud scales persistent.
Leaves oblong, oblanceolate or oblong-obovate; **lamina** thick, somewhat rigid, 6–13.5 cm long, 2–4.2 cm broad; **upper surface** dark green, shiny, somewhat rough, eglandular, midrib with long branched hairs; **underside** somewhat shiny, glabrous, punctulate with minute glands, rough, midrib hairy with long, coarse rusty hairs.
Petiole densely hairy with very long rusty hairs.
Inflorescence in trusses of 6–12 flowers.
Pedicel hairy.
Calyx glandular.
Corolla campanulate, 3.5–4.5 cm long, pink, rose or rosy-red, with deep crimson blotch.
W Sichuan. 2,300–3,500 m (7,500–11,500 ft).
Epithet: With long scales.   Hardiness 3.   May.

R. maculiferum Franch.

Shrub or **small tree** 1–10 m (3–33 ft) high; **branchlets** floccose or glabrous, eglandular.
Leaves oblong-oval, oblong, obovate or elliptic; **lamina** 5–11.6 cm long, 2.3–4.5 cm broad; **upper surface** olive-green, matt, glabrous; **underside** pale silvery green, shiny, glabrous, midrib prominent, densely or moderately hairy one-half to two-thirds its length or its entire length.
Inflorescence in trusses of 6–10 flowers.
Calyx floccose, eglandular.
Corolla campanulate, 2.6–3.5 cm long, white, pink or white suffused with pink, with or without deep purple blotch, with or without dark spots.
Sichuan, W and NW Hubei, Guizhou. 1,600–3,300 m (5,300–11,000 ft).
Epithet: Spotted.   Hardiness 3.   April–May.

Very hairy stems with a thick petiole and midrib base (hairy with long, coarse, rusty hairs) and persistent perulae.

Branchlets densely hairy with very long, coarse, shaggy, branched, rusty hairs.

Calyx glandular.

Corolla pink, rose or rosy red.

R. longesquamatum at Deer Dell

Less hairy than *R. pachytrichum* and with broader leaves that are pale shiny silvery-green beneath. Midrib densely or moderately hairy one-half to two-thirds it length or its entire length.

Calyx and ovary floccose.

Corolla white, pink or white flushed with pink.

R. maculiferum at Deer Dell

Ponticum
(Maculifera)

R. morii Hayata

(*R. nakotaisanense* Hayata)

Shrub or **tree** 1.5–7.6 m (5–25 ft) high; **branchlets** glabrous or floccose, eglandular or glandular.

Leaves lanceolate or oblong-lanceolate; **lamina** 6.5–15 cm long, 1.9–4.1 cm broad; **upper surface** dark green or pale green, glabrous; **underside** pale green, somewhat shiny, glabrous, midrib floccose, glandular.

Inflorescence in trusses of 5–15 flowers.

Pedicel glandular.

Corolla campanulate, 3.5–5 cm long white or white suffused with rose, with crimson blotch and crimson spots.

Ovary hairy, glandular.

Taiwan. 2,000–3,200 m (6,500–10,800 ft).

AM 1956 (Collingwood Ingram), flowers white, with crimson blotches and spots.

Epithet: After U. Mori, collector in Taiwan.   Hardiness 3.   April–May.

R. ochraceum Rehd. & E.H.Wilson

Shrub or small **tree** up to 3 m, **branchlets** setulose-glandular.

Leaves oblanceolate or lanceolate; **lamina** 4–9.5 cm long, 0.9–2.5 cm broad, apex acuminate; **upper surface** matt, glabrous or with vestiges of hairs, midrib grooved, hairy or with vestiges of hairs, glandular with medium or long-stalked glands or eglandular; **underside** with somewhat thick, woolly, continuous, brown, unistrate indumentum of hairs, primary veins concealed.

Petiole 0.9–2 cm long, setulose-glandular.

Inflorescence 6–12 flowers.

Pedicel moderately or rather densely glandular.

Corolla broadly campanulate or tubular-campanulate, 2.5–4 cm long, crimson, unspotted, 5 nectar pouches at the base.

Stamens 10–12.

Ovary densely glandular with long-stalked strigose glands, hairy with strigose hairs or glabrous, calyx lobes persistent.

S Sichuan, N Yunnan. 1,700–3,000 m (5,500–10,000 ft).

Epithet: Yellowish.   Hardiness 3?   April–May.

Morphologically, this species fits better into Subsection Neriiflora*, where Davidian placed it, but geographically it is closer to Subsection* Maculifera*, where more recent authorities have placed it.*

Superficially similar to *R. irroratum* but has a floccose glandular midrib.

Corolla white or white tinged with rose, with crimson blotch and crimson spots.

R. morii RV73-100 at Deer Dell

Dense growing, has small pointed leaves with a matt upper surface, and has a continuous, brown or fawn indumentum, primary veins concealed.

Corolla scarlet-crimson with 5 nectar pouches.

R. ochraceum AC1056 at Deer Dell

Ponticum
(Maculifera)

R. pachysanthum Hayata

We give first a shortened form of Hayata's original 1913 description:

Branchlets tomentose at first, later glabrescent.
Leaves oblong, 8.9 cm long, 3.5 cm wide, apex acute; **upper surface** glabrous, **underside** whitish brown tomentose beneath (with close packed bristly hairs), becoming glabrous, margin recurved.
Petioles 1.5 cm long, tomentose above.
Inflorescence in trusses of 11–20.
Pedicel pubescent, glandular.
Calyx very short, ciliate glandular at the edges.
Corolla widely campanulate, 4 cm long with sides and dorsal lobe densely spotted, glabrous.
Stamens 10.
Ovary densely glandular.
Style glabrous.
Capsule densely hairy.

Davidian's description, below, appears to be from cultivated plants:

Compact rounded **shrub** 1–1.2 m (3–4 ft) high; **branchlets** densely tomentose, eglandular.
Leaves ovate, ovate-lanceolate or lanceolate; **lamina** 5.8–11.8 cm long, 2.5–4 cm broad; **upper surface** dark green, sometimes becoming glabrous (in **young leaves**, densely white tomentose); margin recurved; **underside** covered with a thick, woolly, cinnamon-coloured or rusty-brown, continuous, bistrate indumentum of hairs, without pellicle.
Petiole densely tomentose.
Inflorescence in trusses of 8–10 or sometimes up to 20 flowers.
Pedicel hairy, glandular.
Calyx minute, 1 mm long, hairy, glandular.
Corolla campanulate, 3.1–4 cm long, white with crimson spots.
Stamens 10.
Ovary rather densely glandular.
Style sparsely glandular or eglandular.
Taiwan. 3,000–3,200 m (10,000–10,500 ft).
Epithet: With thick flowers.   Hardiness 3.   April–May.

*Well known and widely cultivated. Although quite variable in the colour and persistence of the leaf tomentum, **the species in cultivation does not match Hayata's 1913 description**. It will therefore require a different name. It may be significant that Stevenson, basing his judgement on Hayata's description, includes R. pachysanthum under R. morii.*

Upper surface of young leaves is densely silvery white tomentose becoming browner, the tomentum often persisting.

Leaf margin recurved; underside covered with a thick, woolly, cinnamon-coloured or rusty-brown, continuous, persistent indumentum.

Leaves usually ovate-lanceolate.

Pedicel glandular.

Calyx glandular.

Corolla white with crimson spots.

R. pachysanthum at Deer Dell

Young foliage of *R. pachysanthum* at Deer Dell

R. pachytrichum var. pachytrichum

R. pachytrichum Franch.

Shrub 1–6 m (4–20 ft) high; **branchlets** rather densely with long shaggy, curly, branched, brownish hairs, eglandular.

Leaves oblanceolate, oblong, lanceolate or oblong-lanceolate; **lamina** 6–13.6 cm long, 2–4 cm broad; **upper surface** bright green, glabrous; **underside** pale green, shiny glabrous, midrib rather densely or moderately hairy with long, shaggy, curly, branched browny hairs.

Petiole rather densely hairy with long curly hairs, eglandular.

Inflorescence in trusses of 6–15 flowers.

Pedicel densely hairy.

Corolla campanulate, 2.8–4.2 cm long, white, pale rose, deep purple or red, with or without a deep purple blotch.

W Sichuan. 2,300–3,600 m (7,500–12,000 ft).

AM 1963 (Bodnant) 'Sesame', flowers white tinged with purple.

Epithet: With thick hairs.   Hardiness 3.   March–April.

Very variable in flower colour.

R. pachytrichum var. monosematum D.F.Chamb.

R. monosematum Hutch.

Shrub 1.5–6 m (5–20 ft) high; **branchlets** moderately or rather densely setulose-glandular and often rather densely setulose.

Leaves lanceolate, oblong-lanceolate or oblong; **lamina** 12.1 cm long, 2–4.3 cm broad, apex acuminate; **upper surface** dark green, glabrous; **underside** pale green, glabrous, midrib hairy or not hairy, often setulose at the base.

Petiole moderately or rather densely setulose-glandular.

Inflorescence in trusses of 6–12 flowers.

Pedicel glandular.

Calyx glandular.

Corolla tubular-campanulate, 2.8–5.2 cm long, white, pink, reddish-purple or deep red, with purple blotch.

W Sichuan. 2,000 m (6,500 ft).

Epithet: With one blotch.   Hardiness 3.   March–April.

Most forms have distinctive long curled brownish branched hairs on the branchlets, petiole, and leaf underside midrib.

Corolla white, rose, purple or red.

R. pachytrichum var. *pachytrichum* at Benmore, Scotland

Branchlets and petiole usually setulose rather than setose (i.e. with shorter glandular hairs than *R. pachytrichum* var. *pachytrichum*). In cultivation, usually has white or pinkish flowers and broader leaves.

Leaf apex acuminate.
Pedicel glandular.
Calyx glandular.
Corolla white or pink in cultivation so far.

R. pachytrichum var. *monosematum* in the Valley Gardens

R. pseudochrysanthum Hayata

Shrub 0.3–3 m (1–10 ft) high; **branchlets** floccose, glandular.

Leaves ovate, elliptic, oblong-lanceolate or ovate-lanceolate; **lamina** thick, coriaceous, rigid, 2–8 cm long, 1.3–3.8 cm broad; **upper surface** green, shiny, glabrous, eglandular (in young leaves, floccose, glandular); margin recurved; **underside** pale green, shiny, glabrous, eglandular, midrib floccose, glandular.

Petiole floccose, glandular.

Inflorescence in trusses of 8–10 flowers.

Pedicel glandular.

Corolla campanulate, 3–4.3 cm long, white, white suffused with pink, or pink with deeper rose lines outside, with crimson spots.

Taiwan. 1,800–4,000 m (6,000–13,000 ft).

AM 1956 (Exbury), flowers white, flushed pink, spotted crimson.

Epithet: False *R. chrysanthum* (*R. aureum*).   Hardiness 3.   April–May.

R. strigillosum Franch.

Shrub or **tree** 2–7 m (6–23 ft) high; **branchlets** rather densely or sparsely bristly-glandular, and often bristly, bristles long, stiff.

Leaves oblong-lanceolate or oblanceolate; **lamina** 7.5–17.5 cm long, 1.8–4.5 cm broad, apex long acuminate; **upper surface** olive-green or dark green, somewhat shiny, glabrous, eglandular; **underside** pale green, rather densely setulose, midrib bristly and rather densely tomentose.

Petiole rather densely bristly; **leaf-buds** sticky.

Inflorescence in trusses of 8–12 flowers.

Pedicel setulose- or setose-glandular; **flower buds** sticky.

Calyx setulose-glandular.

Corolla tubular-campanulate, 4–6 cm long, crimson, scarlet, rose-pink or rarely white.

W Sichuan. 2,300–3,350 m (7,500–11,000 ft).

AM 1923 (Bodnant), flowers crimson.

Epithet: With short bristles.   Hardiness 3.   February–April.

Small stiff leaves, shiny beneath with a pale floccose midrib.

Branchlets floccose, glandular.

Petiole floccose, glandular.

Corolla white or pale pink.

R. pseudochrysanthum 'White Edge' at Deer Dell

New foliage of *R. pseudochrysanthum* 'White Edge' at Deer Dell

Bristly rather than hairy.
Flowers usually red in cultivation.

Leaf buds sticky.

Leaf apex long acuminate, underside pale green, rather densely setulose, midrib bristly and rather densely tomentose.

Flower buds sticky.

Calyx setulose-glandular.

R. strigillosum at Deer Dell

Ponticum
(Maculifera)

Subgenus *Hymenanthes*
Section *Ponticum* G.Don
Subsection *Neriiflora* Sleumer
Neriiflorum Series

Prostrate shrubs to small trees.
Branchlets usually thin, tomentose.
Bark often peeling, brownish.
Inflorescence often loose, hanging between the leaves.
Corolla with 5 nectar pouches, often fleshy.
Stamens 10.

The four Subseries used by Davidian are helpful in identifying plants. The unofficial term 'Group' used here is similar to the previous Balfourian classification.

Forrestii Group. *Dwarf creeping or compact shrubs not over 70 cm high. Leaves, small, glabrous or with a thin discontinuous indumentum.*

> R. chamaethomsonii (var. chamaethomsonii, var. chamaethauma and var. chamaedoron), R. forrestii and R. trilectorum.

Haematodes Group. *Thick woolly brown or cinnamon indumentum. Leaves usually broadly or oblong obovate.*

> R. beanianum, R. catacosmum, R. chionanthum, R. coelicum, R. haematodes (ssp. haematodes and ssp. chaetomallum), R. × hemigynum, R. mallotum, R. piercei, R. pocophorum (var. pocophorum and var. hemidartum).

Neriiflorum Group. *Ovary slender, usually tapered into the style. Leaves small, oblong to lanceolate, mostly glaucous papillate beneath. Shrubs or small trees.*

> R. albertsenianum, R. floccigerum, R. neriiflorum (ssp. neriiflorum, ssp. agetum, var. appropinquans and ssp. phaedropum), R. sperabile (var. sperabile and var. weihsiense) and R. sperabiloides.

Sanguineum Group. *Ovary impressed, i.e. not tapered onto the style. All species have thinner indumentum than the Haematodes Group in which the leaf veins are not or are barely visible through the thick woolly indumentum.*

Further grouping within Sanguineum may be helpful:-

Sanguineum Group (Part A). *Moderately thick indumentum, but leaf veins visible.*

> R. citriniflorum var. citriniflorum, R. horaeum and R. horaeum var. rubens.

Sanguineum Group (Part B). *Indumentum appears continuous and is adpressed, whitish or fawn.*

> R. dichroanthum (ssp. dichroanthum, ssp. apodectum, ssp. scyphocalyx and ssp. septentrionale), R. eudoxum (var. eudoxum, var. brunneifolium and var. mesopolium), R. microgynum, R. parmulatum, and R. sanguineum (ssp. sanguineum [including var. cloiophorum, var. didymoides, var. haemaleum and var. himertum] and ssp. didymum).

Sanguineum Group (Part C). *Leaves glabrous or almost glabrous.*

> R. aperantum and R. temenium (var. temenium and var. gilvum).

All authorities agree that many more field studies are needed to distinguish the stable species in the Sanguineum Group and it should be recorded here that, when asked about the purity of species in this Group, Davidian admitted that there is a considerable amount of hybridity in the wild and thus in cultivation.

The species most commonly grown are R. aperantum, R. citriniflorum var. citriniflorum, R. horaeum, R. dichroanthum, R. parmulatum, R. sanguineum (ssp. sanguineum and ssp. didymum, and var. haemaleum).

Subsection *Neriiflora* Sleumer
Forrestii Group

Dwarf creeping or compact shrubs, less than 70 cm high.
Leaves, small, glabrous or with a thin discontinuous indumentum.

Leaves of the Forrestii and Haematodes Groups of Subsect. Neriiflora
at Deer Dell. Top row (left to right): **Forrestii Group**
R. forrestii (4 leaves) and *R. chamaethomsonii* (2 leaves);
Haematodes Group *R. haematodes* (1 leaf). Middle row (left to
right): **Haematodes Group** *R. chaetomallum*, *R. coelicum* and
R. pocophorum. Bottom row (left to right): **Haematodes Group**
R. catacosmum, *R. piercei* (1 leaf of each) and *R. beanianum* (2 leaves).
Bar = 5 cm

Ponticum
(Neriiflora/Forrestii Group)

R. chamaethomsonii var. chamaethomsonii

Forrestii Group

R. chamaethomsonii (Tagg & Forrest) Cowan & Davidian

Shrub, broadly upright or compact, 30–90 cm (1–3 ft) high; **branchlets** glandular or eglandular.

Leaves obovate, oblong-obovate or oblong; **lamina** rigid, 3–9 cm long, 2–4 cm broad; **upper surface** dark green, glabrous; **underside** pale glaucous green or pale green, glabrous or with a thin discontinuous indumentum of hairs.

Inflorescence 1–5 flowers.

Corolla tubular-campanulate, deep crimson or crimson, 5 nectar pouches.

E and SE Tibet, NW Yunnan, N Burma. 3,300–4,600 m (11,000–15,000 ft).

Epithet: Dwarf *R. thomsonii.*   Hardiness 3.   March–May.

Note the epithet: from this, it is clear that KW21073 is the true species. Many of the other plants in cultivation are natural hybrids of R. forrestii.

R. chamaethomsonii var. chamaethauma

(Tagg) Cowan & Davidian

Forrestii Group

The variety differs from *R. chamaethomsonii* var. *chamaethomsonii* in its smaller leaves; **lamina** 1.6–5 cm long, 1–3 cm broad.

The flower colour is very variable, deep crimson, crimson, deep crimson-rose, deep scarlet, rose-red, rose or pale pink.

SE Tibet, NW Yunnan. 3,400–4,600 m (11,000–15,000 ft).

1932 (Bodnant) as *R. repens* var. *chamaedoxa*, flowers crimson.

Epithet: Marvellous dwarf plant.   Hardiness 3.   March–May.

*The only recorded stable population in the wild has **pale to deep pink flowers**. Other colours may be hybrids.*

R. chamaethomsonii var. chamaedoron

(Tagg & Forrest) D.F.Chamb.

Forrestii Group

R. chamaethomsonii var. *chamaethauma* (Tagg) Cowan & Davidan

Eglandular branchlets and petioles with a thin discontinuous indumentium on the leaf underside.

Likely to be a natural hybrid. A form of R. chamaethomsonii *var.* chamaethauma *with thin indumentum and glandular below the leaf.*

More upright
growing than
R. forrestii with
somewhat larger
leaves: the habit is
not particularly
compact.

Slightly rugulose
upper leaf surface;
leaf underside
almost glabrous,
pale glaucous green
or pale green.

Corolla crimson.

R. chamaethomsonii var. *chamaethomsonii* KW21073 at
Deer Dell

Smaller leaves than
R. chamaethomsonii
var. *chamaethomsonii*.

Corolla deep
crimson-rose or
pale pink.

R. chamaethomsonii var. *chamaethauma* 'Exbury Pink' at Deer Dell

Ponticum
(Neriiflora/Forrestii Group)

R. forrestii Balf.f. ex Diels.   Forrestii Group

Creeping or prostrate **shrub** 3–45 cm (1–1.5 ft) high; **branchlets** glandular;
   leaf-bud scales persistent.
Leaves obovate, orbicular, elliptic or oblong-elliptic; **lamina** rigid, 0.6–5.5 cm
   long, 0.5–3.4 cm broad; **upper surface** dark green, semi-bullate; **underside**
   deep purple-red, glabrous, glandular with sessile or short-stalked glands.
Petiole floccose, glandular.
Inflorescence 1–2 flowers.
Pedicel hairy or glabrous, glandular.
Calyx outside often glandular, margin glandular.
Corolla tubular-campanulate, 2.5–4 cm long, crimson, scarlet or carmine; 5 nectar
   pouches at the base.

SE and E Tibet, NW Yunnan, NE Upper Burma. 3,000–4,400 m (10,000–14,500 ft).
Epithet: After George Forrest (1873–1932).   Hardiness 3.   April–May.

*Many rhododendrons have juvenile leaves with purple undersides. The leaf
underside of this species may also be green, so Davidian's description is misleading
in this respect. Some growers still distinguish between **Repens Group** (prostrate)
and **Tumescens Group** (mound-forming).*

R. pyrrhoanthum, *described by Davidian but not by more recent authors,
originated as a rogue in a batch of* R. forrestii *collected by Forrest. It was
rediscovered by Rock subsequently, and may therefore be a well-defined species.
It has **larger leaves than** R. forrestii (5.5–8 cm long) and a **4–5-flowered
crimson truss**. It is very rare in cultivation.*

R. trilectorum Cowan   Forrestii Group

Prostrate **shrub** 20–30 cm (9–12 inches) high; **branchlets** reddish brown,
   glabrous: **leaf-bud scales** persistent.
Leaves subsessile, obovate; **lamina** rigid, leathery, 0.6–3.2 cm long, 0.6–1.8 cm
   broad; **upper surface** dark green, semi-bullate; **underside** pale green,
   glabrous, eglandular.
Petiole very short.
Inflorescence usually with 3 flowers.
Pedicel glabrous.
Calyx glabrous.
Corolla funnel or tubular-campanulate, 3–4 cm long, pale yellow flushed pink.

SE Tibet, Arunachal Pradesh. 3,700–4,300 m (12,000–14,000 ft).
Epithet: Three collectors.   Hardiness 3?   March–May.

Introduced in 2003 from the Arunachal Pradesh by P.A. and K.N.E. Cox.

Small round rugose leaves and low growth habit.

Inflorescence 1–2-flowered, pedicel glandular.

Corolla tubular-campanulate, crimson, scarlet or carmine.

R. forrestii 'Tumescens' at RBG Edinburgh (Davidian)

Very similar to *R. forestii* in leaf, but the leaves are subsessile and more finely veined. Corolla pale yellow, flushed pink.

Branchlets reddish brown, glabrous.

Leaf underside pale green, eglandular.

Inflorescence usually 3-flowered.

Pedicel glabrous.

New foliage of *R. trilectorum* HECC10056 at Deer Dell

Ponticum
(Neriiflora/Forrestii Group)

Subsection *Neriiflora* Sleumer
Haematodes Group

**Thick woolly brown or cinnamon indumentum.
Leaves usually broadly or oblong obovate.**

R. beanianum Cowan   Haematodes Group

Lax **shrub** 1–3 m (3–10 ft) high; **branchlets** moderately or densely bristly and
often bristly-glandular.

Leaves oblong-obovate or oblong; **lamina** 4.8–10.3 cm long, 2.3–4.6 cm broad;
upper surface dark green, rugulose; **underside** covered with woolly, continuous,
brown or cinnamon-brown, bistrate or unistrate indumentum of hairs.

Petiole bristly and often bristly glandular.

Inflorescence in trusses of 4–10 flowers.

Corolla tubular-campanulate, fleshy, 3–4.5 cm long, deep crimson or crimson; 5
nectar pouches.

Upper Burma. 3,000–3,400 m (10,000–11,000 ft).

AM 1953 (Minterne), flowers crimson (KW6805).

Epithet: After W.J. Bean (1863–1947), former Curator, Royal Botanic Gardens,
Kew.   Hardiness 3.   March–May.

R. beanianum (red) from KW6805 at Deer Dell

Straggly growth. Similar to *R. piercei*, but has smaller dark green and very rugulose leaves and bristly branchlets and petioles. Indumentum woolly, continuous, brown or cinnamon-brown.

Corolla crimson, red, or occasionally pink.

New foliage of *R. beanianum* KW6805 at Deer Dell

R. beanianum (pink) form KW6805 in the Valley Gardens

Ponticum
(Neriiflora/Haematodes Group)

R. catacosmum Balf.f. ex Tagg **Haematodes Group**

Rounded **shrub** 1–2.8 m (4–9 ft) high; **branchlets** densely tomentose, not bristly.

Leaves obovate; **lamina** 5–11 cm long, 2.5–5.3 cm broad; **upper surface** green, matt; **underside** covered with a thick, woolly, continuous, cinnamon-brown, dark brown or brown, bistrate indumentum of hairs.

Petiole densely tomentose.

Inflorescence in trusses of 4–9 flowers.

Pedicel densely tomentose, not bristly.

Calyx large, 1.6–2.5 cm long, red or crimson, cup-shaped, irregularly lobed.

Corolla broadly campanulate, fleshy, 4–4.8 cm long, crimson, crimson-rose or pink, 5 nectar pouches.

E and SE Tibet. 4,000–4,300 m (13,000–14,000 ft).

Epithet: Adorned, referring to the calyx.   Hardiness 3.   April–May.

R. chionanthum Tagg & Forrest **Haematodes Group**

Indumentum thick woolly, interrupted, in shreds or patches, dark brown.
Corolla white.

Somewhat rounded **shrub** 60–90 cm (2–3 ft) high; **branchlets** bristly.

Leaves obovate or oblanceolate; **lamina** 5–7.6 cm long, 2.3–3 cm broad; **upper surface** dark green, glabrous; **underside** with a thick woolly, interrupted, in shreds and patches, dark brown, unistrate indumentum of hairs.

Petiole tomentose, bristly.

Inflorescence in trusses of 4–6 flowers.

Calyx 3–9 mm long, lobes unequal.

Corolla campanulate, 3.4–4 cm long, white; 5 nectar pouches.

NW Yunnan, NE Upper Burma. 4,000–4,300 m (13,000–14,000 ft).

Epithet: With snowy flowers.   Hardiness 3.   April–May.

Possibly lost to cultivation but once grown at the RBG Edinburgh. No longer recorded in The Rhododendron Handbook 1998.

Can be confused with
R. chaetomallum,
but has rounded
leaves and a large
petaloid calyx.

Leaves obovate.
Indumentum
cinnamon-brown,
dark brown or brown.

Calyx large, red or
crimson, cup-shaped,
irregularly lobed.

R. catacosmum F21727 at Deer Dell

New foliage of *R. catacosmum* F21727 at Deer Dell

R. coelicum Balf.f. & Farrer

Haematodes Group

Shrub 1–1.8 m (3–6 ft) high; **branchlets** eglandular, often glabrous.
Leaves obovate or oblong-obovate; **lamina** 5.4–9.5 cm long, 2.7–4.7 cm broad; **upper surface** dark green, shiny; **underside** covered with a thick, woolly, continuous brown, unistrate indumentum of hairs.
Petiole eglandular, often glabrous.
Inflorescence in trusses of 6–15 flowers.
Pedicel glandular.
Calyx eglandular.
Corolla tubular-campanulate, 3.4–4.5 cm long, deep scarlet or deep crimson; 5 nectar pouches.
NE Upper Burma, NW Yunnan. 2,750–4,300 m (9,000–14,000 ft).
AM 1955 (Minterne), flowers deep scarlet.
Epithet: Heavenly.   Hardiness 3.   April–May.

R. haematodes ssp. haematodes

Haematodes Group
R. haematodes Franch.

Shrub often compact spreading, 0.3–1.2 m (1–4 ft) high, or sometimes up to 1.8–3 m (6–10 ft) high; **branchlets** densely tomentose, eglandular.
Leaves oblong, oblong-obovate or obovate; **lamina** thick, leathery, 3.5–9.6 cm long, 1.6–4.5 cm broad, **upper surface** dark green, shiny, glabrous; **underside** covered with a very thick woolly spongy, continuous, dark brown, brown or cinnamon-brown, bistrate indumentum of hairs.
Petiole densely tomentose.
Inflorescence in trusses of 2–8 flowers.
Pedicel often densely tomentose.
Corolla tubular-campanulate, fleshy, 3–5 cm, deep crimson, crimson, scarlet, or rarely rose; 5 nectar pouches.
W Yunnan. 3,350–4,000 m (11,000–13,000 ft).
FCC 1926 (Werrington), flowers bright scarlet.
Epithet: Blood-like.   Hardiness 3.   May–June.

Similar to
R. pocophorum,
but with a smaller,
rounder (obovate
or oblong-obovate)
leaf.

The young growth
has crimson
perulae (absent
in *R. pocophorum*)
and eglandular
petioles.

Indumentum brown.

Corolla deep scarlet or
deep crimson.

R. coelicum at RBG Edinburgh

New foliage of *R. coelicum* F25647 and F25625 at Deer Dell

Usually dense
growing: densely
tomentose
branchlets.

Smooth dark
moderately shiny
leaves.

Indumentum very
thick woolly spongy,
continuous, dark
brown, brown or
cinnamon-brown.

Petiole densely
tomentose.

Corolla deep crimson,
crimson, scarlet.

R. haematodes ssp. *haematodes* F6773 at Deer Dell

R. haematodes ssp. chaetomallum D.F.Chamb.

Haematodes Group

R. chaetomallum Balf.f. & Forrest
(*R. hillieri* Davidian)

Shrub 0.6–3 m (2–10 ft) high; **branchlets** tomentose, densely to moderately bristly, eglandular.

Leaves oblong-obovate or obovate; **lamina** 4.2–12.5 cm long, 2–5.5 cm broad; **upper surface** olive-green or pale green, matt, glabrous; **underside** with a thick, woolly, continuous or sometimes interrupted, brown or dark brown, bistrate or unistrate indumentum of hairs.

Petiole tomentose, moderately or densely bristly, eglandular.

Inflorescence in trusses of 5–7 flowers.

Pedicel moderately or densely bristly, eglandular.

Corolla tubular-campanulate, 3.2–5.3 cm long, deep crimson, crimson, scarlet, deep rose or pink; 5 nectar pouches.

SE and E Tibet, NE Upper Burma, NW Yunnan. 2,750–4,600 m (9,000–15,000 ft).

AM 1959 (Exbury), flowers red (F25601).

Epithet: With woolly hair.   Hardiness 3.   March–May.

*An extreme form (F21077) used to be called **var. glaucescens**.*

R. × hemigynum D.F.Chamb.   Haematodes Group

R. chaetomallum var. *hemigynum* Tagg & Forrest

The variety differs from R. haematodes *ssp.* chaetomallum *in that the indumentum on the underside of the leaves consists of closely or widely scattered hairs or a thin veil of hairs (rarely glabrous), and usually in having bristly glandular branchlets, petioles, pedicels and ovaries.*

SE Tibet, NE Upper Burma. 3,350–4,300 m (11,000–14,000 ft).

AM 1957 (Tower Court) Orient red (F25605).

Epithet: Half glabrous leaf.   Hardiness 3.   April–May.

Obovate leaves with a sometimes interrupted indumentum beneath.

Less compact than *R. haemadotes* ssp. *haematodes*.

Leaf uppersides appearing paler and rougher because of a covering of detersile fawn hairs in some forms. Bristly branchlets.

Corolla crimson, scarlet, rose or pink.

R. haematodes ssp. *chaetomallum* KW21077 at Deer Dell

New foliage of *R. haematodes* ssp. *chaetomallum* KW21077 at Deer Dell

In cultivation, the leaf shape is close to *R. haematodes* ssp. *chaetomallum*. This taxon is a hybrid.

Can be mistaken for *R. pocophorum* var. *hemidartum* in leaf.

Indumentum of scattered hairs or a thin veil of hairs.

R. × hemigynum at Deer Dell

R. mallotum Balf.f. & Ward   Haematodes Group
(*R. aemulorum* Balf.f.)

Shrub or small **tree** 1.5–4.6 m (5–15 ft) high; **branchlets** densely cinnamon-brown tomentose.

Leaves obovate; **lamina** very thick, leathery, stiff, 7–15.8 cm long, 3.5–8 cm broad; **upper surface** dark green, markedly rugulose, rough, glabrous; **underside** covered with a thick woolly continuous cinnamon-brown bistrate indumentum of hairs.

Petiole densely tomentose.

Inflorescence very compact racemose umbel of 10–20 flowers; **flower-bud** large globose, densely tomentose.

Pedicel stout, densely tomentose.

Corolla tubular-campanulate, 2.8–4.3 cm long, deep crimson, crimson or scarlet-crimson; 5 nectar pouches.

E Upper Burma, W Yunnan. 3,000–3,700 m (10,000–12,000 ft).

AM 1933 (Borde Hill), flowers crimson.

AM 1973 (Windsor), as a foliage plant.

Epithet: Woolly.   Hardiness 3.   March–April.

This species has some similarities to plants of Subsection Falconera *and is different from all the other species in Subsection* Neriiflora.

R. piercei Davidian   Haematodes Group
(*R. beanianum* var. *compactum* Cowan)

Compact spreading **shrub** 1.2–1.5 m (4–5 ft) high; **branchlets** densely hairy, not bristly, not bristly-glandular.

Leaves oblong-oval, oblong-obovate or oblong; **lamina** 6–11 cm long, 2.7–5.2 cm broad, apex rounded; **upper surface** dark green, shiny, glabrous; **underside** covered with a thick woolly, brown, continuous, bistrate indumentum.

Petiole densely hairy, not bristly, not bristly glandular.

Inflorescence in trusses of 6–8 flowers.

Pedicel hairy, eglandular.

Calyx glabrous, eglandular.

Corolla tubular-campanulate 2.8–3.6 cm long, crimson; 5 nectar pouches.

Tibet. 3,700–4,000m (12,000–13,000 ft).

Epithet: After Mr. & Mrs. Lawrence J. Pierce, Washington, USA.   Hardiness 3. March–May.

Once known as R. beanianum *var.* compactum, *which may assist identification, but it can still be confused with* R. beanianum.

Tree-like: the leaves alone should allow identification because of their size.

Leaves obovate, lamina very thick, upper surface markedly rugulose, rough, glabrous; underside covered with a thick woolly continuous cinnamon-brown.

Inflorescence very compact, 10–20 flowers.

Corolla deep crimson, crimson or scarlet-crimson.

R. mallotum Far815 at Deer Dell

Young growth tomentose, becoming somewhat shiny at maturity often with traces of tomentum on the upper leaf surface.

Petiole densely hairy, not bristly, not bristly glandular.

Indumentum thick woolly, brown.

Corolla crimson.

R. piercei KW8254 at Deer Dell

Ponticum
(Neriiflora/Haematodes Group)

R. pocophorum var. pocophorum Balf.f. ex Tagg

Haematodes Group

R. pocophorum Balf.f. ex Tagg

Shrub 1–3 m (3–10 ft) high; **branchlets** rather densely glandular, not tomentose, not bristly.

Leaves oblong-obovate, obovate or oblong; **lamina** thick, leathery, 6–16 cm long, 2.4–6.8 cm broad; **upper surface** dark green, glabrous, shiny; **underside** covered with a thick woolly, continuous, brown or dark brown, unistrate indumentum.

Petiole glandular.

Inflorescence in trusses of 6–20 flowers.

Pedicel rather densely glandular.

Calyx 0.3–1.4 cm long.

Corolla tubular-campanulate, 3.6–5 cm long, deep crimson, crimson, crimson-scarlet or red; 5 nectar pouches.

SE and E Tibet, NE Upper Burma, Assam. 3,700–4,600 m (12,000–15,000 ft).

AM 1971 (Nymans) 'Cecil Nice', flowers deep red (KW8289).

Epithet: Wool-bearing.   Hardiness 3.   March–April.

R. pocophorum var. hemidartum (Tagg) D.F.Chamb.

Haematodes Group

R. hemidartum Balf.f. ex Tagg

Shrub 0.6–1.8 m (2–6 ft) high; **branchlets** rather densely or moderately glandular.

Leaves oblong, obovate or oblong-obovate; **lamina** 6–14 cm long, 2.4–5.2 cm broad; **upper surface** green with a glaucous bloom, matt, glabrous; **underside** with a thick, woolly, interrupted, in shreds and patches, brown, unistrate indumentum.

Petiole rather densely or moderately glandular.

Inflorescence in trusses of 5–10 flowers.

Calyx 0.3–1.1 cm long.

Corolla tubular-campanulate, 3.8–5.3 cm long, crimson, scarlet or light red; 5 nectar pouches.

E and SE Tibet. 3,400–4,250 m (11,000–14,000 ft).

Epithet: Half-flayed.   Hardiness 3.   April–May.

Obovate shiny leaves with a continuous brown or dark brown indumentum.

Branchlets rather densely glandular, not tomentose, not bristly.

Petiole glandular.

Pedicel rather densely glandular.

Corolla crimson, scarlet or red.

R. pocophorum var. *pocophorum* KW8289 'Cecil Nice' AM at Deer Dell

Very similar to *R. pocophorum*, but note the epithet.

Leaf underside with a thick, woolly, interrupted, indumentum in shreds and patches.

Branchlets rather densely or moderately glandular.

Leaf upper surface green with a glaucous bloom.

Petiole rather densely or moderately glandular.

Corolla crimson, scarlet or light red.

R. pocophorum var. *hemidartum* F20028 in the Valley Gardens

Ponticum
(Neriiflora/Haematodes Group)

Subgenus *Hymenanthes*
Section *Ponticum* G.Don
Subsection *Neriiflora* Sleumer
Neriiflorum Group

Shrubs or small trees.
Ovary slender, usually tapered into the style.
Leaves small, oblong to lanceolate.
Leaves mostly glaucous papillate beneath.

R. albertsenianum Forrest   Neriiflorum Group

Broadly upright **shrub** 1–2.1 m (4–7 ft) high; **branchlets** whitish-tomentose,
   eglandular.
Leaves oblong-lanceolate or oblong; **lamina** 6–10 cm long, 1.5–3.3 cm broad;
   underside covered with a woolly, continuous, brown, bistrate indumentum
   of hairs, the top layer partially detersile, epapillate.
Petiole whitish-tomentose, eglandular.
Inflorescence in trusses of 5–6 flowers.
Calyx eglandular.
Corolla campanulate, 2.8–3.5 cm long, bright crimson-rose or scarlet-crimson.
SE Tibet. 3,750–3,950 m (12,500–13,000 ft).
Epithet: After M.O. Albertson, Chinese Maritime Customs.   Hardiness 3.
   April–May.

Extremely rare in cultivation. Many plants labelled R. albertsenianum *are, in fact,*
R. sperabiloides. *This confusion is in part caused by the herbarium specimen of*
F14195 (the type form collector's number), collected in W or NW Yunnan. This
herbarium specimen is R. sperabiloides. *An adjacent herbarium sheet at Kew is*
the correct R. albertsenianum. *This is a collection by* **Rock from SE Tibet: R21994.**

Leaves of the Neriiflorum Group of Subsect. *Neriiflora*. Top to bottom: left, *R. albertsenianum* (2 leaves); middle, *R. floccigerum* (2 leaves), *R. sperabile* var. *weihsiense* (1 leaf) and *R. neriiflorum* ssp. *neriiflorum* (2 leaves); right, *R. sperabile* var. *sperabile* (2 leaves), *R. sperabiloides* (2 leaves) and *R. neriiflorum* var. *appropinquans* (1 leaf) at Deer Dell. Bar = 5 cm

Wider flatter leaf than *R. sperabile* or *R. sperabiloides*, and the leaf surface under the indumentum is not waxy and is **without the papillae** of those species.

Leaf underside covered with a woolly, continuous, brown indumentum, the top layer partially detersile.

Calyx eglandular.
Corolla crimson or scarlet.

The type form is R21994 from SE Tibet.

A probable *R. albertsenianum* at Deer Dell

R. floccigerum Franch.   Neriiflorum Group

Shrub 0.6–2.4 m (2–8 ft) high; **branchlets** moderately or densely floccose, eglandular.

Leaves lanceolate or oblong; **lamina** 4–12 cm long, 0.9–2.8 cm broad; **upper surface** dark green, glabrous; **underside** with a thick, woolly, interrupted, brown or cinnamon-brown, unistrate indumentum of hairs.

Petiole floccose, eglandular.

Inflorescence in trusses of 4–8 flowers.

Pedicel moderately or densely floccose, eglandular.

Corolla tubular-campanulate, 2.5–4.3 cm long, crimson, scarlet, or bright rose; 5 nectar pouches at the base.

NW Yunnan, SE Tibet. 2,300–4,300 m (7,500–14,000 ft).

Epithet: Bearing flecks of wool.   Hardiness 3.   April–May.

Flowers are variable, some forms can be called 'muddy'.

R. neriiflorum ssp. neriiflorum

Neriiflorum Group

R. neriiflorum Franch.
(*R. phoenicodum* Balf.f. & Forrest)

Shrub 0.6–3 m (2–10 ft) high; **branchlets** with a thin fawn or white tomentum or glabrous, not bristly, eglandular.

Leaves oblong, oblong-lanceolate or lanceolate; **lamina** 3–12.3 cm long, 1.3–3.9 cm broad; **upper surface** dark green, glabrous; **underside** glabrous, white, covered with waxy papillae.

Petiole with a thin fawn or white tomentum or glabrous.

Inflorescence with 4–12 flowers.

Pedicel tomentose, not bristly, eglandular.

Calyx 0.1–1.8 cm long, hairy or glabrous, eglandular.

Corolla tubular-campanulate, 2.8–4.5 cm long, deep crimson, crimson, scarlet, carmine or deep rose; 5 nectar pouches.

Mid-W and W Yunnan, S Tibet, N Burma. 2,750–3,700 m (9,000–12,000 ft).

Epithet: With flowers like *Nerium oleander*.   Hardiness 3.   April–May.

R. neriiflorum var. *euchaites* Davidian

Tall upright **shrub** or **tree** usually 2.4–6 m (8–20 ft) high.

W and mid-W Yunnan, NE Upper Burma, SE Tibet. 2,500–3,400 m (8,000–11,000 ft).

AM 1929 (Bodnant), flowers rich ruby-red.

Epithet: With beautiful hairs.   Hardiness 3.   April–May.

Differs from R. neriiflorum *ssp.* neriiflorum *only in its stature, so should not have a separate name.*

Leaf underside
with a partial
interrupted floccose
indumentum, and
having waxy papillae
underneath.
The leaf shape
(usually appearing
elliptic because of
the recurved margins)
distinguishes this
species from
R. sperabiloides.

Branchlets moderately
or densely floccose,
eglandular.
Leaves lanceolate or
oblong.
Corolla crimson, scarlet
or rose.

R. floccigerum R18467 in the Valley Gardens

R. floccigerum F25831, a bullate form, at Borde Hill

Oblong leaves with
a smooth brightly
glaucous (almost
white), underside
covered with
waxy papillae.

Corolla deep crimson,
crimson, scarlet,
carmine or deep rose.
Pedicel tomentose,
eglandular.
Ovary eglandular.

R. neriiflorum ssp. *neriiflorum* F6780 at Deer Dell

This striking form
appears to have
retained deep
purple juvenile
characteristics.

R. neriiflorum ssp. *neriiflorum* 'Rosevallon' at Deer Dell

R. neriiflorum ssp. agetum

Neriiflorum Group

R. neriiflorum var. agetum (Balf.f. & Forrest) Davidian

W Yunnan, Burma–Tibet Frontier. 2,500–3,700 m (8,000–12,000 ft).
Epithet: Wondrous.   Hardiness 3.   April–May.

Leaves smaller than *R. neriiflorum* ssp. *neriiflorum*; **lamina** 3.1–6 cm long and
1–1.5 cm broad, narrowly oblong or lanceolate; with characteristic cavities or
depressions (alveoli) on the underside.

Probably not in culivation.

R. neriiflorum var. appropinquans W.K.Hu

Neriiflorum Group

R. floccigerum var. appropinquans Tagg & Forrest

This variety differs from *R. floccigerum* in that the underside of the **leaves** is
glabrous except the floccose midrib, and the **branchlets** and **petioles** are
setulose glandular.
Epithet: Nearly glabrous.   Hardiness 3.   March–May.

Although the glandular branchlets and the leaf shape place this variety with
R. floccigerum, *Hu, in the* 'Flora of China', *sensibly suggests it is closer to*
R. neriiflorum. *In cultivation, it has most attractive bi-coloured flowers.*

R. neriiflorum ssp. phaedropum Tagg

Neriiflorum Group

R. phaedropum Balf.f. & Farrer

Shrub or **tree** 0.6–6 m (2–20 ft) high; **branchlets** tomentose with a thin fawn
   tomentum, or densely or moderately tomentose, glandular with long-stalked
   glands or eglandular.
Leaves lanceolate, oblong or oblong-lanceolate; **lamina** 4–14.3 cm long,
   0.8–4.5 cm broad; **upper surface** pale or dark green, glabrous; **underside**
   glabrous, white, covered with waxy papillae.
Inflorescence in trusses of 4–8 flowers.
Pedicel rather densely or moderately glandular or eglandular.
Calyx glandular or eglandular.
Corolla tubular-campanulate, 2.6–4.5 cm long, crimson, scarlet, salmon-rose,
   tawny-orange or straw-yellow; 5 nectar pouches.
Ovary moderately or densely glandular or eglandular.
NE Upper Burma, S Tibet, Burma–Tibet Frontier, Assam, Bhutan. 2,300–3,700 m
   (7,500–12,000 ft).
Epithet: Of bright appearance.   Hardiness 3.   March–May.

Leaf underside
glabrous except
the floccose midrib.
Most attractive
bicoloured flowers –
often yellow
flushed pink.

R. neriiflorum var. *appropinquans*, formerly *R. floccigerum* var.
appropinquans, at Deer Dell

Longer more
rugulose leaves than
spp. *neriiflorum*.
Pedicel usually
rather densely or
moderately
glandular.

Leaves lanceolate,
oblong or oblong-
lanceolate.
Corolla red shades,
tawny orange or straw
yellow.
Ovary often glandular.

R. neriiflorum ssp. *phaedropum* KW6857 at Deer Dell

R. sperabile var. sperabile Balf.f. & Farrer

Neriiflorum Group

R. sperabile Balf.f. & Farrer

Shrub 1–2.4 m (3–8 ft) high; **branchlets** setulose-glandular, densely tomentose with a whitish or fawn tomentum.

Leaves lanceolate, oblong-lanceolate or oblong; **lamina** 3.8–10.5 cm long, 1–3.3 cm broad; **upper surface** dark green, glabrous; **underside** covered with a thick, woolly, continuous yellowish-fawn, fawn or brown, unistrate indumentum.

Petiole setulose-glandular, tomentose.

Inflorescence in trusses of 3–5 flowers.

Pedicel hairy, glandular with long-stalked glands.

Calyx glandular or eglandular.

Corolla tubular-campanulate, 2.6–4.5 cm long, scarlet, crimson or deep-crimson; 5 nectar pouches.

Ovary tomentose, densely or moderately glandular.

NE Upper Burma. 3,000–3,700 m (10,000–12,000 ft).

AM 1925 (Exbury), flowers scarlet (Farrer888).

Epithet: To be hoped for.   Hardiness 3.   April–May.

Rare in cultivation.

R. sperabile var. weihsiense Tagg & Forrest

Neriiflorum Group

NW Yunnan, NE Upper Burma. 3,200–4,300 m (10,600–14,000 ft).

Epithet: From Weihsi, China.   Hardiness 3.   April–May.

This variety differs from var. sperabile *in that the indumentum on the lower surface of the leaves is less dense, white or rarely pale fawn and the leaves are larger, up to 14 cm long and 5 cm broad.*

Common in cultivation. **Recently, it has been sensibly suggested in** 'The Flora of China' **that this taxon should have specific status.**

Continuous yellowish-fawn indumentum, papillate beneath.

Branchlets setulose-glandular, densely tomentose.

Petiole setulose-glandular, tomentose.

Corolla scarlet, crimson or deep-crimson.

R. sperabile var. *sperabile* at Deer Dell

Whitish (or rarely pale fawn) indumentum.

Larger and flatter leaves than var. *sperabile* (up to 15 cm long), also longer petioles. Many more flowers in the truss.

Corolla scarlet to crimson.

R. sperabile var. *weihsiense* KW7124 'Rouge et Noir' at Deer Dell

Ponticum
(Neriiflora/Neriiflorum Group)

R. sperabiloides Tagg & Forrest   Neriiflorum Group

Shrub 0.6–1.2 m (2–4 ft) high; **branchlets** with a whitish tomentum, eglandular.
Leaves oblanceolate, lanceolate or oblong-elliptic; **lamina** 3–7.7 cm long,
1.4–2.6 cm broad; **upper surface** green, glabrous; **underside** with a thin,
interrupted, whitish or fawn, unistrate indumentum of hairs, covered with
waxy papillae.
Inflorescence in trusses of 4–11 flowers.
Pedicel floccose, eglandular.
Corolla tubular-campanulate, 2.5–3.8 cm long, deep to light crimson; 5
nectar pouches.
E and SE Tibet. 3,700–4,000 m (12,000–13,000 ft).
AM 1933 (Exbury), flowers deep crimson.
Epithet: Like *R. sperabile*.   Hardiness 3.   April–May.

Plants cultivated as R. albertsenianum *may often be found to have the waxy
papillae, giving a pale appearance to the leaf surface under the indumentum,
and are therefore referable to this species.*

Subgenus *Hymenanthes*
Section *Ponticum* G.Don
Subsection *Neriiflora* Sleumer
Sanguineum Group

Ovary impressed, i.e. not tapered into the style.
All species have thinner indumentum than in Haematodes Group.

Further grouping within Sanguineum may be helpful:-

Sanguineum Group (Part A).
Moderately thick indumentum, but leaf veins visible.

Sanguineum Group (Part B).
*Indumentum appears continuous and is adpressed and
whitish or fawn.*

Sanguineum Group (Part C).
Leaves glabrous or almost glabrous.

Leaf shape like *R. sperabile.*

The lower leaf surface is glaucous and waxy under the partial indumentum, which is thick, interrupted, whitish or fawn.

Corolla deep to light crimson.

R. sperabiloides at Deer Dell. This species is often wrongly labelled *R. albertsenianum*

Leaves of the *Sanguineum* Group of Subsect. *Neriiflora* at Deer Dell. Top to bottom (all 2 leaves each): left, *R. dichroanthum* ssp. *septentrionale*, *R. temenium* var. *temenium* and *R. eudoxum* var. *eudoxum*; middle, *R. dichroanthum* ssp. *dichroanthum*, *R. aperantum* and *R. dichroanthum* ssp. *apodectum*; right, *R. citriniflorum* var. *citriniflorum*, *R. parmulatum* and *R. sanguineum* ssp. *didymum*. Bar = 5 cm

R. citriniflorum var. citriniflorum

Sanguineum Group (Part A)

R. citriniflorum Balf.f. & Forrest

Shrub 0.3–1.5 m (1–5 ft) high; **branchlets** with a thin tomentum, eglandular.

Leaves obovate, oblong-obovate, oblong or oblong-lanceolate; **lamina** 2.8–8.5 cm long, 1.2–2.8 cm broad, base decurrent on the petiole; **upper surface** dark or pale green; **underside** covered with a thick woolly, continuous, brown, fawn or whitish, bistrate indumentum of hairs.

Petiole margins narrowly winged or ridged, eglandular.

Inflorescence 3–6 or sometimes up to 10 flowers.

Calyx 0.2–1.4 cm long.

Corolla tubular-campanulate, 2.3–4.5 cm long, bright lemon-yellow, yellow, yellow margined rose, or yellowish suffused with rose; 5 nectar pouches.

NW Yunnan, SE Tibet. 4,000–4,600 m (13,000–15,000 ft).

Epithet: With lemon-yellow flowers.   Hardiness 3.   April–May.

R. horaeum Balf.f. & Forrest

Sanguineum Group (Part A)

R. citriniflorum var. *horaeum* D.F.Chamb.

Shrub 0.15–1.5 m (0.5–5 ft) high; **branchlets** with a thin white tomentum; **leaf-bud scales** persistent.

Leaves obovate or oblong-obovate; **lamina** 2.3–6 cm long, 1.2–2.4 cm broad, base decurrent on the petiole; **upper surface** dark green, semi-bullate, glabrous; **underside** covered with a thick, woolly continuous, bistrate indumentum of hairs.

Petiole margins narrowly winged.

Inflorescence in trusses of 2–4 flowers.

Pedicel hairy, often setulose and setulose-glandular.

Calyx 0.2–2 cm long, coloured like the corolla.

Corolla tubular-campanulate, 2.6–3.8 cm long, deep crimson, rose-crimson, orange-crimson, or yellow heavily margined crimson; 5 nectar pouches.

E and SE Tibet, NW Yunnan. 4,000–4,600 m (13,000–15,000 ft).

Epithet: Beautiful.   Hardiness 3.   May–June.

With its persistent perulae, this taxon should have specific status. The orange-flowered form is very distinctive.

Indumentum continuous, brown, fawn or whitish, **bistrate**, variable in thickness from woolly to (rarely) thinly plastered: leaf clearly visible. Corolla yellow.

Leaf base decurrent on the petiole.

R. citriniflorum var. *citriniflorum* R108 at Deer Dell

Thick woolly persistent brown bistrate indumentum. **Persistent perulae**. Corolla red or orange.

Leaf upper surface dark green, semi-bullate.

R. horaeum F21850 at Deer Dell

R. horaeum var. rubens (Cowan) Davidian

Sanguineum Group (Part A)

Epithet: Reddish.   Hardiness 3.   May–June.

The var. rubens form is a more compact plant with red or crimson flowers; it has the thick indumentum of R. horaeum, but non-persistent perulae, making its status less than certain. Plants labelled R. sanguineum ssp. consanguineum may be referable to this taxon, but if these have persistent perulae they are simply R. horaeum.

R. dichroanthum

Sanguineum Group (Part B)

Four subspecies:

ssp. **dichroanthum** *has a compacted silvery or sometimes fawn indumentum;*

ssp. **apodectum** *has the shortest leaves – up to 2.4 times as long as broad, and usually oval with a rounded base; leaves have a shiny upper surface, usually with longitudinal ribbing and a somewhat thicker silvery to fawn indumentum;*

ssp. **scyphocalyx** *has matt obovate leaves and a distinctive cup-shaped calyx;*

ssp. **septentrionale** *has the longest leaves (more than 3 times as long as broad).*

R. dichroanthum ssp. dichroanthum

Sanguineum Group (Part B)

R. dichroanthum Diels.

Shrub 0.6–2.4 m (2–8 ft) high; **branchlets** with a thin white tomentum.

Leaves oblong, oblanceolate, obovate or oblong-obovate; **lamina** 4–10 cm long, 1.3–4 cm broad; **upper surface** dark green; **underside** covered with a thin, plastered, continuous, white or sometimes fawn, unistrate indumentum of hairs.

Petiole with a thin, white tomentum, eglandular.

Inflorescence in trusses of 3–8 flowers.

Pedicel brown-hairy, eglandular.

Calyx 0.4–2.5 cm long, cup-shaped, divided to the base or to about the middle.

Corolla tubular-campanulate, 3.2–4.5 cm long, yellowish-rose, orange, orange-red or pinkish-red; 5 nectar pouches.

W Yunnan. 2,750–3,700 m (9,000–12,000 ft).

AM 1923 (Bodnant), flowers brick red.

Epithet: With flowers of two colours.   Hardiness 3.   May–June.

Leaf-bud scales deciduous.

Ovary densely setulose-glandular.

Calyx usually small, margin moderately setulose-glandular.

Corolla crimson to red.

R. horaeum var. *rubens* at Nymans (Davidian)

Upper leaf surface not noticeably shining.

Indumentum compacted, usually silvery.

Pedicel brown-hairy, eglandular.

Calyx 0.4–2.5 cm long, cup-shaped.

Corolla yellowish rose, orange, orange-red to pinkish red.

R. dichroanthum ssp. *dichroanthum* at Deer Dell

R. dichroanthum ssp. apodectum Cowan

Sanguineum Group (Part B)

R. apodectum Balf.f. & W.W.Sm.

Shrub 0.3–2.4 m (1–8 ft) high; **branchlets** with a thin white tomentum.

Leaves oval, obovate, elliptic or oblanceolate; **lamina** 3.5–8.5 cm long, 2–3.8 cm broad; **upper surface** dark or pale green, with or without longitudinal ribbing, glabrous; **underside** covered with a somewhat thick somewhat woolly or sometimes thin plastered, continuous unistrate indumentum of rosulate hairs.

Inflorescence in trusses of 2–5 flowers.

Corolla tubular-campanulate, 2–4.3 cm long, deep crimson, crimson, deep rose flushed orange, or deep orange suffused with rose.

W and mid-W Yunnan, North Triangle, North Burma. 3,000–3,700 m (10,000–12,000 ft).

Epithet: Acceptable.   Hardiness 3.   May–June.

R. dichroanthum ssp. scyphocalyx Cowan

Sanguineum Group (Part B)

R. scyphocalyx Balf.f. & Forrest
(*R. herpesticum* Balf.f. & Forrest)

Shrub 1–1.5 m (3–5 ft) high; **branchlets** with a thin, white or fawn tomentum.

Leaves obovate or oblong-obovate; **lamina** 3.5–9.5 cm long, 2–4 cm broad, base decurrent on the petiole; **upper surface** dark or pale green, glabrous, **underside** covered with a thin, plastered, continuous, white or fawn, unistrate indumentum of hairs.

Petiole margins narrowly winged or ridged.

Inflorescence in trusses of 3–5 flowers.

Pedicel rather densely setulose-glandular.

Calyx 0.4–1 cm long, cup-shaped, divided to about the middle.

Corolla tubular-campanulate, 2.5–4.3 cm long, rose-orange, orange, deep crimson or orange suffused with crimson; 5 nectar pouches.

NE Upper Burma, mid-W Yunnan. 3,000–4,300 m (10,000–14,000 ft).

Epithet: Cup-shaped calyx.   Hardiness 3.   May–June.

R. herpesticum *Balf.f. & Kingdon Ward* *is a more densely mounded form of* R. dichroanthum *ssp.* scyphocalyx *and is distinctively different in cultivation.*

Upper leaf surface somewhat shining, and usually with longitudinal ribbing.

Leaves oval or obovate.

Somewhat smaller leaves than ssp. *dichroantum* or ssp. *scyphocalyx*.

Indumentum sometimes thicker than those of the other subspecies.

Corolla orange to orange flushed rose to crimson.

R. dichroanthum ssp. *apodectum* at Deer Dell

Leaves matt, similar to those of ssp. *dichroanthum*, obovate or oblong-obovate, indumentum thin. Pedicel rather densely setulose-glandular.

Calyx 0.4–1 cm long, cup-shaped.

Corolla orange, pinkish or crimson orange to deep crimson.

R. dichroanthum ssp. *scyphocalyx* at Deer Dell

R. dichroanthum ssp. septentrionale Cowan

Sanguineum Group (Part B)

R. scyphocalyx var. *septentrionale* Tagg ex Davidian

NE Upper Burma, NW Yunnan. 3,700–4,300 m (12,000–14,000 ft).
Epithet: Northern.   Hardiness 3.   May–July.

R. eudoxum and its varieties

***var.* brunneifolium** *is an eglandular expression of R. eudoxum with a larger corolla, a tiny calyx and a longer pedicel.*

***var.* eudoxum** *has thin leathery leaves and a glandular ovary.*

***var.* mesopolium** *is eglandular, but without the larger corolla or the tiny calyx.*

var. brunneifolium *and var.* mesopolium *are very rare in cultivation. More collections from the wild are needed to ascertain their status as they may well be hybrids.*

R. eudoxum var. eudoxum

Sanguineum Group (Part B)

R. eudoxum Balf.f. & Forrest
(*R. trichomiscum* [syn: *R. herpesticum* Balf.f. & Forrest])

Shrub 0.3–1.8 m (1–6 ft) high; **branchlets** often floccose, not bristly, often glandular.
Leaves oblong, obovate or oblong-elliptic; **lamina** thinly leathery, 3–9 cm long, 1.5–3.3 cm broad; **upper surface** dark green, glabrous; **underside** with a thin veil of white or fawn unistrate hairs.
Petiole floccose, often glandular.
Inflorescence in trusses of 3–6 flowers.
Pedicel often hairy, often glandular.
Calyx 2–8 mm long.
Corolla campanulate or tubular-campanulate, 2.4–4 cm long, crimson-rose, rose, white suffused with rose or dark red; 5 nectar pouches at the base.
Ovary hairy and glandular.
NW Yunnan, SE Tibet, Assam. 3,400–4,300 m (11,000–14,000 ft).
AM 1960 (Glendoick), flowers purple.
Epithet: Of good report.   Hardiness 3.   April–May.

Late flowering (May–July).

Narrow oblanceolate leaves, longer than those of the type.

Pedicel, ovary and calyx usually eglandular.

R. dichroanthum ssp. *septentrionale* F25577 at Deer Dell

Thin leaves.

Related to *R. temenium*, but that species is glaucous papillate. Ovary hairy and glandular.

Leaf underside with a thin veil of white or fawn unistrate hairs.

Petiole floccose, often glandular.

Corolla crimson-rose, rose, white-flushed rose or dark red.

R. eudoxum var. *eudoxum* (as *R. trichomiscum*) at Blackhills, Scotland

R. eudoxum var. *eudoxum* (S. Hootman)

R. eudoxum var. brunneifolium D.F.Chamb.

Sanguineum Group (Part B)

R. brunneifolium Balf.f. & Forrest

Shrub 1–1.2 m (3–4 ft) high; **branchlets** floccose or glabrous.

Leaves oblong, oblong-elliptic or oblong-obovate; **lamina** 5.3–8.5 cm long, 2.3–3.2 cm broad; **upper surface** dark green, glabrous; **underside** with a thin veil of white or fawn rosulate hairs.

Petiole glabrous or floccose.

Inflorescence in trusses of 3–4 flowers.

Pedicel 2–3 cm long, hairy, eglandular.

Calyx 1–2 mm long, outside and margin hairy, eglandular.

Corolla tubular-campanulate, 3.5–4 cm long, rose-crimson; 5 nectar pouches.

Ovary densely hairy, eglandular.

SE Tibet, NW Yunnan. 3,700–4,300 m (12,000–14,000 ft).

Epithet: With brown leaves.   Hardiness 3.   April–May.

Extremely rare in cultivation. Likely to be part of a hybrid swarm.

R. eudoxum var. mesopolium D.F.Chamb.

Sanguineum Group (Part B)

R. mesopolium Balf.f. & Forrest

Somewhat compact **shrub** 0.45–1.5 m (1.5–5 ft) high; **branchlets** not bristly, eglandular.

Leaves oblong, oblong-obovate or oblong; **lamina** 3–6.5 cm long, 1.2–2.8 cm broad, base decurrent on the petiole; **upper surface** dark green, glabrous; **underside** with a thin veil of white or fawn, unistrate hairs.

Petiole margins narrowly winged or ridged, not bristly, eglandular.

Inflorescence in trusses of 2–4 or sometimes up to 10 flowers.

Pedicel eglandular.

Calyx 0.1–1 cm long, eglandular.

Corolla tubular-campanulate, pale rose or rose, or rose-margined and lined a deeper shade, bright pink or white; 5 nectar pouches.

SE Tibet, NW Yunnan. 3,700–4,300 m (12,000–14,000 ft).

Epithet: Grey in middle.   Hardiness 3.   April–May.

Extremely rare in cultivation. There are two types of hairs in the unistrate indumentum, indicating that this is likely to be a hybrid.

An eglandular
expression of
R. eudoxum with
a larger corolla
and a tiny calyx.

**Indumentum thin,
white or fawn
(note the misleading
epithet).**

**Ovary densely hairy,
eglandular.**

Pedicel long,
eglandular.

Corolla crimson-rose.

R. eudoxum var. *brunneifolium* at Werrington, Cornwall

**Corolla smaller than
var. *eudoxum*.**

Ovary hairy, eglandular.
Pedicel and calyx
eglandular.

Corolla pale rose or
rose, or rose-margined
and lined a deeper
shade, bright pink
or white.

R. eudoxum var. *mesopolium* at RBG Edinburgh

R. microgynum Balf.f. & Forrest Sanguineum Group (Part B)

(*R. perulatum* Balf.f. & Forrest)
(*R. gymnocarpum* Balf.f. ex Tagg)

Shrub 1–1.2 m (3–4 ft) high; **branchlets** with a thin tomentum, not setulose.

Leaves lanceolate or oblanceolate; **lamina** 4–9 cm, long, 1.5–2.5 cm broad; **upper surface** dark green, glabrous; margin recurved; **underside** covered with a thin, continuous, brown, unistrate indumentum of hairs.

Petiole with a thin tomentum, not setulose.

Inflorescence in trusses of 5–6 flowers.

Pedicel glandular.

Corolla broadly campanulate, 2.2–3 cm long, dull soft rose with faint crimson markings; 5 nectar pouches.

Style short, half the length of the corolla or less.

SE Tibet. 3,660 m (12,000 ft).

Epithet: With small ovary.   Hardiness 3.   April–May.

*Very rare in cultivation, and aberrant in the Sanguineum Group because of its leaf size and **rugose upper leaf surface**. More suited to the* Neriiflorum *Group, which contains species with rugose upper leaf surfaces.*

R. gymnocarpum Balf.f. ex Tagg

Corolla deep claret crimson.

Only separated from R. microgynum *by the flower colour and size, so a separate name is not justified, although this form is much more common in cultivation.*

R. parmulatum Cowan Sanguineum Group (Part B)

Shrub 0.6–1.2 m (2–4 ft) high, **branchlets** floccose or glabrous, eglandular.

Leaves oblong-oval, oblong or oblong-obovate; **lamina** 3.5–8.6 cm long, 1.5–4.5 cm broad; **upper surface** green, glabrous; **underside** glabrous or hairy on the lateral veins, covered with waxy papillae, midrib floccose.

Petiole eglandular.

Inflorescence in trusses of 3–5 flowers.

Pedicel glabrous, eglandular.

Calyx 3–7 mm long, cup-shaped or disc-like, glabrous.

Corolla tubular-campanulate, 3.3–5 cm long, white, pale yellow, whitish-pink, or white suffused with crimson, with numerous crimson spots; 5 nectar pouches.

SE Tibet. 3,400–4,300 m (11,000–14,000 ft).

AM 1977 (Tremeer) 'Ocelot', flowers yellowish-green, heavily spotted with purple.

Epithet: With a small round shield.   Hardiness 3.   April–May.

Rugulose upper leaf surface and crimson to rose flowers.

Leaf lanceolate or oblanceolate, margin recurved.

Indumentum thin, felted, continuous brown.

Pedicel glandular.

R. microgynum at Darts Hill, Oregon (M.L.A. Robinson)

R. microgynum F16687 (as *R. gymnocarpum*) at Deer Dell

Rugulose leaves with only vestiges of indumentum on the lateral veins beneath.

Lamina papillate and appearing glaucous.

Corolla white, pale yellow, whitish-pink, or white-flushed crimson, with numerous crimson spots.

Leaves oblong-oval, oblong or oblong-obovate.

Calyx 3–7 mm long, cup-shaped or disc-like.

R. parmulatum KW5875 at Borde Hill

R. sanguineum, its subspecies and varieties

Petiole narrowly winged.

Only **R. sanguineum *ssp*. didymum** *and the related* **R. sanguineum *ssp*. sanguineum *var*. didymoides** *show significant botanical differences. The others differ primarily in flower colour, and some may be hybrids.*

Very difficult to separate out of flower.

R. sanguineum ssp. sanguineum
Sanguineum Group (Part B)

R. sanguineum Franch.

Shrub 0.3–1.2 m (1–4 ft) or rarely up to 2.4 m (8 ft) high; **branchlets** with a thin, white tomentum, eglandular; **leaf-bud scales** deciduous.

Leaves obovate, oblong-obovate or oblong, base decurrent on the petiole, **upper surface** dark green, glabrous; **underside** covered with a thin plastered, continuous, unistrate indumentum.

Petiole margins narrowly winged or ridged, eglandular.

Inflorescence in trusses of 3–5 flowers.

Pedicel hairy, eglandular.

Corolla tubular-campanulate, 2.4–4 cm long, deep crimson, crimson, rose-crimson, carmine or scarlet; 5 nectar pouches.

Ovary densely hairy, eglandular.

SE Tibet, NW Yunnan. 3,400–4,400 m (11,000–14,500 ft).

Epithet: Blood red.   Hardiness 3.   March–May.

R. sanguineum ssp. didymum Cowan
Sanguineum Group (Part B)

R. didymum Balf.f. & Forrest

Broadly upright **shrub** 30–90 cm (1–3 ft) high; **branchlets** with a thin white tomentum, setulose-glandular. **Leaf-bud** scales more or less persistent.

Leaves obovate or oblong-obovate; **lamina** rigid, 2–6 cm long, 1–1.8 cm broad, upper surface dark green; **underside** covered with a thin or somewhat thick, continuous, white or fawn unistrate or bistrate indumentum.

Petiole with a thin tomentum, setulose-glandular.

Inflorescence in trusses of 3–6 flowers.

Pedicel rather densely setulose-glandular.

Calyx setulose-glandular.

Corolla tubular-campanulate, 2.3–3 cm long, black-crimson or dark crimson; 5 nectar pouches.

Ovary densely glandular.

E and SE Tibet, Yunnan–Tibet border. 4,250–4,600 m (14,000–15,000 ft).

Epithet: Two-fold.   Hardiness 3.   June–July.

A glandular expression of R. sanguineum *var.* haemaleum.

Thin white or brown indumentum, somewhat compacted.

In cultivation, flowers usually bright crimson.

Leaf-bud scales deciduous.

Ovary densely hair, eglandular

R. sanguineum ssp. *sanguineum* at RBG Edinburgh

June–July flowering with dark black-crimson flowers.

Branchlets setulose-glandular.

Leaf-bud scales more or less persistent.

Small rigid shiny leaves, usually a thinish fawn indumentum.

Petiole thinly tomentose, setulose glandular.

Calyx setulose glandular. Ovary densely glandular.

Indumentum thin or somewhat thick, continuous, white or fawn.

R. sanguineum ssp. *didymum* at Deer Dell

Ponticum
(Neriiflora/Sanguineum Group)

R. sanguineum ssp. sanguineum var. haemaleum D.F. Chamb.

Sanguineum Group (Part B)

R. haemaleum Balf.f. & Forrest
(*R. sanguineum* ssp. *mesaeum* [Balf.f.] Cowan)

Shrub 0.3–1.8 m (1–6 ft) high; **branchlets** with a thin, white tomentum, eglandular.

Leaves oblanceolate, oblong or oblong-obovate; **lamina** 2.6–9.5 cm, long, 1.3–3.8 cm broad; **upper surface** dark green, glabrous; **underside** covered with a thin plastered or thin suede-like, continuous, white, fawn or brown, unistrate indumentum of hairs.

Inflorescence in trusses of 2–6 flowers.

Calyx eglandular.

Corolla tubular-campanulate, 2.3–4 cm long, black-crimson, black-carmine, dark carmine or nearly black; 5 nectar pouches.

SE Tibet, NW Yunnan, Assam. 3,050–4,450 m (10,000–14,500 ft).

AM 1973 (Nymans).

Epithet: Blood-red.   Hardiness 3.   April–May.

R. haemaleum var. atrorubrum Cowan

Davidian described black crimson flowers, and a densely glandular pedicel and ovary, but no persistent perulae. Of uncertain status.

R. sanguineum spp. sanguineum var. cloiophorum D.F.Chamb.

Sanguineum Group (Part B)

R. cloiophorum Balf.f. & Forrest

Shrub 0.3–1.5 m (1–5 ft) high; **branchlets** with a thin tomentum, esetulose, eglandular.

Leaves oblong-obovate, obovate or oblanceolate; **lamina** 2.8–7.5 cm long, 1.3–2.6 cm broad, base decurrent on the petiole; **underside** covered with a thin, plastered continuous, white or fawn, unistrate indumentum.

Petiole margins narrowly winged.

Inflorescence in trusses of 2–6 flowers.

Calyx 0.1–1 cm long.

Corolla tubular-campanulate, 2.5–3.7 cm long, rose-crimson, yellowish suffused with rose, or white margined rose or orange-red; 5 nectar pouches.

NW Yunnan, SE Tibet. 3,400–4,300 m (11,000–14,000 ft).

Epithet: Wearing a collar, referring to the calyx.   Hardiness 3.   April–May.

Rare in cultivation. Status uncertain.

R. cloiphorum *var.* leaucopetalum *has white flowers.*

April–May flowering usually with black crimson flowers.

Leaf bud scales deciduous.

Eglandular petiole.

Pedicel hairy, eglandular.

Calyx eglandular, the same colour as the corolla.

Indumentum thin plastered or thin suede-like, continuous, usually white or fawn.

R. sanguineum ssp. *sanguineum* var. *haemaleum* at Benmore

Corolla rose-crimson, yellowish flushed rose, or white margined rose or orange-red, 2.5–3.7 cm long.

Calyx lobes usually deflexed.

Ovary eglandular. Distinguishable only in flower.

Leaves 2.8–7.5 cm long. Indumentum thin, plastered, continuous, white or fawn.

R. sanguineum ssp. *sanguineum* var. *cloiophorum* in the UBC Botanic Garden (M.L.A. Robinson)

Ponticum
(Neriiflora/Sanguineum Group)

R. sanguineum ssp. sanguineum var. didymoides Tagg & Forrest   Sanguineum Group (Part B)

(*R. roseotinctum* Balf.f. & Forrest, *R. sanguineum* ssp. *consanguineum* Cowan)
R. cliophorum Balf.f. & Forrest var. *roseotinctum*

Leaves 1.8–4 cm long, indumentum thin, plastered, continuous, white, fawn or brown.
Epithet: Resembling *R. sanguineum* ssp. *didymum*.   Hardiness 3.   April–May.

Very rare in cultivation.

R. sanguineum ssp. sanguineum var. himertum D.F.Chamb.   Sanguineum Group (Part B)

R. himertum Balf.f. & Forrest
(*R. nebrites* Balf.f. & Forrest)

Leaves oblong-obovate, oblong or obovate; **lamina** 3–6.7 cm long, 1–2.3 cm broad, base decurrent on the petiole; **underside** covered with a thin plastered, continuous, white or fawn unistrate indumentum.
Inflorescence in trusses of 3–7 flowers.
Corolla campanulate, 2.5–3.6 cm long, yellow, lemon-yellow or bright yellow; 5 nectar pouches.
SE Tibet, NW Yunnan. 3,700–4,000 m (12,000–13,000 ft).
Epithet: Lovely.   Hardiness 3.   April–May.

Extremely rare in cultivation. Similar to R. citriniflorum *but has a thin plastered unistrate indumentum.*

R. aperantum Balf.f. & Kingdon Ward

Sanguineum Group (Part C)

Compact rounded or a low spreading **shrub** 8–60 cm (0.25–2 ft) high, sometimes 90 cm (3 ft) or rarely 1.5–1.8 m (5–6 ft) high, with usually short annual growth; **leaf-bud scales** persistent.
Leaves almost sessile in close whorls at the ends of the branchlets, oblanceolate or obovate, rarely oval; **lamina** 2–6.4 cm long, 1–2.8 cm broad, base decurrent on the petiole; **underside** white or whitish, covered with waxy papillae, glabrous or sparsely floccose.
Petiole margins winged.
Inflorescence in trusses of 2–6 flowers.
Corolla tubular-campanulate, 2.8–4.8 cm long, deep crimson, crimson, rose, pink, yellow, orange or white.
Ovary densely hairy.
NE Upper Burma, NW Yunnan. 3,000–4,500 m (10,000–14,600 ft).
AM 1931 (Marquess of Headfort, Kells), flowers crimson.
Epithet: Boundless.   Hardiness 3.   April–May.

Corolla rose, yellow margined rose, or, orange red, or, rarely, red flowers, 2–2.6 cm long.

Ovary somewhat glandular.

Perulae often persistent.

***R. sanguineum* ssp. *sanguineum* var. *didymoides*
(as *R. roseotinctum*) R10903 at Deer Dell**

***R. aperantum* F27075 at Deer Dell**

Dwarf spreading shrub.

Glaucous leaves in whorls and **persistent perulae**.

Leaf underside white or whitish, covered with waxy papillae, glabrous or sparsely floccose.

Corolla usually crimson or pink in cultivation.

***R. aperantum* at RBG Edinburgh (Davidian)**

R. temenium var. temenium

Sanguineum Group (Part C)

R. temenium Balf.f. & Forrest
(*R. pothinum* Balf.f. & Forrest)

Shrub 0.3–1.2 m (1–4 ft) high; **branchlets** floccose or glabrous, often setulose.

Leaves oblong, obovate or oblong-obovate; **lamina** 2–7.5 cm long, 0.6–2,5 cm broad, base decurrent on the petiole; **upper surface** dark green, glabrous; **underside** glabrous; petiole margins narrowly winged, often setulose.

Inflorescence in trusses of 3–10 flowers.

Pedicel hairy.

Corolla tubular-campanulate, 2–4.3 cm long, deep crimson, crimson, purplish-crimson or purplish-red; 5 nectar pouches.

SE and E Tibet, NW Yunnan. 3,400–4,400 m (11,000–14,500 ft).

Epithet: From a sacred place near Doker La, Tsarong, in E Tibet.   Hardiness 3. April–May.

R. temenium var. gilvum D.F.Chamb.

Sanguineum Group (Part C)

R. temenium Balf.f. & Forrest ssp. *gilvum*
R. temenium ssp. *chrysanthemum* Cowan

Branchlets bristly.

Leaf underside papillate.

Corolla yellow.

Calyx glandular.

The name of this taxon has been changed a few times. Davidian gives it as **R. fulvastrum** *Balf.f. & Forrest, and it received awards as* **R. chrysanthemum**:

AM 1958 (Ardrishaig) 'Cruachan'.

FCC 1964 (Ardrishaig) 'Cruachan'.

Hardiness 3.   April–May.

R. temenium *var.* **dealbatum** *D.F.Chamb., formerly R. glaphyrum var. dealbatum (Cowan) Davidian, has poorly developed papillae on the leaf underside, and white flowers. Its branchlets are not bristly.*

R. glaphyrum *Balf.f. & Forrest also with poorly developed papillae, but with pale or deep rose or pale yellow flushed rose flowers has been included here. Its branchlets are sometimes bristly. Both are of uncertain status.*

Smooth, not rugulose, oblong to obovate leaves, glabrous beneath. Corolla usually purplish red. Calyx eglandular.

Closely related to *R. sanguineum*. Petiole often setulose.

R. temenium var. *temenium* R104 at Deer Dell

Bristly branchlets. Leaf underside papillate. Corolla yellow. Calyx glandular.

R. temenium var. *gilvum* R22272 'Cruachan' FCC at Glendoick

Ponticum
(Neriiflora/Sanguineum Group)

Subgenus *Hymenanthes*
Section *Ponticum* G.Don
Subsection *Parishia* Sleumer
Parishii Series

Large shrubs or small trees.

Branchlets, petioles, pedicels and ovaries with stellate hairs: a microscope is needed to distinguish these.

Mature leaves very shiny beneath.

Indumentum, if any, is detersile except near the petiole.

Corolla with nectar pouches.

Mostly red late-flowering species.

R. elliottii, R. facetum *and* R. sikangense *are the only species of this Subsection commonly found in cultivation.*

R. venator KW6285 at Deer Dell

Leaves of Subsect. *Parishia* at Deer Dell. Top to bottom (2 leaves each): *R. venator*, *R. sikangense* var. *sikangense*, *R. facetum* and *R. schistocalyx*. Bar = 5 cm

R. agapetum Balf.f. & Ward

Shrub or **tree** 1.8–11 m (6–35 ft) high; **branchlets** hairy with stellate hairs or glabrous, setulose-glandular or eglandular.

Leaves oblong or oblong-oval, 10–22.8 cm long, 4.5–8.3 cm broad; **upper surface** dark green, glabrous (in young leaves, with stellate hairs); **underside** at first moderately or densely hairy with a discontinuous indumentum of stellate hairs, which ultimately remain or partially or almost completely fall off.

Inflorescence in trusses of 9–11 flowers.

Corolla campanulate, 3.8–5.3 cm long, crimson-scarlet or deep crimson; 5 nectar pouches.

Ovary densely tomentose with stellate hairs.

Style hairy at the base with stellate hairs, glandular at the base or rarely eglandular.

E Upper Burma, W Yunnan. 1,800–3,000 m (6,000–10,000 ft).

Epithet: Delightful.   Hardiness 1–3.   May–June.

Very rare in cultivation.

R. elliottii Watt ex Brandis

Shrub or **tree**, 2.4–4.6 m (8–15 ft) high; **branchlets** hairy with stellate hairs, setulose-glandular or eglandular.

Leaves oblong-elliptic, oblong-oval or oblong-lanceolate; **lamina** 6–15 cm long, 2.6–6 cm broad; **upper surface** dark green, shiny, glabrous (in young leaves, hairy with stellate hairs); **underside** paler, glabrous (in young leaves, hairy with stellate hairs).

Inflorescence in trusses of 9–15 flowers; **pedicel** glabrous, glandular.

Calyx glabrous, glandular.

Corolla tubular-campanulate, 4.5–5.5 cm long; crimson, scarlet or deep rose; 5 nectar pouches at the base.

Style glandular throughout to the tip.

Naga Hills, Assam. 2,400–2,750 m (8,000–9,000 ft).

AM 1934 (Embley Park), flowers crimson (KW7725).

FCC 1937 (Clyne Castle), flowers deep scarlet (KW7725).

Epithet: After Mr. Elliott, a friend of its discoverer, Sir George Watt.   Hardiness 1–3.   May–July.

Submerged into
R. kyawii but is
earlier flowering and
has shorter petioles.
Style usually
glandular, but only
at the base.

Indumentum almost
completely detersile.

Corolla crimson-scarlet
or deep crimson.

R. agapetum at Borde Hill

Hard to distinguish
from *R. facetum* out
of flower.

Calyx and pedicel
glandular, not hairy.

R. facetum usually
has more persistent
hairs on the leaves,
and usually larger
leaves.

Leaf lamina 6–15 cm
long, 2.6–6 cm broad.

Style glandular
throughout to the tip.

Corolla usually scarlet.

R. elliottii from KW7725 at Deer Dell

R. facetum Balf.f. & Ward

(*R. eriogynum* Balf.f. & W.W.Sm.)

Shrub or **tree** 1.5–12 m (5–40 ft) high; **branchlets** glabrous or hairy with stellate hairs.

Leaves oblong-elliptic, oblong-obovate or oblong-lanceolate; **lamina** 9–25 cm long, 3–7.8 cm broad; **upper surface** dark green, matt, glabrous (in young leaves, mealy (i.e. hairy) with stellate hairs); **underside** paler, somewhat shiny, glabrous or sometimes with a few small patches of stellate hairs (in young leaves, densely or moderately mealy with stellate hairs).

Inflorescence in trusses of 8–16 flowers.

Pedicel hairy with stellate hairs.

Calyx hairy with stellate hairs.

Corolla tubular-campanulate, 2.5–5 cm long; crimson, scarlet, crimson-rose or deep rose; 5 nectar pouches at the base.

Ovary densely tomentose with stellate hairs.

E and NE Upper Burma, mid-W and W Yunnan. 2,300–3,700 m (7,500–12,000 ft).
AM 1924 (Sunninghill), flowers reddish.
AM 1938 (Clyne Castle), flowers scarlet.
Epithet: Elegant.   Hardiness 1–3.   June–July.

R. kyawii Lace & W.W.Sm.

(*R. prophantum* Balf.f. & Forrest)

Shrub or **tree** 3–7.6 m (10–25 ft) high; **branchlets** at first hairy with stellate hairs, setulose-glandular, finally glabrous, eglandular.

Leaves oblong, oblong-obovate or oblong-oval; **lamina** 14–28 cm long, 3–11 cm broad; **upper surface** dark green, at first sparsely hairy with stellate hairs later glabrous; **underside** at first moderately or rather densely sprinkled (discontinuous) with stellate hairs, which ultimately remain or partially or completely fall off leaving the underside partly or completely glabrous.

Inflorescence in trusses of 12–16 flowers.

Pedicel setulose-glandular, glabrous or tomentose.

Calyx glandular, glabrous or tomentose.

Corolla tubular-campanulate, 4.5–6 cm long, crimson or crimson-scarlet, outside hairy with stellate hairs; 5 nectar pouches.

Ovary densely tomentose with stellate hairs.

Style wholly or partly hairy with stellate hairs, glandular throughout to the tip or sometimes eglandular.

NE Upper Burma, NW and mid-W Yunnan. 1,350–3,000 m (6,000–12,000 ft).
Epithet: After Maung Kyaw, a Burmese plant collector.   Hardiness 1–3.
June–August.

Tender, and therefore rare in the UK.

Hard to distinguish from *R. elliottii* out of flower.

Pedicel hairy with stellate hairs, eglandular.

Calyx hairy with stellate hairs, eglandular.

Leaf lamina 9–25 cm long, 3–7.8 cm broad. Style sometimes glandular.

Corolla usually scarlet or crimson scarlet.

R. facetum F15917 at Deer Dell

Larger leaves than *R. elliottii* and *R. facetum*, and even later flowering.

Pedicel setulose-glandular, glabrous or tomentose.

Calyx glandular, glabrous or tomentose.

Leaf lamina 14–28 cm long, 3–11 cm broad.

Corolla crimson or crimson-scarlet, outside hairy with stellate hairs.

R. kyawii (S. Hootman)

Ponticum
(Parishia)

R. parishii C.B.Clarke

Shrub or **tree** 6–8 m (20–26 ft) high; **branchlets** hairy with stellate hairs or glabrous.

Leaves elliptic or obovate, broad; **lamina** 6–12.3 cm, long, 3.4–6.5 cm broad; **upper surface** dark green, matt, glabrous; **underside** paler, somewhat shiny, glabrous (in young leaves, mealy (i.e. hairy) with brown stellate hairs), midrib hairy with stellate hairs or glabrous.

Petiole hairy with stellate hairs.

Inflorescence in trusses of 8–12 flowers.

Pedicel glandular.

Calyx glandular.

Corolla tubular-campanulate, 3–3.5 cm long, deep red with darker bands along the petals; 5 nectar pouches.

Ovary densely tomentose with stellate hairs.

SE Lower Burma. 1,800–1,900 m (6,000–6,200 ft).

Epithet: After Rev. C.S. Parish (1822–1897), chaplain at Moulmein.   Hardiness 1–2.   March–April.

Similar to R. elliottii *but has broader leaves. Probably not in cultivation.*

R. schistocalyx Balf.f. & Forrest

Shrub 1.8–6 m (6–20 ft) high; **branchlets** hairy with stellate hairs, eglandular.

Leaves oblanceolate or oblong; **lamina** 10–15.5 cm long, 2.3–5 cm broad; **upper surface** dark green, matt, glabrous (in young leaves, hairy with stellate hairs); **underside** paler, glabrous (in young leaves, hairy with stellate hairs).

Petiole hairy with stellate hairs.

Inflorescence in trusses of 8–10 flowers.

Pedicel densely hairy, hairs stellate.

Calyx large, 0.9–2 cm long, reddish, split on one side.

Corolla tubular-campanulate, 4–5 cm long, rose-crimson or crimson; 5 nectar pouches; outside glabrous.

Ovary densely tomentose with stellate hairs.

W Yunnan. 2,700–3,300 m (9,000–11,000 ft).

Epithet: With split calyx.   Hardiness 2–3.   April–May.

Rare in cultivation.

Large petaloid calyx, but the species is hard to distinguish from *R. facetum* out of flower, save that the indumentum is much less detersile.

Branchlets eglandular (glandular in *R. elliottii* and *R. kyawii*, and sometimes glandular in *R. facetum*).

Corolla rose crimson, or crimson.

R. schistocalyx F17637 at Deer Dell

R. schistocalyx F17637 detail at Deer Dell (M.L.A. Robinson)

Ponticum
(Parishia)

R. sikangense var. sikangense

R. sikangense Fang

Shrub or small **tree** 3–5 m (10–16 ft), high; **stem** and **branches** with rough
bark; **branchlets** at first whitish-floccose, later glabrous.

Leaves lanceolate or oblong-lanceolate; **lamina** 8–10 cm long, 2.5–3 cm broad;
upper surface dark green, glabrous; **underside** pale green, glabrous.

Petiole red-brown tomentose when young with stellate hairs.

Inflorescence in trusses of 10 flowers.

Pedicel white-tomentose with stellate hairs.

Calyx outside hairy with stellate hairs.

Corolla campanulate, 4 cm long, purple, with a red blotch and with purple spots.

W Sichuan. 3,700–4,550 m (12,000–15,000 ft).

Epithet: From Sikang.   Hardiness 3.   May–July.

Chamberlain places this species in Subsection Maculifera, *but out of flower it fits
better into Subsection* Parishia *for identification purposes because of the
minimal persistent indumentum on the midrib. The corolla is without nectar
pouches and not indicative of either Subsection. Forms with any persistent
indumentum are almost certainly hybrids, and this may include the suggested
R. sikangense var.* exquisitum.

R. sikangense var. sikangense

R. cookeanum Davidian

Shrub or **tree** 1.2–8 m (4–26 ft) high; **stem** and **branches** with smooth,
brown, flaking bark; **branchlets** with a thin whitish or fawn tomentum.

Leaves oblong-lanceolate or oblong-elliptic; **lamina** 6.5–15 cm long,
2.4–6.5 cm broad; **upper surface** dark green, glabrous; **underside**
glabrous (in young leaves, sparsely hairy with stellate hairs).

Inflorescence in trusses of 8–15 flowers.

Pedicel densely hairy with stellate hairs.

Calyx densely or moderately hairy with stellate hairs.

Corolla campanulate, 3–5 cm long, usually white, with or without a
crimson blotch.

Ovary densely tomentose with stellate hairs.

SW Sichuan. 3,700–4,450 m (12,000–15,000 ft).

Epithet: After R.B. Cooke (1880–1973), rhododendron grower, Corbridge,
Northumberland.   Hardiness 3.   May–July.

Submerged into R. sikangense, *but the leaves of the earlier collections are
distinctively broader and the flowers are white or pink. Two recent collections,
at least, of* R. sikangense *have white flowers and may be referable to*
R. cookeanum.

The species is glabrous with a shiny lower leaf surface.

Stem and branches with rough bark.

Corolla campanulate, **purple**.

Leaves usually lanceolate.

Petiole red-brown tomentose when young.

Pedicel tomentose with stellate hairs.

Calyx hairy with stellate hairs.

R. sikangense var. *sikangense* at Kilbryde, Northumberland (Davidian)

Stem and branches with smooth, brown, flaking bark.

Leaves oblong-lanceolate or oblong-elliptic and broader.

Corolla campanulate, usually **white**, with or without a crimson blotch.

'*R. cookeanum*' AC879 at Deer Dell

Ponticum
(Parishia)

R. venator Tagg

Shrub 1.5–3 m (5–10 ft) high; **branchlets** hairy with stellate hairs, setulose-glandular.

Leaves oblong-lanceolate, lanceolate or oblanceolate; **lamina** somewhat recurved, 6–14 cm long, 1.7–4 cm broad; **upper surface** dark green, somewhat rugulose, glabrous or hairy with stellate hairs; **underside** with widely or closely scattered white stellate hairs, eglandular, papillate.

Petiole with stellate hairs.

Inflorescence in trusses of 5–10 flowers.

Pedicel moderately or densely hairy with stellate hairs, often setulose-glandular.

Corolla tubular-campanulate, 2.8–3.8 cm long, scarlet or deep crimson; 5 nectar pouches.

SE Tibet. 2,000–2,450 m (7,000–8,000 ft).

AM 1933 (Bodnant), flowers reddish orange (KW6285).

Epithet: Hunter, alluding to the scarlet colour of the flowers.   Hardiness 3. May–June.

Formerly separated from Subsection Parishia, it has been restored in the 'Flora of China'. *A related species with more indumentum has recently been introduced by P.A. and K.N.E Cox under KC0104.*

Subgenus *Hymenanthes*
Section *Ponticum* G.Don
Subsection *Pontica* Sleumer
Ponticum Series

Shrubs or small trees.
Leaves with a fairly smooth, not rugulose, surface.
Rhachis usually long, giving rise to a candelabroid inflorescence.
Calyx white tomentose.
Pedicels noticeably long when in fruit.
Corolla deeply lobed to half its length.
Stamens 10 (14 in *R. degronianum* var. *heptamerum*).

Group A. Leaf underside always glabrous:
> *R. aureum, R. catawbiense, R. hyperythrum, R. macrophyllum* and *R. ponticum.*

Group B1. Leaf underside not glabrous, indumentum thin:
> *R. brachycarpum* (ssp. *brachycarpum*, Tigerstedtii Group and ssp. *fauriei*), *R. caucasicum* and *R. maximum.*

Group B2. Leaf underside not glabrous, indumentum thick:
> *R. degronianum* (ssp. *degronianum*, ssp. *heptamerum* [including var. *heptamerum*, var. *hondoense* and var. *kyomaruense*] and spp. *yakushimanum*), *R. makinoi, R. smirnowii* and *R. ungernii.*

A shrub of straggly growth with relatively small irroratum-type rugulose matt leaves, with a somewhat detersile indumentum of stellate hairs beneath.

The red glandular hairs on the branchlets of the new leaves are an outstanding feature.

Vivid 'Hunting Scarlet' flower colour.

R. venator KW6285 at Deer Dell

New foliage of *R. venator* KW6285 at Deer Dell

Leaf types of Subsect. *Pontica* at Deer Dell. Top to bottom: left, *R. degronianum* ssp. *heptamerum* var. *kyomaniense* (1 leaf), *R. degronianum* ssp. *heptamerum* 'Ho Emma' FCC (2 leaves) and *R. hyperythrum* (2 leaves); right (2 leaves each), *R. makinoi* narrow leaf form, *R. makinoi* and *R. ungernii*. Bar = 5 cm

Ponticum
(Parishia/Pontica)

R. aureum Georgi

Pontica Gp A

(*R. chrysanthum* Pallas)

Small, prostrate or semi-prostrate or compact spreading **shrub**, 10–60 cm
(4 inches–2 ft) high; annual growth short; **leaf-bud scales** persistent.
Leaves obovate, oblanceolate, oblong-elliptic or ovate; **lamina** 2.3–9.5 cm long,
0.8–4 cm broad; **upper surface** dark green or olive-green; **underside**
glabrous, under magnification punctulate with minute hairs.
Inflorescence a candelabroid umbel of 3–8 flowers, flower-bud scales persistent.
Corolla widely funnel-shaped, 2–3 cm long, yellow, with or without purplish spots.

Extends from the Altai mountains in W Siberia to the mountains of Mongolia,
Manchuria, Kamtschatka, Sakhalin Island, Kurile Islands, Korea and Japan.
1,500–2,700 m (5,000–6,000 ft).
Epithet: Golden.   Hardiness 3.   April–May.

Often chlorotic in cultivation.

R. catawbiense Michaux

Pontica Gp A

Shrub 1–3 m (3–10 ft) or rarely 6 m (20 ft) high; **branchlets** floccose,
eglandular.
Leaves oblong-oval, oval, elliptic or oblong-elliptic; **lamina** 6.5–15 cm long,
2.6–3 cm broad; **upper surface** dark green, shiny, glabrous; **underside** pale
whitish-green, glabrous.
Inflorescence a candelabroid umbel of 8–20 flowers.
Corolla funnel-campanulate, 3.5–4.3 cm long, rose, lilac-purple, pink or white,
with olive-green spots.

N Carolina, Virginia, USA. 1,200–2,000 m (3,950–6,500 ft).
Epithet: After the Catawba river, N Carolina, USA.   Hardiness 3.   May–June.

Generally a prostrate shrub.

Annual growth short.

Leaf-bud scales persistent.

Smaller leaves than its relatives.

Flower-bud scales persistent.

Corolla yellow.

R. aureum at Deer Dell

Rounded fairly dense-growing shrub.

Convex leaves, glabrous but whitish beneath, with rounded or obtuse bases.

Corolla rose, lilac, pink or white.

R. catawbiense at Borde Hill

Ponticum
(Pontica Group A)

R. hyperythrum Hayata   Pontica Gp A

Shrub 1–2.4 m (3–8 ft) high; **branchlets** glabrous or floccose.
Leaves lanceolate, oblong-lanceolate or oblong; **lamina** 6.8–11.5 cm long,
    1.6–5 cm broad, apex acute or shortly acuminate; **upper surface** dark green,
    glabrous; **underside** pale green, glabrous, under magnification with reddish
    or brownish punctulations.
Inflorescence candelabroid umbel of 7–12 flowers.
Pedicel glandular.
Calyx glandular.
Corolla funnel-campanulate, 3–5 cm long, white or pink, with or without purple
    spots.
Taiwan. 900–1,200 m (3,000–4,000 ft).
AM 1976 (Collingwood Ingram) 'Omo', flowers white.
Epithet: Reddish below.   Hardiness 3.   April–May.

This species often has very recurved marginal leaves in cultivation.

R. macrophyllum D.Don ex G.Don   Pontica Gp A

Shrub 1.8–3.7 m (6–12 ft) or sometimes 9 m (30 ft) high; **branchlets** glabrous.
Leaves elliptic, oblong-elliptic or oblong-lanceolate; **lamina** 6.5–17 cm long,
    3–6 cm broad; **upper surface** dark green, glabrous; **underside** pale
    green, glabrous.
Petiole glabrous, eglandular.
Inflorescence candelabroid umbel of 9–20 flowers.
Calyx minute, 1 mm long.
Corolla widely funnel-shaped, 2.8–4 cm long, white, pink, rose or rose-purple,
    with reddish-brown or yellowish spots.
California to British Columbia. Sea level up to 1,200 m (4,000 ft).
Epithet: With big leaves.   Hardiness 3.   May–June.

Recurved leaves, appearing narrow and similar to *R. degronianum*, but with no indumentum: underside pale green, glabrous.

Almost always white flowers.

Pedicel glandular.
Calyx glandular.
Corolla funnel-campanulate, white or pink.

R. hyperythrum at Deer Dell

Resembles *R. ponticum*, but the leaves are usually more oblong and larger.

Flat usually elliptic or oblong leaves, underside pale green, glabrous.

Ovary tomentose.

Corolla usually rose or rose-purple and always with reddish-brown or yellowish spots.

R. macrophyllum in the Valley Gardens

Ponticum
(Pontica Group A)

R. ponticum Linn.   Pontica Gp A

Shrub 1–4.6 m (3–15 ft) or **tree** up to 7.6 m (25 ft) high; **branchlets** glabrous.

Leaves lanceolate, oblong-lanceolate, oblanceolate or oblong; **lamina** 6.5–21 cm long, 2.1–6 cm broad; **upper surface** dark green, shiny, glabrous; **underside** pale green, glabrous.

Petiole glabrous.

Inflorescence a candelabroid umbel; 6–19 flowers.

Calyx 1–2 mm long, glabrous.

Corolla widely funnel-shaped, 3–5 cm long, pinkish-purple, purple, pink, lilac-purple, lavender or mauve, with greenish-yellow spots.

Pontic ranges of North Anatolia, Turkey, Caucasus, Lebanon, SE Bulgaria, Portugal, Spain. Sea-level up to 1,200–1,800 m (4,000–6,000 ft).

Epithet: From the Pontus, Asia Minor.   Hardiness 4.   June–July.

There are quite a few selected extreme and atypical forms of R. ponticum in cultivation: these are probably hybrids as this species has been promiscuously crossing with garden rhododendrons. White-flowered forms are also likely to be hybrids.

R. brachycarpum ssp. brachycarpum

R. brachycarpum D.Don ex G.Don   **Pontica Gp B1**

Rounded, occasionally, lax **shrub**, 1.2–3 m (4–10 ft) high; **branchlets** often tomentose with a thin fawn tomentum.

Leaves oblong-elliptic, oblong or oblong-obovate; **lamina** 7–15 cm long, 2.6–7 cm broad; **upper surface** bright green, glabrous; **underside** covered with a thin, brown or fawn, continuous unistrate indumentum of radiate hairs.

Petiole glabrous.

Inflorescence a candelabroid umbel of 8–21 flowers.

Corolla widely funnel-shaped, 2.5–3 cm long, white, white flushed pink or creamy-white flushed pink along the middle of the petals, with green spots.

North and Central Japan, Korea. 1,700–2,300 m (5,500–7,500 ft).

Epithet: With short fruit.   Hardiness 3.   June–July.

Leaf underside pale green.
Glabrous in all its parts.

R. ponticum from seed collected in Turkey at Barnhourie Mill

R. ponticum at Deer Dell

Pointed growth buds.

Leaves have a pale yellow petiole and midrib.

Persistent thin pale fawn indumentum on leaf underside.

Late-flowering.

Corolla white flushed pink or creamy-white flushed pink.

R. brachycarpum ssp. **brachycarpum** at Deer Dell

Leaves of R. brachycarpum ssp. **brachycarpum (two forms) at Deer Dell.** Bar = 5 cm

Ponticum
(Pontica Group A/B1)

R. brachycarpum Tigerstedtii Group   Pontica Gp B1

R. brachycarpum var. *tigerstedtii* (Nitzelius) Davidian

This variety differs from the species type in that the **leaves** are longer, 15–25 cm in length, and the **flowers** are broader, 7 cm in breadth.

Korea. 200–900 m (650–3,000 ft).

Epithet: After Dr Tigerstedt, the former owner of Mustila Arboretum in south Finland.   Hardiness 3.   June–July.

R. brachycarpum ssp. fauriei D.F.Chamb.

R. fauriei Franch.   Pontica Gp B1

Rounded compact **shrub** 1–3 m (3–10 ft) high; **branchlets** with a thin, whitish tomentum, eglandular.

Leaves oblong-obovate or oblong-elliptic; **lamina** 6.6–10.3 cm long, 3–5 cm broad; **upper surface** bright green, glabrous; **underside** pale green, glabrous.

Inflorescence a candelabroid umbel of 12–20 flowers.

Pedicel rather densely hairy.

Calyx minute, 1 mm long.

Corolla widely funnel-shaped, 2.1–2.5 cm long, white or yellowish, flushed pink along the middle of the petals, with green spots, outside glabrous, eglandular.

N Japan, Korea. About 2,500 m (8,000 ft).

Epithet: After Pere L.F. Faurie, French Foreign Missions, China.   Hardiness 3. June–July.

R. caucasicum Pallas   Pontica Gp B1

Shrub 0.3–1 m (1–3 ft) high; **branchlets** with a thin tomentum; **leaf-bud scales** persistent.

Leaves oblong, oblong-obovate or obovate; **lamina** 4–14 cm long, 1.5–4.5 cm broad; **upper surface** dark green, glabrous; **underside** covered with a thin, fawn, brown or yellowish, continuous unistrate indumentum of hairs.

Inflorescence a candelabroid umbel of 3–14 flowers; flower-bud scales persistent.

Corolla widely funnel-shaped, 2.3–3.5 cm long, pale cream, white flushed lemon-yellow, white flushed pink, very pale lemon flushed pink, or pale lemon, with or without green spots.

NE Turkey, the Caucasus, and the adjacent parts of Russia. 1,800–3,000 m (6,000–10,000 ft).

Epithet: From the Caucasus.   Hardiness 3.   April–May.

Leaves longer and corolla broader than those of the subspecies type.

R. brachycarpum Tigerstedii Group at the Rhododendron Species Foundation (M.L.A. Robinson)

Small compact shrub.
Leaf underside glabrous at maturity.
Corolla white to yellowish.

R. brachycarpum ssp. *fauriei* in the Valley Gardens

Indumentum continuous, thin, fawn, brown or yellowish.
Leaf-bud scales persistent; flower-bud scales persistent.
Similar to, but larger in all its parts than, *R. aureum*, which is glabrous beneath.

Corolla pale cream or pale yellow sometimes pink flushed or white flushed pink.

R. caucasicum at Borde Hill

Ponticum
(Pontica Group B1)

R. maximum Linn.

Pontica Gp B1

Shrub 1.2–3.7 m (4–12 ft) or sometimes a **tree** up to 12 m (40 ft) high; **branchlets** floccose, moderately or rather densely glandular.

Leaves lanceolate or oblong-obovate; **lamina** 9–20 cm long, 2.5–7 cm broad; **upper surface** dark green, glabrous; **underside** with a thin film of hairs, or covered with a thin, continuous unistrate indumentum of hairs, or glabrous.

Petiole often glandular.

Inflorescence a candelabroid umbel of 12–30 flowers.

Pedicel rather densely glandular.

Calyx glandular.

Corolla campanulate, 2.3–3.1 cm long, white, pink, rose or purplish, with yellowish-green spots.

Ovary densely glandular.

Ontario, Quebec, Nova Scotia southward along the Appalachian Mountains to Georgia. From nearly sea level to about 900 m (3,000 ft).

AM 1974 (Windsor) 'Summer Time', flowers white with yellow-green spots.

Epithet: Maximum.   Hardiness 3.   July.

A form with smaller leaves than those of the species type, with deeply waved margins and globular flower buds, is known as **R. maximum var. leachii** *Harkness.*

R. degronianum ssp. degronianum Carriere

R. degronianum Carriere **Pontica Gp B2**
(*R. japonicum* [Blume] Schneider, *R. degronianum* ssp. *pentamerum* Matsumura)

Shrub 1–2 m (3–6 ft) high; **branchlets** with a thin, whitish tomentum.

Leaves oblong, obovate or oblong-lanceolate; **lamina** 8–18 cm long, 2.2–4.6 cm broad; **upper surface** dark green or olive-green, shiny, glabrous; **underside** covered with a thick or somewhat thin, felty, fawn or brown, continuous, bistrate indumentum of hairs, eglandular.

Inflorescence candelabroid umbel of 6–15 flowers.

Corolla campanulate, 2.8–4.3 cm long, pink, rose, deep rose, reddish or rarely white, with or without deep pink lines along the middle of the petals; lobes 5.

Stamens 10.

Ovary 5-celled.

Central and Southern Japan. Up to 1,800 m (6,000 ft).

AM 1974 (Wakehurst) 'Gerald Loder'.

Epithet: After M. Degron, Director of the French Posts in Yokohama in 1869.   Hardiness 3.   April–May.

Branchlets floccose, moderately or rather densely glandular.

Large leaves.

The greyish indumentum may be partly shed at maturity.

Upright compact truss.

Late-flowering. Pedicel rather densely glandular.

Calyx glandular.

Corolla white, pink, rose or purplish.

R. maximum in the Valley Gardens (Davidian)

Narrow recurved leaves similar to, but larger than, those of the well-known *R. yakushimanum*.

Indumentum felty, fawn or brown, continuous.

Corolla 5-lobed with 10 stamens, usually pink to rose, sometimes white.

R. degronianum ssp. *degronianum* at Deer Dell

R. degronianum ssp. heptamerum var. heptamerum D.F.Chamb. & F.Doleshy Pontica Gp B2

R. degronianum var. heptamerum (Maxim.) Sealy
(*R. metternichii* Seibold & Zuccherini)

AM 1976 (Borde Hill) 'Ho Emma'.
FCC 1982 (Borde Hill).
Epithet: 7-lobed.   Hardiness 3.   April–May

Formerly R. metternichii. Fairly common in cultivation.

R. degronianum ssp. heptamerum var. hondoense Pontica Gp B2

Merges with R. degronianum ssp. heptamerum var. heptamerum in the wild, and probably of no botanical significance. Rare in cultivation.

R. degronianum ssp. heptamerum var. kyomaruense Pontica Gp B2

Superficially very similar to R. degronianum ssp. degronianum, but from a geologically different area of Japan, and likely to have evolved separately. Rare in cultivation. The designation 'heptamerum' is misleading as the corolla has 5 lobes.

Corolla has 7 or
sometimes 6 lobes.
Stamens 14.
Ovary is 7–8-celled.

Indumentum thin,
plastered.
Corolla 7-lobed.

R. degronianum ssp. *heptamerum* var. *heptamerum*
(*R. metternichii*) 'Ho Emma' AM FCC at Deer Dell

Indumentum thin,
plastered.
Corolla 5-lobed.

R. degronianum ssp. *heptamerum* var. *kyomuruense* at Deer Dell

R. degronianum ssp. yakushimanum Hara

R. yakushimanum Nakai   **Pontica Gp B2**

Very compact or lax **shrub**, 0.3–2.5 m (1–8 ft) high; **branchlets** densely or
moderately whitish woolly.

Leaves lanceolate, oblanceolate or oblong-obovate; **lamina** 5–10 cm long,
3–8 cm broad; **upper surface** dark green, markedly convex or flat,
glabrous (in young leaves, densely whitish woolly), margin recurved or flat;
underside covered with a thick, woolly, fawn, continuous, bistrate
indumentum of hairs.

Petiole densely whitish woolly.

Inflorescence a candelabroid umbel of 8–12 flowers.

Corolla campanulate, 3.2–3.6 cm long, flower-buds rich pink, finally corolla
pure white or very pale pink.

Yakushima Island, Japan. 500–2,000 m (1,600–6,500 ft).

FCC 1947 (RHS Wisley) 'Koichiro Wada', flowers pink in bud, opening white.

Epithet: From Yakushima Island, Japan.   Hardiness 3.   May.

The laxer growing (lower altitude) forms are rare in cultivation.

R. makinoi Tagg   **Pontica Gp B2**

Shrub 1–2.4 m (3–8 ft) high; **branchlets** densely brown woolly, eglandular;
leaf-bud scales persistent.

Leaves lanceolate or narrowly lanceolate; **lamina** 5–17 cm long, 0.8–2.5 cm
broad; **upper surface** dark green or bright green, glabrous (in young leaves,
whitish woolly); margin recurved; **underside** covered with a thick, woolly,
brown, continuous bistrate indumentum.

Petiole densely brown woolly, eglandular.

Inflorescence a candelabroid umbel of 5–8 flowers.

Calyx 1–7 mm long, densely hairy, eglandular.

Corolla funnel-campanulate, 3–4 cm long, pink or rose, with or without
crimson spots.

Central Honshu, Japan. 450–550 m (1,500–1,800 ft).

Epithet: After T. Makino, a Japanese botanist.   Hardiness 3.   May–June.

In cultivation, nearly always a compact shrub.

Leaves recurved: young leaves densely whitish woolly.

Indumentum thick, woolly, fawn.

Corolla white or very pale pink.

R. degronianun ssp. *yakushimanum* at Deer Dell

R. degronianum ssp. *yakushimanum* in the Valley Gardens

Long, narrow or very narrow, leaves strongly recurved in two dimensions.

Branchlets densely brown woolly.

Indumentum thick, woolly, brown.

Corolla rose or pink.

R. makinoi narrow leaf form at Deer Dell

R. smirnowii Trautvetter   Pontica Gp B2

Shrub or **tree**, 0.5–6 m (1.8–20 ft) high; **branchlets** rather densely woolly with whitish or fawn wool, glandular.

Leaves oblong-obovate or oblanceolate; **lamina** 6.5–16 cm long, 1.8–4.2 cm broad; **upper surface** olive-green or dark green, shiny, glabrous (in young leaves, densely whitish woolly); **underside** covered with a thick, woolly, fawn or brown, continuous bistrate indumentum of hairs.

Petiole densely whitish or fawn woolly, glandular.

Inflorescence a candelabroid umbel of 6–12 flowers.

Pedicel floccose, glandular.

Corolla funnel-campanulate, 3.1–4.8 cm long, pink, deep pink, rose-red or rose-purple, with or without spots.

NE Turkey, Caucasus, Georgia, Russia. 800–2,800 m (2,600–9,000 ft).

AM 1991 (Exbury) 'Vodka'.

Epithet: After M. Smirnow, a friend of its discoverer, Baron Ungern-Sternberg. Hardiness 3–4.   May–June.

R. ungernii Trautvetter   Pontica Gp B2

Shrub or **tree**, 1–6 m (3–20 ft) high; **branchlets** with whitish tomentum, glandular.

Leaves oblong-obovate, oblanceolate or oblong-oblanceolate; **lamina** 9–25.5 cm long, 2.8–7.5 cm broad; **upper surface** dark green, glabrous; **underside** covered with a thick, woolly, whitish or fawn, continuous, bistrate indumentum of hairs.

Inflorescence a candelabroid umbel of 12–30 flowers.

Pedicel moderately or rather densely glandular.

Calyx outside glandular.

Corolla campanulate, 3–4.5 cm long, pale rose, pink or white, with or without pale green spots, outside of tube pubescent, glandular.

Ovary densely glandular.

NE Turkey, Georgia, Russia. 850–2,000 m (2,600–6,500 ft).

AM 1973 (Bodnant), flowers white suffused pink.

Epithet: After Baron F. von Ungern-Sternberg (1800–1868), Professor at Dorpat. Hardiness 3.   June–July.

Branchlets rather densely woolly.

Dark strongly recurved leaves, broader than those of *R. degronianum* ssp. *yakushimanum*.

Thick woolly fawn or brown indumentum.

Resembles *R. ungernii* but has a tomentose ovary.

Petiole densely whitish or fawn woolly.

Corolla pink to rose red or rose purple.

R. smirnowii in the Valley Gardens

Leaves similar to, but larger, flatter, and broader than those of *R. smirnowii*.

Indumentum thick, woolly, whitish or fawn.

Late-flowering.

Ovary densely glandular.

Corolla pale rose, pink or white.

R. ungernii at Deer Dell

Subgenus *Hymenanthes*
Section *Ponticum* G.Don
Subsection *Selensia* Sleumer

Species within this Subsection are placed in one of two series:

Subsection *Selensia* Sleumer
Thomsonii Series, Selense Subseries

Shrubs or small trees.
Branchlets bristly glandular.
Leaves rounded, obovate to elliptic.
Leaf underside glabrous or with a thin indumentum.
Inflorescence lax.
Corolla 5-lobed, lacking nectar pouches.
Stamens 10.

R. bainbridgeanum, *R.* × *erythrocalyx*, *R. esetulosum*, *R. eurysiphon*, *R. hirtipes*, *R. martinianum*, *R. selense* (ssp. *selense* and ssp. *dasycladum*, ssp. *jucundum* [which includes var. *probum*] and ssp. *setiferum*).

R. martinianum KW21557 at Deer Dell

Leaves of Subsect. *Selensia* at Deer Dell. From top left to bottom right (2 leaves each): *R. hirtipes, R. dasycladum, R. setiferum, R. bainbridgeanum* and *R. selense* ssp. *jucundum* var. *probum.* Bar = 5 cm

R. bainbridgeanum Tagg & Forrest

Shrub 0.6–2.8 m (2–9 ft) high; **branchlets** rather densely or moderately setulose-glandular.

Leaves oblong-elliptic, obovate or oblong; **lamina** 5.5–12 cm long, 2–4.8 cm broad; **upper surface** dark green, glabrous; **underside** with a thin discontinuous or continuous indumentum of woolly hairs, or with a thin veil of hairs.

Petiole rather densely or moderately setulose-glandular.

Inflorescence in trusses of 5–8 flowers.

Corolla campanulate, 2.8–4 cm long, creamy-yellow, white, reddish-purple or pink, with or without a crimson blotch, with or without crimson spots.

SE Tibet, NW Yunnan. 3,000–4,400 m (10,000–14,500 ft).

Epithet: After Mr. Bainbridge, a friend of George Forrest.   Hardiness 3.
March–April.

R. × erythrocalyx

R. erythrocalyx Balf.f. & Forrest

Shrub 1–2.4 m (3–8 ft) high; **branchlets** often glandular.

Leaves elliptic, oblong-elliptic or oblong-oval; **lamina** thin, chartaceous, 3–10.6 cm long, 2.1–5.3 cm broad; **upper surface** dark green, glabrous; **underside** pale glaucous green, glabrous or minutely hairy.

Petiole often glandular.

Inflorescence 4–10 flowers.

Pedicel glandular.

Calyx glandular.

Corolla campanulate, 2.8–4.5 cm long, white, white suffused with rose, rose or creamy-yellow, with or without a crimson blotch.

Ovary densely glandular.

NW Yunnan, SE Tibet, NE Upper Burma. 3,400–4,000 m (11,000–13,000 ft).

Epithet: With a red calyx.   Hardiness 3.   April–May.

This is probably a natural hybrid of R. selense.

Leaves usually oblong.

The thin 'scurfy' indumentum is unusual in this Subsection.

The form with reddish-purple flowers is noteworthy.

Branchlets rather densely or moderately setulose-glandular.

Petiole rather densely or moderately setulose-glandular.

Corolla white, creamy yellow, pink or reddish purple.

R. bainbridgeanum F21834 at Deer Dell

Leaf underside pale glaucous green, glabrous or minutely hairy.

Pedicel glandular.

Calyx glandular.

Corolla usually white.

Ovary densely glandular.

R. × erythrocalyx at Deer Dell

Ponticum
(Selensia)

R. esetulosum Balf.f. & Forrest

(*R. manopeplum* Balf.f. & Forrest)

Shrub 1.2–1.8 m (4–6 ft) high; **branchlets** glandular.
Leaves oblong, elliptic or oblong-ovate; **lamina** thick, coriaceous, 5–9.4 cm long, 2.3–5.1 cm broad; **upper surface** dark green, glabrous; **underside** with a thin veil of hairs, midrib glandular.
Petiole often glandular.
Inflorescence in trusses of 8–10 flowers.
Pedicel glandular.
Calyx margin fringed with glands.
Corolla campanulate, 3.1–5 cm long, creamy-white or white, often flushed rose or purplish, often spotted crimson.
Ovary densely glandular.
NW and mid-W Yunnan, SE Tibet. 3,000–4,300 m (10,000–14,000 ft).
Epithet: Hairless.   Hardiness 3.   April–May.

Very rare in cultivation. Possibly a hybrid. Suffers badly from powdery mildew. The thick coriaceous leaf is unusual in this Subsection. Indumentum is very thin.

R. eurysiphon Tagg & Forrest

Shrub 1–1.5 m (3–5 ft) high; **branchlets** often glandular.
Leaves oblong, oblong-oval or oblong-elliptic; **lamina** leathery rigid, 2.6–7.5 cm long, 1.3–2.8 cm broad; **upper surface** dark green, glabrous; **underside** pale glaucous green, glabrous, often minutely glandular.
Petiole often glandular.
Inflorescence 3–5 flowers.
Pedicel glandular.
Calyx glandular.
Corolla campanulate, 3–4 cm long, creamy-white or very pale rose, flushed deep magenta-rose, copiously spotted with crimson.
Ovary densely glandular.
SE Tibet. 4,000 m (13,000 ft).
Epithet: With a broad tube.   Hardiness 3.   May.

Some plants so labelled are forms of R. stewartianum. *It can also be confused with* R. martinianum. R. eurysiphon *has a broad-based campanulate corolla with copious spotting, whereas* R. martinianum *has a funnel-campanulate corolla with or without spotting.*

R. eurysiphon F13949A at Deer Dell showing broad-based campanulate corolla

Very short petioles. Corolla campanulate, pale cream or rose, copiously spotted with crimson.

Leaf lamina leathery rigid, underside pale glaucous green, glabrous, often minutely glandular.

Calyx glandular.

R. eurysiphon F21694 at Deer Dell

New foliage of *R. eurysiphon* at Deer Dell

R. hirtipes Tagg

Shrub or **tree**, 1–7.6 m (3–25 ft) high; **stem** and **branches** with rough bark; **branchlets** moderately or rather densely brown setose-glandular.

Leaves oval, obovate, elliptic or oblong-oval; **lamina** 6–12 cm long, 3–7 cm broad; **upper surface** dark green, shiny, glabrous, midrib setulose-glandular in the lower half; **underside** not setose, punctulate with vesicular hairs and short-stalked glands, midrib setulose-glandular in its entire length or in the lower half.

Petiole setose-glandular.

Inflorescence in lax trusses of 3–5 flowers, flower-bud sticky.

Pedicel setulose-glandular.

Calyx glandular.

Corolla campanulate, 3.5–4.8 cm long, rose-pink, pink or white flushed pink, with or without purple blotch and spots.

SE Tibet. 3,000–4,000 m (10,000–13,500 ft).

AM 1965 (Glenarn) 'Ita', flowers phlox pink (LS&T3624).

Epithet: With a hairy foot.   Hardiness 3.   April–May.

With similarities to Subsection Glischra, *this species does not fit well into that Subsection or into Subsection* Selensia.

R. martinianum Balf.f. & Forrest

Shrub 0.6–1.8 m (2–6 ft) high; **branchlets** often setulose-glandular, nodular, thin.

Leaves oblong, oblong-oval or oblong-elliptic; **lamina** thinly leathery, rigid, 1.9–4 cm long, 0.8–2.5 cm broad; **upper surface** green, glabrous; **underside** pale glaucous green, glabrous, punctulate with minute glands.

Petiole glandular.

Inflorescence 1–3 flowers.

Pedicel glandular.

Calyx usually glandular.

Corolla funnel-campanulate, 2.5–4 cm long, pale rose, white or creamy-white faintly flushed pale rose, with or without crimson spots.

Ovary densely glandular.

W Yunnan, Upper Burma, E and SE Tibet. 2,750–4,300 m (9,000–14,000 ft).

Epithet: After J. Martin, gardener at Caerhays, Cornwall.   Hardiness 3. April–May.

Similar to R. eurysiphon – *differences are described under* R. eurysiphon (p. 286).

The leaf is usually oblong-oval and is shiny above in some forms.

Copious setose glands on the petiole.

Branchlets brown setose-glandular.

Leaf midrib setulose-glandular in its entire length or in the lower half.

Calyx glandular.

Corolla rose pink to pale pink.

R. hirtipes at Eckford, Scotland

Rigid oval shiny leaf. Few (1–3) flowers in the truss.

Branchlets often setulose-glandular.

Petiole glandular.

Pedicel glandular.

Corolla usually pale rose, white or creamy-white.

R. martinianum KW6795 at Deer Dell showing the funnel-campanulate corolla

New foliage of *R. martinianum* KW6795 at Deer Dell

Ponticum
(Selensia)

R. selense ssp. selense

R. selense Franch.
(*R. nanothamnum* Balf.f. & Forrest)

Shrub 0.6–1.8 m (2–6 ft) or sometimes 2.4–3 m (8–10 ft) high; **branchlets** often glandular, hairy or glabrous.

Leaves oblong, obovate, oblong-oval or elliptic; **lamina** 2.6–8.2 cm long, 1.5–3.9 cm broad; **upper surface** dark green or green, glabrous; **underside** pale glaucous green, glabrous or minutely hairy or with a thin veil of hairs.

Inflorescence 3–8 flowers.

Pedicel glandular.

Calyx outside glandular, margin gland-fringed.

Corolla funnel-campanulate, 2.2–4 cm long, pink, rose, white flushed rose, or rarely reddish-purple, with or without a crimson blotch.

Ovary densely glandular.

NW Yunnan, SE Tibet, SW Sichuan. 3,000–4,400 m (10,000–14,500 ft).

Epithet: From Sie-La, W Yunnan.   Hardiness 3.   April–May.

R. selense var. *pagophilum* Cowan & Davidian

Tibet–Yunnan frontier. NW Yunnan, SE Tibet. 3,700–4,900 m (12,000–16,000 ft).

Epithet: Rock-loving.   Hardiness 3.   April–May.

R. selense ssp. dasycladum D.F.Chamb.

R. dasycladum Balf.f. & W.W.Sm.
(*R. rhaibocarpum* Balf.f. & W.W.Sm.)

Shrub or **tree**, 1–4.5 m (3–15 ft) high; **branchlets** moderately or densely setulose-glandular.

Leaves oblong, oblong-oval, oval or elliptic; **lamina** 3–12.5 cm long, 1.5–4.7 cm broad; **upper surface** dark green; **underside** glabrous or with a thin veil of hairs.

Petiole often setulose-glandular.

Inflorescence in trusses of 5–10 flowers.

Pedicel glandular.

Calyx outside and margin moderately or densely glandular.

Corolla funnel-campanulate, 2.5–3.6 cm long, rose, pink, white or white suffused with rose, with or without a crimson blotch.

NW and mid-W Yunnan. 3,000–4,000 m (10,000–13,000 ft).

Epithet: With hairy boughs.   Hardiness 3.   April–May.

The petioles are long in relation to the size of the leaf (also in ssp. *jucundum*).
Leaf underside green or pale glaucous green.

Leaves oblong, obovate, or oblong-oval, undersides glabrous or minutely hairy.

Petiole glandular or eglandular.

Pedicel glandular.

Calyx glandular, margin gland-fringed.

Corolla usually very pale rose, rose or pink.

R. selense ssp. *selense* R11127 in the Valley Gardens

Small flowers and small leaves.
Corolla often dark rose to dark crimson.

R. selense var. *pagophilum* F21743 at Deer Dell

Branchlets moderately or densely setulose-glandular.
Leaves usually oblong; underside not glaucous.
More flowers to the truss than in the type of *R. selense*.

Leaves oblong, oblong-oval, or elliptic.

Undersides glabrous or minutely hairy.

Pedicel glandular.

Calyx and margin glandular.

Corolla white to rose, or pink.

R. selense ssp. *dasycladum* CNW384 at Deer Dell

Ponticum
(Selensia)

R. selense ssp. jucundum

R. jucundum Balf.f. & W.W.Sm.

Shrub 60 cm–3 m (2–10 ft) high or **tree** up to 6.10 m (20 ft) high; **branchlets** not glandular, not setulose-glandular or sometimes setulose-glandular, often warty.
Leaves elliptic, oblong or oblong-elliptic; **lamina** 3.8–7.4 cm long, 2–4 cm broad; **upper surface** dark green, glabrous; **underside** markedly glaucous, glabrous or minutely hairy.
Petiole not glandular, not setulose-glandular.
Inflorescence 5–8 flowers.
Pedicel glandular.
Calyx 1–6 mm long, margin glandular.
Corolla funnel-campanulate, 3–4 cm long, pale rose, rose, pink or white suffused with rose, with or without crimson blotch.
Ovary densely glandular.
W Yunnan. 3,000–3,700 m (10,000–12,000 ft).
Epithet: Pleasant.　　Hardiness 3.　　May–June.

R. selense ssp. jucundum var. probum

R. selense var. *probum* Cowan & Davidian

In this variety, the **flowers** are white without spots, with or without a crimson blotch at the base, and the underside of the **leaves** is often glaucous.
NW Yunnan, SE Tibet. 3,000–4,000 m (10,000–13,000 ft).
Epithet: Excellent.　　Hardiness 3.　　April–May.

R. selense ssp. setiferum

R. setiferum Balf.f. & Forrest

Shrub 1.5–2.8 m (5–9 ft) high; **branchlets** setulose-glandular and glandular with shorter-stalked glands, often hairy.
Leaves oblong; **lamina** thick, coriaceous, 5–9 cm long, 2–3.5 cm broad; **upper surface** dark green or green, glabrous; **underside** with a thin veil of hairs and short-stalked or sessile glands, midrib glandular.
Petiole setulose-glandular and glandular with shorter-stalked glands.
Inflorescence 6–10 flowers.
Pedicel glandular.
Corolla funnel-campanulate, 3.1–3.5 cm long, creamy-white.
Ovary densely glandular.
NW Yunnan. 3,700–4,000 m (12,000–13,000 ft).
Epithet: Bearing bristles.　　Hardiness 3.　　April–May.

The leaves are very similar to those of *R. selense* ssp. *selense*, but have a brightly glaucous underside.

Leaves elliptic-oblong, oblong or elliptic.
Petiole eglandular.
Pedicel glandular.
Calyx margin glandular.
Corolla white flushed rose to rose, or pink.

R. selense ssp. *jucundum* at Deer Dell

A white-flowered ssp. *jucundum*.

R. selense ssp. *jucundum* var. *probum* R.10929 at Deer Dell

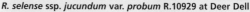

Within this Subsection, only *R. esetulosum* and ssp. *setiferum* have a coriaceous leaf.

Leaf underside with a thin veil of hairs and glands.

Branchlets setulose-glandular.
Petiole setulose-glandular.
Pedicel glandular.
Calyx glandular.
Corolla creamy white.

R. selense ssp. *setiferum* at Deer Dell

Ponticum
(Selensia)

Subgenus *Hymenanthes*
Section *Ponticum* G.Don
Sherriffii Series

Shrubs or small trees.
Leaves rounded to oblong.
Inflorescence lax.
Calyx crimson or carmine.
Corolla crimson or carmine, with 5 nectar pouches.

Sufficiently distinct from Subection Thomsonia to be separated for identification purposes.

R. lopsangianum Cowan

Shrub 0.6–1.8 m (2–6 ft) high; **branchlets** glandular or eglandular.
Leaves elliptic, oblong-elliptic, oval or almost orbicular; **lamina** 2.7–6 cm long, 1.3–4.5 cm broad; **upper surface** dark green; **underside** papillate, glabrous.
Petiole glandular or eglandular.
Inflorescence 3–5 flowers.
Calyx 2–3 mm long, crimson.
Corolla tubular-campanulate, 3–4.2 cm long, crimson or dark crimson; 5 nectar pouches.
S and SE Tibet. 2,600–4,300 m (8,500–14,000 ft).
Epithet: After Juga-Wang Lopsang Tup-Den Gyatso, the late Dalai Lama of Tibet. Hardiness 3.   April.

Like a glabrous-leaved R. sherriffii *and closely similar to* R. sherriffii *in the number of flowers, flower colour, the calyces, leaf shape and size, the habit of growth, and geographical origins. In fact, some plants marked* R. lopsangianum *are* R. sherriffii *hybrids.*

R. lopsangianum LS&T6561 at RBG Edinburgh

Leaves elliptic,
oblong-elliptic, oval
or almost orbicular,
underside papillate,
glabrous.

Calyx 2–3 mm long,
crimson.

Corolla crimson or
dark crimson.

R. miniatum Cowan

Shrub or small **tree**, 1.5–4.6 m (5–15 ft) high; **branchlets** floccose, eglandular.

Leaves oblong-ovate, oval, elliptic, obovate or oblong; **lamina** 3.3–5.5 cm long, 1.5–3 cm broad; **upper surface** glabrous, midrib grooved, moderately or sparsely hairy; **underside** covered with a thick woolly, brown or fawn, continuous, unistrate indumentum of fasciculate hairs.

Inflorescence 4–6 flowers.

Pedicel 5–6 mm long, glabrous or sparsely hairy.

Calyx 2 mm–1.4 cm long, lobes unequal, fleshy, crimson, glabrous.

Corolla campanulate or tubular-campanulate, 2.8–3 cm long, fleshy, deep crimson or very deep rose, dark nectar pouches.

S Tibet. 3,200–3,700 m (10,500–12,000 ft).

Epithet: Flame-scarlet.

Closely allied to R. sherriffii. Both have a thick woolly, unistrate, continuous indumentum, a lax 3–6-flowered inflorescence, a crimson corolla and a glabrous ovary. The main distinctions between them are that the indumentum is brown or fawn and the pedicels are 5–6 mm long in R. miniatum whereas, in R. sherriffii, the indumentum is cinnamon-brown and the pedicels are 0.8–2 cm long.

R. sherriffii Cowan

Shrub or **tree**, 1.5–6 m (5–20 ft) high; **stem** and **branches** with smooth, brown flaking bark; **branchlets** sparsely hairy or glabrous, sparsely glandular or eglandular.

Leaves obovate, oval or oblong-elliptic; **lamina** 4.5–6 cm long, 2.5–4 cm broad; **upper surface** dark green, glabrous; **underside** covered with a thick woolly cinnamon-brown, continuous, unistrate indumentum of long-rayed hairs.

Inflorescence in trusses of 3–6 flowers.

Calyx 3–5 mm long, carmine, glaucous.

Corolla tubular-campanulate, 3.5–4 cm long, rich deep carmine; 5 nectar pouches.

S Tibet. 3,500–3,800 m (11,500–12,500 ft).

AM 1966 (Windsor), flowers deep carmine (L&S2751).

Epithet: After Major G. Sherriff (1898–1967), notable plant collector.
Hardiness 3.   March–April.

Indumentum brown or fawn.
Pedicel 5–6 mm long.

Leaves oblong-ovate, oval, elliptic, obovate or oblong.
Calyx 0.2–1.4 cm long, unequal, fleshy, crimson.
Corolla 2.8–3 cm long, fleshy, deep crimson or very deep rose.

Foliage of *R. miniatum* at Deer Dell

Indumentum almost chocolate brown.
Pedicels 0.8–2 cm long.
Stem and branches with smooth, brown flaking bark.

Leaves small, oval, flat and glabrous above.
Calyx 3–5 mm long, carmine, glaucous.
Corolla 3.5–4 cm long, rich deep carmine.

R. sherriffii L&S2751 AM at Deer Dell

New foliage of
R. sherriffii L&S2751
AM at Deer Dell

Ponticum
(Sherriffii Series)

Subgenus *Hymenanthes*
Section *Ponticum* G.Don
Subsection *Taliensia* Sleumer
Taliense Series

Slow-growing shrubs or small trees.

Bark usually rough.

Leaves almost always with an indumentum, which may be woolly, compacted or felty, and is sometimes covered with a pellicle.

Inflorescence usually compact, white or pink, often with markings.

Corolla 5-lobed except for R. clementinae, always without nectar pouches.

Stamens 10 or rarely up to 14.

The division of this large Subsection into four groups is not presented in any taxonomic manner. It is done to simplify the task of identifying a species of Taliensia for the non-specialist.

The Subsection has been arranged using the general shape of the leaf together with indumentum characters as seen by the naked eye.

It must be noted that the indumentum colour changes with age in most Taliensia *species, that young plants may take some time to develop their typical indumentum, and that the colour of the indumentum on herbarium specimens can be deceptive.*

For a more accurate determination of a plant, it may be necessary to consult the detailed taxonomic publications. A microscope will be required for detailed investigation of the character of the indumentum hairs and the presence of flowers is often necessary to ensure a correct identification.

Group A (Roxieanum Group). Leaves linear/elliptic:
R. iodes, R. proteoides, R. roxieanum (var. *roxieanum*, var. *cucullatum*, *R. roxieanum* Globigerum Group and var. *oreonastes*), *R. russotinctum*, *R. triplonaevium* and *R. triplonaevium* Tritifolium Group.

Group B. Leaves oblong to more or less ovate:
R. aganniphum (var. *aganniphum* Doshongense Group, var. *aganniphum* Glaucopeplum Group and var. *flavorufum*), *R. balangense, R. bhutanense, R. clementinae, R. faberi* ssp. *prattii, R. mimetes* (var. *mimetes* and var. *simulans*), *R. przewalskii, R. rufum, R. rufum* Weldianum Group, *R. sphaeroblastum, R. sphaeroblastum* var. *wumengense, R. wasonii* and *R. wasonii* (Rhododactylum Group).

Group C. Leaves lanceolate/oblanceolate:
R. adenogynum, R. alutaceum, R. balfourianum, R. balfourianum var. *aganniphoides, R. bureavii, R. bureavioides, R. elegantulum, R. faberi* ssp. *faberi, R. nigroglandulosum, R. phaeochrysum* (var. *phaeochrysum*, var. *agglutinatum* and var. *levistratum*), *R. pronum, R. taliense, R. traillianum* (var. *traillianum* and var. *dictyotum*), *R. vellereum* and *R. wiltonii*.

The varieties of R. phaeochrysum *are very variable, and could also be placed in Group B.*

Group D. (Lacteum Group).
These species have a thin greyish or brown 'wash' of hairs that is distinctive to the naked eye, and that gives members of this group a different appearance when compared with other species in Subsection *Taliensia*
R. beesianum (including the submerged *R. colletum*), *R. dignabile, R. lacteum, R. nakotiltum* and *R. wightii*.

Taliensia hybrids are described at the end of the Subsection:
R. bathyphyllum, R. comisteum, R. detonsum, R. paradoxum, R. roxieanum var. *parvum* and *R. wightii* (cultivated hybrid).

Subsection *Taliensia* Sleumer
Group A. (Roxieanum Group)

Leaves linear/elliptic.
Shiny leaves mostly with a fawn to rufous indumentum.
Most species of the old Roxieanum Subseries fit here.

R. iodes, R. proteoides, R. roxieanum (var. roxieanum, var. cucullatum,
R. roxieanum Globigerum Group and var. oreonastes), R. russotinctum, R. triplonaevium
and R. triplonaevium Tritifolium Group.

It might be worth noting here that R. iodes, R. triplonaevium, and R. triplonaevium
Tritifolium Group have aggregations of special indumentum hairs that are not long-
rayed. These hairs are in 'clumps' on the lamina and each have multiple wide and
diaphanous stems with 'rotors' on each one. The clumps on mature leaves have mostly
crimson centres and are of a most unusual character when seen under a microscope.
Although submerged in recent revisions, these three species are maintained here on
this account.

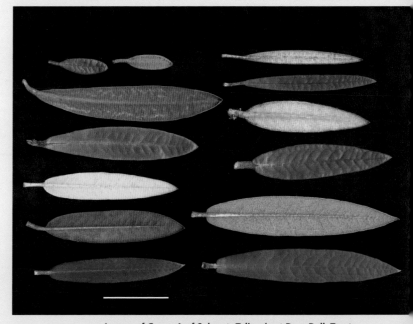

Leaves of Group A of Subsect. *Taliensia* at Deer Dell. Top to bottom: left column, *R. proteoides* (2 leaves), *R. roxieanum* (Globigerum Group) (2 leaves) and *R. iodes* (3 leaves: underside of 1-year-old leaf, underside of 2-year-old leaf and upperside); right column (2 leaves each) *R. roxieanum* var. *oreonastes*, *R. roxieanum* var. *roxieanum* and *R. triplonaevium*. Bar = 5 cm

R. iodes Balf.f. & Forrest   Taliensia Gp A

Shrub or rarely tree, 0.6–5 m (2–15 ft) high; **branchlets** densely rust-coloured or brown tomentose, eglandular.

Leaves lanceolate, oblanceolate or oblong-lanceolate; **lamina** 5–12 cm long, 1–3.3 cm broad; **upper surface** dark green, shiny, glabrous; **underside** covered with a somewhat thick, compact felty, rust or cinnamon-coloured or brown (in young leaves, greenish-yellow) continuous, bistrate indumentum of hairs, upper layer not detersile.

Petiole densely or moderately rust-coloured or brown tomentose, eglandular.

Inflorescence in trusses of 8–18 flowers.

Pedicel densely or moderately floccose, eglandular.

Corolla campanulate, 2.5–3.5 cm long, white or white suffused with rose.

SE Tibet, NW Yunnan. 3,350–4,300 m (11,000–14,000 ft).

AM 1978 (Borde Hill) 'White Plains', flowers white with red-purple spots.

AM 1980 (Borde Hill) 'Easter Island', flowers white with dorsal red spots.

Epithet: Rust-coloured.   Hardiness 3.   April–May.

R. proteoides Balf.f. & W.W.Sm.   Taliensia Gp A

(*R. lamropeplum* Balf.f. & Forrest)

Shrub, compact, rounded or spreading, 0.3–1 m (1–3 ft), or rarely creeping 8–15 cm (3–6 inches) high; annual growths and internodes very short; **branchlets** densely woolly with cinnamon or rust-coloured wool; **leaf-bud scales** persistent.

Leaves close-set, narrowly oblong or narrowly oblanceolate; **lamina** short, narrow, 1.3–4 cm long, 4–8 mm or rarely 1.1–1.2 cm broad; margin markedly recurved; **underside** covered with a thick, woolly, rust- or cinnamon-coloured, continuous, unistrate indumentum of hairs.

Petiole short, densely woolly.

Inflorescence in compact trusses of 6–12 flowers.

Pedicel densely woolly.

Corolla funnel-campanulate or campanulate, 1.9–3.8 cm long, creamy-white, creamy-yellow, white or white flushed rose, with crimson spots.

NW Yunnan, SE Tibet. 3,700–4,600 m (12,000–15,000 ft).

Epithet: Resembling a *Protea*.   Hardiness 3.   April–May.

Leaves quite narrow: young leaves with a distinctively yellowish indumentum.

Indumentum continuous, somewhat thick, felty, rust or cinnamon-coloured or brown on older leaves.

Corolla white or white flushed rose.

New foliage of *R. iodes* at Deer Dell

R. iodes at Deer Dell

Slow-growing compact, rounded or spreading shrub.

Leaves close-set, narrowly oblong or narrowly oblanceolate, margin markedly recurved. Indumentum: unistrate, continuous thick, woolly, rust or cinnamon-coloured.

Leaf-bud scales persistent.

Corolla white to creamy yellow often flushed rose, with spots.

R. proteoides at Deer Dell

New foliage of *R. proteoides* (J.C. Birck)

Ponticum
(Taliensia Group A)

R. roxieanum var. roxieanum   Taliensia Gp A

R. roxieanum Forrest

Shrub 1–3 m (3–9 ft) or rarely a tree 4.5 m (15 ft) high; annual growths and internodes short; **branchlets** densely or moderately woolly often with cinnamon or rust-coloured wool.

Leaves lanceolate, oblanceolate or oblong-lanceolate; **lamina** 5–12 cm long, 1–2.3 cm broad; **upper surface** dark green, shiny, glabrous; margin moderately or slightly recurved; **underside** covered with a thick, woolly, rust or cinnamon-coloured, brown or fawn, continuous, bistrate indumentum of hairs.

Petiole woolly.

Inflorescence in trusses of 8–20 flowers.

Corolla funnel-campanulate, 2.1–3.8 cm long, creamy-white, white, white suffused with rose, pink or rose, with or without crimson spots.

NW Yunnan, SW Sichuan, E Tibet. 3,000–4,900 m (10,000–16,000 ft).

Epithet: After Mrs. Roxie Hanna of Tali-fu, China, friend of George Forrest. Hardiness 3.   April–May.

R. roxieanum var. cucullatum D.F.Chamb.

R. cucullatum Balf.f. & Forrest   **Taliensia Gp A**

Shrub 0.6–1.5 m (2–5 ft) high; annual growths and internodes short; **branchlets** densely woolly, glandular; **leaf-bud scales** often persistent.

Leaves oblong-lanceolate, oblanceolate or lanceolate; **lamina** 6–13 cm long, 2–6 cm broad; **upper surface** dark green; **underside** covered with a thick, loose woolly, rust or cinnamon-coloured, continuous, bistrate indumentum of hairs.

Petiole short and broad, densely woolly with cinnamon, rust-coloured or brown wool.

Inflorescence in compact trusses of 10–20 flowers.

Pedicel glandular.

Corolla campanulate, 2–3 cm long, white or white suffused with rose, with or without crimson spots.

SW Sichuan, NW Yunnan. 3,700–4,800 m (12,000–15,700 ft).

Epithet: Hooded.   Hardiness 3.   April–May.

Plants under this designation are very variable, suggesting possible hybridity.

Thick bistrate woolly indumentum with white understratum showing through gaps in the upper layer (when gaps are present).

Leaves not especially narrow and not hooded.

Annual growths and internodes short. Branchlets densely or moderately woolly.
Petiole woolly.
Corolla usually white or white flushed rose.

R. roxieanum var. *roxieanum* (a form with a matt upper surface of the leaf and buff indumentum) at Deer Dell

Leaves wider than the other variations of *R. roxieanum*. Cucullate (hooded) leaf apex, not always pronounced.

Annual growths and internodes short.
Branchlets densely woolly.
Indumentum continuous, thick, loosely woolly, rust or cinnamon-coloured.
Petiole densely woolly.
Corolla usually white or white flushed rose.

R. roxieanum var. *cucullatum* R10920 at RBG Edinburgh

R. roxieanum Globigerum Group

Talensia Gp A

Shrub 1–1.8 m (3–6 ft) high; **branchlets** densely woolly, eglandular.
Leaves oblanceolate, lanceolate or oblong-lanceolate; **lamina** 4.5–10.5 cm long,
1.5–3 cm broad; **upper surface** dark green, glabrous; margin recurved;
underside covered with a thick woolly, brown or rust-coloured, continuous,
bistrate indumentum of hairs.
Petiole densely woolly, eglandular.
Inflorescence in compact trusses of 12–15 flowers.
Pedicel densely or moderately tomentose, eglandular.
Corolla campanulate, 2–4 cm long, white with crimson spots.
SW Sichuan, NW Yunnan. 3,400–4,300 m (11,000–14,000 ft).
Epithet: Bearing a globe.   Hardiness 3.   April–May.

*Collections by Rock in cultivation are fairly uniform and have the roxiaenum type
of indumentum 'in spades'.*

R. roxieanum var. oreonastes (Balf.f. & Forrest)
Davidian   Taliensia Gp A

NW and mid-W Yunnan, SW Sichuan. 3,700–4,500 m (12,000–13,000 ft).
AM 1973 (Crown Estate Commissioners Windsor), flowers white with darker spots.
Epithet: Mountain-dwelling.   Hardiness 3.   April–May.

R. roxieanum var. recurvum Davidian

The variety differs from the species type in that the leaves are narrowly
lanceolate or narrowly oblanceolate, usually 0.6–1.2 cm broad, and markedly
recurved at the margins.

NW and mid-W Yunnan, SE Tibet. 3,400–4,500 m (11,000–14,500 ft).
Epithet: Recurved.   Hardiness 3.   April–May.

Probably just an intermediate form between R. roxieanum *Globigerum Group
and* R. roxieanum *var.* roxieanum: *there is an almost continuous variation in
leaf width in the wild.*

On the forms in cultivation, the indumentum is deep rust coloured and thicker than that of the other varieties.

The white understrata shows through in patches.

Branchlets densely woolly.

Indumentum continuous, thick woolly, brown or rust-coloured.

Petiole densely woolly.

Corolla white with crimson spots.

R. roxieanum Globigerum Group (Rock 1930 expedition) at Deer Dell

Very narrow leaves with markedly recurved margins.

Corolla usually white.

R. roxieanum var. *oreonastes* at Deer Dell

R. russotinctum Balf.f. & Forrest   Taliensia Gp A

R. alutaceum var. *russotinctum* D.F.Chamb.

Shrub 2–2.5 m (5–8 ft) high; **branchlets** brown floccose, glandular.

Leaves oblong or lanceolate; **lamina** 5–11 cm long, 1.5–3 cm broad; **upper surface** dark green, matt, glabrous or almost glabrous; margin slightly recurved; **underside** covered with a thin or somewhat thick woolly, rust-coloured, continuous, bistrate indumentum of hairs, upper layer detersile, falling off in large patches.

Petiole floccose, glandular.

Inflorescence lax, about 8 flowers.

Pedicel sparsely floccose, glandular.

Calyx glabrous, densely or moderately glandular.

Corolla campanulate, 2.4–3 cm long, white flushed rose with a few spots.

NW Yunnan. 3,600–3,900 m (12,000–13,000 ft).

Epithet: Tinged with red.   Hardiness 3.   April–May.

There is much confusion in the taxonomy of this species. Morphologically, it is akin to R. triplonaevium *but differs in its indumentum of hairs, the upper layer of which is partly detersile while the remainder is red-tinged on mature leaves.* **This species has no affinities with** R. alutaceum *and it may be found labelled* R. triplonaevium *or* R. iodes.

R. triplonaevium Balf.f. & Forrest   Taliensia Gp A

Shrub 1–4.5 m (3–14 ft) high; annual growths and internodes short; **branchlets** with a thin, rust-coloured or brown tomentum, eglandular.

Leaves lanceolate, oblanceolate or oblong-lanceolate; **lamina** 6–17.5 cm long, 1.8–4.3 cm broad; **upper surface** dark green, glabrous; **underside** covered with a thin or somewhat thick, rust, cinnamon-coloured or brown, continuous, bistrate indumentum of hairs, upper layer detersile, usually falling off in small patches.

Inflorescence in trusses of 10–14 flowers.

Pedicel eglandular.

Calyx eglandular.

Corolla campanulate, 2.7–3.8 cm long, white flushed rose or pure white, with a crimson blotch.

NW Yunnan, SE Tibet. 3,400–3,700 m (11,000–12,000 ft).

Epithet: With triple moles.   Hardiness 3.   April–May.

Close to R. tritifolium.

Closely allied to *R. tritifolium*, but the branchlets, leaf midrib and petiole are glandular, and the ovary is not tomentose.

Indumentum hairs are ramiform.

Foliage of a probable *R. russotinctum* at Deer Dell. Bar = 5 cm

Annual growths and internodes short.

Upper layer of the indumentum detersile, usually falling off in small patches.

Pedicel **eglandular**.

Calyx **eglandular**.

Corolla usually white or rose flushed, with a crimson blotch.

R. triplonaevium at Deer Dell

R. triplonaevium Tritifolium Group Taliensia Gp A

R. tritifolium Balf.f. & Forrest

Shrub 1.2–3.7 m (4–12 ft) high; annual growths and internodes short;
branchlets with a thin, rust-coloured or brown tomentum, eglandular.
Leaves oblanceolate or lanceolate; **lamina** 8–19 cm long, 2–4 cm broad; **upper
surface** dark green, matt or shiny, glabrous; **underside** covered with a thin or
somewhat thick, cinnamon or rust-coloured, continuous, bistrate indumentum
of hairs, upper layer detersile falling off in large or small patches.
Petiole eglandular.
Inflorescence in trusses of 10–15.
Pedicel glandular.
Calyx glandular.
Corolla campanulate, 2.9–4 cm long, white suffused with rose, with or without
a crimson blotch.
Ovary tomentose, densely or moderately glandular.
Capsule often glandular.
NW Yunnan. 3,400–4,000 m (11,000–13,000 ft).
Epithet: With polished leaves.   Hardiness 3.   April–May.

A glandular expression of R. triplonaevium.

Subsection *Taliensia* Sleumer
Group B.
Taliense Series

Leaves oblong to more or less ovate.

R. aganniphum (var. *aganniphum* Doshongense Group, var. *aganniphum*
Glaucopeplum Group and var. *flavorufum*), *R. balangense*, *R. bhutanense*,
R. clementinae, *R. faberi* ssp. *prattii*, *R. mimetes* (var. *mimetes* and var.
simulans), *R. przewalskii*, *R. rufum*, *R. rufum* Weldianum Group,
R. sphaeroblastum, *R. sphaeroblastum* (var. *sphaeroblastum* and var.
wumengense) and *R. wasonii*.

Annual growths and
internodes short.

Upper layer of the
indumentum
detersile, usually
falling off in small
patches.

Pedicel **glandular**.

Calyx **glandular**.

Ovary **densely or
moderately
glandular**.

Corolla white flushed
rose, with or without a
crimson blotch.

R. triplonaevium Tritifolium Group at Deer Dell

**Leaves of Group B of Subsect. Taliensia at Deer Dell. Top to
bottom: left column, *R. aganniphum* var. *flavorufum* (1 leaf)
and *R. wasonii* (3 leaves: 2-year-old underside, 1-year-old
underside and upperside); middle, *R. clementinae* (2 leaves);
right, *R. faberi* ssp. *prattii* (2 leaves) and *R. aganniphum* var.
aganniphum (2 leaves). Bar = 5 cm**

R. aganniphum var. aganniphum

R. aganniphum Balf.f. & Kingdon-Ward　　　　　　　　　　　**Taliensia Gp B**

Shrub 0.3–3 m (1–10 ft) high; **branchlets** glabrous or with a thin tomentum.

Leaves ovate, elliptic, oblong or oblong-lanceolate; **lamina** 3.5–8.5 cm long, 1.5–4 cm broad; **upper surface** dark green or pale green, glabrous; **underside** covered with a thick, spongy, brown, whitish, yellowish or fawn, continuous bistrate indumentum of hairs, with a silky glossy surface pellicle, not splitting.

Inflorescence in trusses of 8–12 flowers.

Calyx eglandular.

Corolla campanulate, small, 2.6–4.5 cm long, white, rose, deep rose or purplish-pink, with crimson spots.

Ovary glabrous, eglandular.

SE Tibet, NW Yunnan, SW Sichuan. 3,400–4,600 m (11,000–15,000 ft).

Epithet: Snowy.　　Hardiness 3.　　May.

Plants with a longitudinally splitting surface indumentum are referable to **Schizopeplum Group**.

Plants with red/brown indumentum at maturity are referable to the **Glaucopeplum Group** (see p. 312).

Plants with a thin somewhat plastered indumentum are referable to the **Doshongense Group**.

R. aganniphum var. aganniphum R11151 in the Valley Gardens (a pink form)

R. *aganniphum* var. *aganniphum* at Deer Dell

Thick and usually very white indumentum.

Indumentum with a silky glossy surface pellicle (i.e. skin), not splitting.

Petiole eglandular or, rarely, glandular.

Corolla white, deep rose or purplish pink, with crimson spots.

Ponticum
(Taliensia Group B)

R. aganniphum var. aganniphum
Glaucopeplum Group
R. glaucopeplum Balf.f. & Forrest   **Taliensia Gp B**

Shrub 1–2.5 m (3–8 ft) high; **branchlets** eglandular.
Leaves ovate, ovate-elliptic, elliptic or oblong-elliptic; **lamina** 4.8–9.5 cm long, 2–3.8 cm broad; **upper surface** dark green, shiny, glabrous; **underside** covered with a somewhat thick, spongy, white, fawn or yellowish, continuous bistrate indumentum of hairs, with a surface pellicle not splitting, often glandular with short-stalked glands.
Petiole often glandular with short-stalked glands.
Inflorescence in trusses of 8–10 flowers.
Corolla campanulate, 3–3.5 cm long, bright rose or white suffused with rose, with crimson spots.
Ovary glabrous, eglandular.
NW Yunnan, SE Tibet. 3,400–4,300 m (11,000–14,000 ft).
Epithet: With a grey covering.   Hardiness 3.   April–May.

The epithet is misleading, though the young leaves are white.

R. aganniphum var. flavorufum D.F.Chamb.
R. flavorufum Balf.f. & Forrest   **Taliensia Gp B**

Shrub 0.3–3 m (1–10 ft) high; **branchlets** glabrous or floccose, eglandular.
Leaves oblong-oval, oblong, elliptic or oblong-elliptic; **lamina** 4.5–12.5 cm long, 1.9–6 cm broad; **upper surface** dark green, shiny or matt, usually glabrous; **underside** covered with a thick, brown, rust-coloured or fawn, continuous, bistrate indumentum of hairs, with a surface pellicle splitting into small or large patches.
Petiole eglandular.
Inflorescence in trusses of 8–15 flowers.
Pedicel eglandular.
Calyx eglandular.
Corolla campanulate, 2.5–5.4 cm long, white, rose, deep rose, white suffused with rose or yellowish-white, usually with crimson spots.
Ovary glabrous, eglandular.
NW Yunnan, SE Tibet, SW Sichuan. 3,400–4,600 m (11,000–15,000 ft).
Epithet: Yellow-red.   Hardiness 3.   April–May.

In cultivation, this has a somewhat thick, spongy rust-coloured indumentum with the surface pellicle not splitting.

Often has a more glandular petiole than its close relatives.

Corolla campanulate, bright rose or white flushed rose.

R. aganniphum var. *aganniphum* Glaucopeplum Group 'Rusty' at Deer Dell

Rust-coloured indumentum **in patches** on a green background.

Ovary glabrous, eglandular.

Corolla white to deep rose or yellowish-white, usually with crimson spots.

R. aganniphum var. *flavorufum* F16488 at RBG Edinburgh

R. balangense W.P.Fang   Taliensia Gp B

Shrub 1–3 m tall; young shoots glabrous; **leaf bud scales** persistent.
Leaves thickly leathery, obovate to elliptic-ovate, 6–10 cm long, 3.5–5 cm broad; base obtuse or subrounded; apex acute, apiculate; **upper surface** glabrous; **underside** with a bistrate indumentum, upper layer yellowish to fawn, loose, detersile, hairs branched, lower layer whitish, adpressed, persistent.
Petiole densely glandular and floccose-tomentose.
Inflorescence 13–15 flowers.
Pedicel densely grey-white to yellowish floccose.
Calyx persistent, densely floccose.
Corolla campanulate, white, 3.5–4 cm long, lobes 5, stamens 10, unequal, filaments densely puberulent in lower half.
Ovary and **style** glabrous.
NW Sichuan. 2,400–3,400 m (8,000–10,100 ft).
Epithet: From Balang.   Hardiness 3?   June.

P.A. and K.N.E. Cox introduced this species recently, placing it tentatively in Subsection Grandia. However, the 5-lobed corolla with 10 stamens fits better into Subsection Taliensia. It is not clear whether the indumentum is bistrate, as described in the 'Flora of China', or unistrate as Peter and Kenneth Cox suggest. It will be easier to decide where this species should be placed when the nature of the indumentum hairs is determined.

R. bhutanense D.G.Long   Taliensia Gp B

Shrub 0.6–3 m; **branchlets** greyish floccose when young.
Leaves broadly elliptic or elliptic-obovate, 6–12.5 cm long, 3–5 cm broad, base subcordate; **upper surface** dark green, glabrous except towards the base of the midrib; **underside** with a slightly granular adpressed felted brown indumentum of radiate hairs.
Petiole floccose, greyish above, brownish beneath.
Inflorescence 10–15 flowers.
Pedicel glabrous or puberulous.
Corolla campanulate, deep pink in bud, opening to pale or deep pink.
Calyx minute, glabrous.
Stamens 10, glabrous.
Ovary glabrous.
Bhutan. 3,500–4,300 m (12,000–14,000 ft).
Epithet: From Bhutan.   Hardiness 4.   March–May.

Recognised as a species in 1989.

Persistent bud scales.
Leaves thickly leathery, obovate to elliptic-ovate.
Indumentum yellowish to fawn. Calyx persistent.
Corolla campanulate, white.

R. balangense at Glendoick (M.L.A. Robinson)

New foliage of *R. balangense* at Glendoick (M.L.A. Robinson)

Broad elliptic-obovate leaves with a pronounced subcordate base, with hairs on the upper leaf midrib.
Stamens and ovary glabrous.
Corolla very deep pink in bud, opening to pale or, often, deep pink.

R. bhutanense KW11586 (as *R. phaeochrysum*) in the Valley Gardens

R. clementinae Forrest

Taliensia Gp B

Shrub or **tree** 1–4.5 m (3–15 ft) high; **branchlets** glabrous, eglandular.

Leaves oval, oblong-oval or elliptic; **lamina** 6–15 cm long, 3.2–8.7 cm broad;
  upper surface dark green or pale green (in young leaves, bluish-green), margin
  recurved; **underside** covered with a thick, spongy, fawn, whitish or silvery-
  white, continuous bistrate indumentum of hairs, with a silky glossy pellicle not
  splitting or splitting into very large patches.

Inflorescence in trusses of 7–15 flowers.

Calyx 6–7-lobed.

Corolla campanulate, 3.6–5 cm long, creamy-white, white, white suffused with
rose, or pink, with or without crimson spots; lobes 6–7.

Stamens 12–14.

NW Yunnan, SW Sichuan. 3,400–4,400 m (11,000–14,600 ft).

Epithet: After Clementine, wife of George Forrest.   Hardiness 3.   April–May.

R. faberi ssp. prattii

Taliensia Gp B

R. prattii Franch.

Shrub or **tree** 1–6.5 m (3–21 ft) high; **branchlets** densely tomentose with a thin,
  whitish or brown tomentum, glandular.

Leaves broadly elliptic, elliptic or oblong-elliptic; **lamina** 6.3–18 cm long,
  2.8–7.8 cm broad, apex acuminate; **upper surface** dark green, shiny, glabrous;
  underside with a thin, fawn or brown veil of hairs, or with a thin, continuous,
  unistrate indumentum of hairs, or with closely scattered tufts of hairs.

Petiole densely thin tomentose.

Inflorescence in trusses of 6–20 flowers.

Calyx glandular.

Corolla campanulate, 3.6–5.6 cm long, white, with red spots.

W Sichuan. 2,600–4,500 m (8,500–14,800 ft).

AM 1967 (Sandling Park) 'Perry Wood', flowers white tinged pink.

Epithet: After A.E. Pratt who discovered the species in W Sichuan.   Hardiness 3.
  April–May.

*The indumentum is often absent from the leaf margin, a characteristic which,
curiously, does not feature in the herbarium specimens.*

Bluish foliage on the young growth.

Smooth convex leaf with spongy fawn, whitish or silvery-white indumentum with a silky glossy pellicle.
Corolla lobes 6–7.

Corolla white, white flushed rose to pink, or yellowish-white.

R. clementinae at Deer Dell

Fawn indumentum, continuous or with closely scattered tufts of hairs.

Larger leaves and a thinner indumentum than the closely related *R. faberi* ssp. *faberi*.

The indumentum is often absent from the leaf margin.

Corolla white, with red spots.

R. faberi ssp. *prattii* 'Perry Wood' AM at Borde Hill

Ponticum
(Taliensia Group B)

R. mimetes var. mimetes Tagg & Forrest

Taliensia Gp B

Leaves elliptic, oblong-elliptic, or oblong lanceolate; base rounded or obtuse.
Indumentum somewhat thick, woolly, brown, continuous, bistrate, upper layer
  worn off in small or large patches revealing a thin whitish or fawn lower layer.
Ovary densely or sparsely hairy.
Corolla white, white flushed rose to pinkish-purple.
Epithet: Imitative.   Hardiness 3?   May?

*Extremely rare in cultivation, one clone has reportedly been found but it may
be a hybrid.*

R. mimetes var. simulans Tagg & Forrest

Taliensia Gp B

Shrub 1–2.2 m (3–7 ft) high; **branchlets** floccose, glandular.
Leaves ovate to oblong-ovate, base cordulate; **lamina** 5–11 cm long, 2.5–5 cm
  broad; **upper surface** dark green, shining, glabrous; **underside** covered with a
  somewhat thick, woolly, brown, continuous, bistrate indumentum of hairs,
  upper layer is worn off in small or large patches revealing a thin whitish or fawn
  lower layer; without pellicle.
Petiole floccose, often glandular.
Inflorescence in trusses of 6–10 flowers.
Calyx 2–7 mm long, often glandular.
Corolla funnel-campanulate, 2.8–4.3 cm long, white, white faintly flushed and
  margined rose, or pinkish-purple, with or without red blotch, and with or
  without crimson spots.
Epithet: Imitative.   Hardiness 3.   May.
SW Sichuan. 3,400–3,700 m (11,000–12,000 ft).

Rare in cultivation. May be found labelled as R. mimetes *var.* mimetes.

The leaves of var. *simulans* have a twisted appearance: they are ovate to oblong-ovate, the base cordulate.

Indumentum thick, brown with upper layer wearing off into very small patches (mottling).

Ovary glabrous.

Corolla white, white flushed rose to pinkish-purple.

R. mimetes var. *simulans* at Sarum, Surrey

R. mimetes var. *simulans* at R.B.G. Edinburgh

Ponticum
(Taliensia Group B)

R. przewalskii Maxim.

Taliensia Gp B

Shrub 1–4.6 m (3–15 ft) high; **branchlets** glabrous, eglandular; young shoots and buds yellow.

Leaves elliptic, oblong-elliptic, ovate-elliptic, ovate or oblong; **lamina** 5–11 cm long, 2–5 cm broad; **upper surface** pale green, glabrous, midrib yellow; **underside** covered with a thin, continuous, unistrate indumentum of hairs, or sometimes with a thin veil of hairs, or rarely with a few scattered hairs.

Inflorescence in trusses of 10–15 flowers.

Pedicel glabrous, eglandular.

Calyx eglandular.

Corolla funnel-campanulate or campanulate, 2.3–3.5 cm long, white or rose-pink, with or without spots.

Gansu, E Tibet, SW Sichuan. 3,000–5,000 m (10,000–16,400 ft).

Epithet: After N.M. Przewalski (1839–88), Russian traveller and geographer. Hardiness 3.   April–May.

R. rufum Batalin

Taliensia Gp B

Shrub or **tree**, 1.2–4 m (4–15 ft) high; **branchlets** glabrous or with a thin, whitish tomentum.

Leaves oblong, oblong-obovate or oblong-lanceolate; **lamina** 5–11 cm long, 2–4 cm broad; **upper surface** green, shiny, glabrous; **underside** covered with a thick, often loosely, brown or rusty-brown, continuous, bistrate indumentum of hairs, without pellicle.

Petiole with whitish tomentum.

Inflorescence in trusses of 5–10 flowers.

Calyx minute, 0.5–1 mm long, tomentose.

Corolla campanulate, 2–3.2 cm long, white, pink or pinkish-purple, with or without crimson spots.

N Sichuan, Gansu. 3,000–3,700 m (10,000–12,000 ft).

Epithet: Red.   Hardiness 3.   April–May.

Pale green upper leaf surface with a yellow midrib.

White to fawn indumentum, but in some examples there is none, and even when present, this has one of the thinnest indumenta in the Subsection.

Branchlets glabrous, eglandular.

Corolla white or rose-pink.

R. przewalskii R13686 in the Valley Gardens

Pale green upper leaf surface.

The brown or rust-brown indumentum is loosely thick.

Corolla white, pink, or pinkish-purple.

R. rufum at the Hillier Arboretum

R. rufum Weldianum Group   Taliensia Gp B

R. weldianum Rehd. & Wils.

Shrub; **branchlets** floccose, eglandular.
Leaves oval, almost orbicular, oblong-elliptic or elliptic; **lamina** 6–13.8 cm long, 2.7–5.9 cm broad; **upper surface** olive-green or pale green, shiny, glabrous; **underside** covered with a thick, often loosely, woolly, brown or rust-coloured (in young leaves, whitish) continuous, bistrate indumentum of hairs, without pellicle.
Petiole tomentose, eglandular.
Inflorescence in trusses of 6–12 flowers.
Calyx minute, 0.5–1 mm long.
Corolla campanulate, 2.2–2.8 cm long, white, white suffused with pink, or pink.
W Sichuan, W and SW Gansu. 2,400–4,000 m (8,000–13,000 ft).
Epithet: After Gerald S.M. Weld, a former President of the Massachusetts Horticultural Society.   Hardiness 3.   April–May.

R. sphaeroblastum Balf.f. & Forrest   Taliensia Gp B

Shrub or **tree** 0.6–6 m (2–20 ft) high; **branchlets** glabrous, eglandular.
Leaves oval, oblong-oval, elliptic, oblong-elliptic or rarely oblong-lanceolate; **lamina** 7–13.2 cm long, 3.5–7 cm broad; **upper surface** dark green, shiny, glabrous; **underside** covered with a thick, woolly, cinnamon, rust-coloured or brown, continuous, bistrate indumentum of hairs, without pellicle.
Petiole glabrous, eglandular.
Inflorescence in trusses of 10–15 flowers.
Pedicel glabrous, eglandular.
Calyx glabrous, eglandular.
Corolla campanulate, 2.9–4 cm long, white, creamy-white, white suffused with rose, or pink, with or without crimson spots.
SW Sichuan, NW Yunnan. 3,000–4,600 m (11,000–15,000 ft).
Epithet: With rounded buds.   Hardiness 3.   April–May.

Some forms of this species have a thinner and somewhat adpressed indumentum. The epithet is misleading.

New foliage of
R. sphaeroblastum
F17110 at Deer Dell

Thicker brown or rust-coloured (in young leaves, whitish) indumentum than *R. rufum*, and more oval leaves.

Corolla white, white flushed pink to pink.

R. rufum Weldianum Group at the Rhododendron Species Foundation (M.L.A. Robinson)

When viewed from above, the growth bud is noticeably **triangular**.

Indumentum thick, woolly, cinnamon, rust-coloured or brown.

Branchlets glabrous eglandular.

Indumentum without pellicle.

Petiole glabrous, eglandular.

Calyx glabrous, eglandular.

Corolla white, creamy or flushed rose, or pink.

R. sphaeroblastum at Benmore

Ponticum
(Taliensia Group B)

R. sphaeroblastum var. wumengense Feng

Taliensia Gp B

Shrub 1–3 m high; **branchlets** glabrous.
Leaves elliptic to oblong-obovate; **lamina** 6–12 cm long, 3.5–6 cm broad;
apex acute; base rounded to cuneate; **upper surface** deep green, glabrous; in
young leaves, bluish; **underside** sparsely clad with soft light yellow-brown
woolly hairs.
Petiole 2–2.5 cm long, glabrous.
Inflorescence 8–16 flowers.
Pedicel 1–2.5 cm long, glabrous.
Calyx glabrous.
Corolla funnel-campanulate, about 4 cm long, white to pink with red spots.
Wumengshan Mts, NE Yunnan. 3,800–4,000 m (13,000 ft).
Epithet: From Wumengshan.   Hardiness 4.   May–June.

Only recently described, but it is in cultivation.

R. wasonii Hemsl. & Wils.   Taliensia Gp B

Shrub 0.6–2 m (2–7 ft) high; **branchlets** densely or moderately brown
tomentose, often glandular.
Leaves ovate, ovate-lanceolate or broadly lanceolate; **lamina** 5.5–10.3 cm long,
2.5–4.5 cm broad, apex acuminate or acute; **upper surface** dark green, shiny;
underside covered with a thick, woolly, brown or cinnamon-coloured (in young
leaves, white), continuous, unistrate indumentum of hairs, without pellicle.
Petiole densely or moderately tomentose.
Inflorescence in trusses of 6–10 flowers.
Pedicel densely tomentose.
Corolla campanulate, 2.6–4 cm long, yellow, creamy-yellow to creamy-white,
with or without reddish spots.
W Sichuan. 2,600–3,600 m (8,500–11,800 ft).
Epithet: After Rear-Admiral C.R. Wason (1874–1941), friend of E.H. Wilson.
Hardiness 3.   April–May.

R. wasonii Rhododactylum Group   Taliensia Gp B

R. wasonii Hemsl. var. *rhododactylum* (Hort.) Davidian

W Sichuan. 2,600–3,000 m (8,500–9,850 ft).
AM 1923 (Bodnant) as *R. rhododactylum*.
AM 1974 (Windsor).
Epithet: Tinged rosy-red finger-like.   Hardiness 3.   April–May.

Yellow-brown indumentum less thick than that of *R. sphaeroblastum* var. *sphaeroblastum*.

Triangular buds.

In some forms, the new young foliage has a bluish hue.

Corolla white to pink, with red spots.

R. sphaeroblastum var. *wumengense* showing the triangular growth bud (M.L.A. Robinson)

Dark green shiny leaves, ovate to ovate-lanceolate.

The yellow flowers distinguish it from the Rhododactylum Group.

Indumentum thick, woolly, brown or cinnamon-coloured (in young leaves, white).

Corolla yellow, creamy-yellow to creamy-white.

R. wasonii at Benmore

Ponticum
(Taliensia Group B)

Subsection *Taliensia* Sleumer
Group C.
Taliense Series

Leaves lanceolate/oblanceolate.

R. adenogynum, R. alutaceum, R. balfourianum, R. balfourianum var.
aganniphoides, R. bureavii, R. bureavioides, R. elegantulum, R. faberi ssp.
faberi, R. nigroglandulosum, R. phaeochrysum (var. *phaeochrysum,* var.
agglutinatum and var. *levistratum), R. pronum, R. taliense, R. traillianum*
(var. *traillianum* and var. *dictyotum), R. vellereum* and *R. wiltonii.*

The varieties of R. phaeochrysum *are very variable, and could also be placed
in Group B.*

Leaves of Group C of Subsect. *Taliensia* **at Deer Dell. Top to bottom (2 leaves each): left,**
***R. balfourianum* and *R. taliense*; right, *R. wiltonii, R. adenogynum* and *R. elegantulum*.**
Bar = 5 cm

Leaves of Group C of Subsect. *Taliensia* at Deer Dell (continued). Left to right: top row, *R. phaeochrysum* var. *levistratum* (syn. *R. dryophyllum*) (1 leaf), *R. phaeochrysum* var. *agglutinatum* (1 leaf) and *R. traillianum* (2 leaves); bottom row (2 leaves each), *R. phaeochrysum* var. *phaeochrysum* and *R. faberi* ssp. *faberi*. Bar = 5 cm

R. adenogynum Diels.   Taliensia Gp C

(*R. adenophorum* Balf.f. & W.W.Sm.)

Shrub or **tree** 0.5–5.5 m (1–18 ft) high; **branchlets** floccose or sometimes glabrous, often glandular.

Leaves lanceolate, oblong-lanceolate or rarely ovate-lanceolate; **lamina** 5–13.5 cm long, 1.3–4 cm broad; **underside** covered with a thick woolly, yellowish, yellowish-brown, fawn or brown, continuous, bistrate indumentum of hairs, sometimes with a shiny pellicle.

Petiole often glandular.

Inflorescence in trusses of 4–12 flowers.

Pedicel densely glandular.

Corolla funnel-campanulate, 3.5–5.3 cm long, rose, reddish-purple, pink or white, with or without crimson spots.

Ovary densely glandular.

NW and mid-W Yunnan, SW Sichuan, SE Tibet. 3,400–4,300 m (11,000–14,000 ft).

AM 1976 (Borde Hill, as *R. adenophorum*) 'Kirsty', flowers white tinged purple with crimson spots.

Epithet: With glandular ovary.   Hardiness 3.   April–May.

R. adenophorum (*synonym*) *is a glandular expression of this species.*

R. alutaceum Balf.f. & W.W.Sm.   Taliensia Gp C

Shrub 1.8–4.5 m (6–14 ft) high; **branchlets** glandular.

Leaves oblong-lanceolate or lanceolate; **lamina** 8–14 cm long, 2–3.9 cm broad; **upper surface** dark green, glabrous; **underside** covered with a thick, woolly, fawn, brown or dark brown, continuous, bistrate indumentum of hairs, without pellicle.

Inflorescence in trusses of 8–12 flowers.

Pedicel moderately or rather densely glandular.

Calyx 0.5–2 mm long.

Corolla funnel-campanulate, 3.5–4.5 cm long, rose or deep rose, with crimson spots.

NW Yunnan. 3,700 m (12,000 ft).

Epithet: Like soft leather.   Hardiness 3.   April–May.

Plants in cultivation derive from the single collection F13098, so whether this is a valid species is uncertain — P.A. and K.N.E. Cox report hybrid swarms of related plants. However the 'Flora of China' extends the range to SE Tibet and W Sichuan, so the status of this species may become more certain when more material has been collected.

Leaves lanceolate, oblong-lanceolate or rarely ovate-lanceolate.

Mustard-coloured indumentum with a shiny pellicle.

Ovary densely glandular. Corolla white to pink or rose, or reddish purple.

R. adenogynum R11471 'Kirsty' AM at Deer Dell

Branchlets glandular.

Indumentum thick, woolly, fawn, brown or dark brown, without pellicle.

Pedicel moderately or rather densely glandular.

Corolla rose or deep rose with crimson spots.

R. alutaceum (ex RBG Edinburgh) at Deer Dell

Ponticum
(Taliensia Group C)

R. balfourianum Diels.   Taliensia Gp C

Shrub 1–4.5 m (3–15 ft) high; **branchlets** glabrous or floccose, eglandular or glandular.

Leaves lanceolate, ovate-lanceolate or oblong-lanceolate; **lamina** coriaceous, 5–13.2 cm long, 1.9–4.8 cm broad, apex acute or acuminate; **upper surface** dark green; **underside** covered with a thin, plastered, brown, fawn or silvery, continuous, unistrate indumentum of large rosulate hairs.

Inflorescence in trusses of 4–10 flowers.

Pedicel rather densely glandular.

Calyx rather densely glandular.

Corolla funnel-campanulate, 3–4.8 cm long, pale or deep rose, pink or white.

Ovary densely glandular.

W and NW Yunnan. 3,400–4,300 m (11,000–14,000 ft).

Epithet: After Sir Isaac Bayley Balfour (1853–1922), former Regius Professor of Botany, Edinburgh.   Hardiness 3.   April–May.

R. balfourianum var. aganniphoides
Tagg & Forrest   Taliensia Gp C

Shrub or sometimes **tree**, 0.6–4 m (2–16 ft) high.

The variety differs from the species type in that the indumentum on the **underside** of the **leaves** is thick, spongy and bistrate; upper layer of ramiform hairs or sometimes very large rosulate hairs, lower layer of rosulate hairs, with a distinct pellicle or skin on the surface.

SW Sichuan, W Yunnan. 3,000–5,400 m (10,000–17,500 ft).

Epithet: Resembling *R. aganniphum*.   Hardiness 3.   April–May.

Always has **unistrate** indumentum.

Leaf reminiscent of *R. arboreum*, but less rugulose.

Ovary densely glandular.
Corolla white, pink, or pale to deep rose.

R. balfourianum at Deer Dell

Indumentum thicker, spongy, **bistrate** with a distinct pellicle.

R. balfourianum var. *aganniphoides* in the Valley Gardens

R. bureavii Franch.   Taliensia Gp C

(*R. cruentum* Levl.)

Shrub or **tree** 1.2–7.6 m (4–25 ft) high; **branchlets** densely woolly with cinnamon-coloured or rusty-red wool, glandular.

Leaves oblong-lanceolate, oblong-elliptic or elliptic; **lamina** 5.5–14 cm long, 2.5–6 cm broad; **upper surface** dark green, slightly rugulose; **underside** covered with a thick, woolly, cinnamon-coloured or rusty-red, continuous, bistrate indumentum of hairs.

Petiole densely woolly with cinnamon-coloured or rusty-red wool, glandular.

Inflorescence in trusses of 6–15 flowers.

Pedicel densely woolly, glandular.

Calyx glandular.

Corolla campanulate, 4–5 cm long, white, white suffused with rose, or rose, with crimson spots.

NW Yunnan, SW Sichuan. 3,200–4,300 m (10,500–14,000 ft).

AM 1939 (Exbury), flowers white with crimson spots.

AM 1972 (Wakehurst), as a foliage plant.

Epithet: After E. Bureau (1830–1918), a French professor.   Hardiness 3.   May.

R. bureavioides Balf.f.   Taliensia Gp C

Shrub 1.2–3 m (4–10 ft) high; **branchlets** densely woolly with pale fawn or rust-coloured wool, sparsely glandular.

Leaves oblong-lanceolate or oblanceolate; **lamina** 10–16.5 cm long, 2–5.8 cm broad; **upper surface** dark green, slightly rugulose or not rugulose; **underside** covered with a thick, woolly, pale fawn, cinnamon-coloured or brown, continuous, bistrate indumentum of hairs.

Petiole densely woolly with pale fawn, rust-coloured or brown wool.

Inflorescence in trusses of 8–15 flowers.

Pedicel densely or moderately woolly, glandular.

Calyx glandular.

Corolla funnel-campanulate, 4–4.5 cm long, rose with a deep red blotch, with crimson spots.

Epithet: Resembling *R. bureavii*.   Hardiness 3.   May.

Plants labelled R. bureavioides *without the above characteristics are likely to be hybrids.*

Leaves usually oblanceolate.

Very woolly rust-coloured or cinnamon indumentum.

Branchlets, petiole and pedicel densely woolly.

Corolla usually white or white flushed rose, with crimson spots.

R. bureavii R11376A at the Valley Gardens

Petiole bent so the leaves appear almost sessile.

Tomentum on the upper leaf surface semi-persistent.

Indumentum loosely woolly, cinnamon or brown.

Branchlets densely woolly.

Leaves oblong-lanceolate or oblanceolate.

Petiole densely woolly.

Corolla rose with a deep red blotch.

Foliage of *R. bureavioides* at Deer Dell (M.L.A. Robinson)

Ponticum
(Taliensia Group C)

R. elegantulum Tagg & Forrest   Taliensia Gp C

Shrub 1–1.8 m (3–6 ft) high; **branchlets** densely woolly with cinnamon-coloured or rusty-red wool.

Leaves oblanceolate or lanceolate; **lamina** 4–9 cm long, 1.3–3.1 cm broad; **upper surface** light green, matt, slightly rugulose; **underside** covered with a thick, felty, somewhat woolly, cinnamon-coloured or rusty-red, continuous, unistrate indumentum of hairs, without pellicle.

Petiole densely woolly with cinnamon-coloured or rusty-red wool.

Inflorescence in trusses of 6–20 flowers.

Calyx glandular.

Corolla funnel-campanulate, 2.6–3.3 cm long, pale purplish-pink with darker spots.

SW Sichuan. 3,700–4,000 m (12,000–13,000 ft).

Epithet: Elegant.   Hardiness 3.   May.

R. faberi ssp. faberi   Taliensia Gp C
R. faberi Hemsley
(*R. faberoides* Balf.f., *R. wuense* Balf.f.)

Shrub 1–6 m (3–20 ft) high; **branchlets** densely woolly with rust-coloured or brown tomentum, glandular.

Leaves elliptic, oblong-elliptic, obovate or lanceolate; **lamina** 4.5–12 cm long, 1.8–4.5 cm broad, apex acute or acuminate; **upper surface** dark green, matt; **underside** covered with thick, woolly rust-coloured or dark brown, continuous, bistrate indumentum of hairs, without pellicle, upper layer is mostly shed at maturity revealing a thin whitish lower layer.

Petiole densely woolly with rust-coloured or brown tomentum.

Inflorescence in trusses of 6–12 flowers.

Pedicel rather densely or moderately glandular.

Calyx glandular.

Corolla campanulate, 3.8–4 cm long, white, often with a blotch at the base, and often with red spots.

W Sichuan. 3,000–3,400 m (10,000–11,000 ft).

Epithet: After Rev. E. Faber who collected in China during 1887–1891.
    Hardiness 3.   May.

Leaves oblanceolate or lanceolate.

Indumentum cinnamon or rust red at maturity.

Branchlets densely woolly with cinnamon-coloured or rusty-red wool.

Petiole densely woolly.

Corolla pale purplish-pink with darker spots.

R. elegantulum at Deer Dell

Leaves usually lanceolate.

In cultivation, the indumentum is usually woolly and light brown to buff; upper layer is mostly shed at maturity revealing a thin whitish lower layer.

Branchlets densely woolly with rust-coloured or brown tomentum.

Petiole densely woolly with rust-coloured or brown tomentum.

Corolla white, usually with a blotch and spots.

R. faberi ssp. *faberi* KR177 at Deer Dell

Ponticum
(Taliensia Group C)

R. nigroglandulosum Nitzelius   Taliensia Gp C

Shrub 2–5 m (6–16 ft) high; **branchlets** densely tomentose, glandular with black short-stalked glands.

Leaves lanceolate, oblanceolate or oblong; **lamina** 12–17 cm long, 3.2–5 cm broad; **upper surface** dark green, glabrous, eglandular; midrib glandular with black short-stalked glands; **underside** covered with a somewhat thick, woolly, reddish-brown or brown, continuous, bistrate indumentum of hairs.

Petiole glandular with black short-stalked glands.

Inflorescence in trusses of 8–10 flowers.

Calyx minute, 1 mm long, eglandular.

Corolla campanulate, 3.2–5 cm long, deep pink at first, later yellowish-pink, with purple spots.

Sikang. 3,500 m (11,500 ft).

Epithet: With black glands.   Hardiness 3.   April–May.

Rare in cultivation. The plants in cultivation may be hybrids, but it is possible that the true species has recently been introduced.

R. phaeochrysum var. phaeochrysum

R. phaeochrysum Balf.f. & W.W.Sm.   Taliensia Gp C
(*R. cupressens* Nitzelius)

Shrub 1–6 m (3–20 ft) high; **branchlets** tomentose with a thin tomentum or glabrous, glandular or eglandular.

Leaves oblong-elliptic, oblong-ovate or oblong-lanceolate; **lamina** 5–14 cm long, 1.8–5.8 cm broad; **upper surface** dark green, glabrous; **underside** covered with a thin, continuous, agglutinate (plastered) or sometimes suede-like, dark brown or fawn, unistrate indumentum of hairs.

Inflorescence in trusses of 8–15 flowers.

Pedicel eglandular.

Corolla funnel-campanulate, 3–5 cm long, white, white suffused with rose, or pinkish, rarely yellow, with or without crimson spots.

NW and mid-W Yunnan, SE Tibet, SW Sichuan. 3,400–5,400 m (11,000–17,500 ft).

Epithet: Dark golden.   Hardiness 4.   April.

This is a very variable species, and unfortunately has a wide range of leaf shapes. The AM clone 'Greenmantle' (Borde Hill) is a hybrid.

Black glands on the leaf upper midrib.

Petiole glandular with black short-stalked glands.

Branchlets glandular with black short-stalked glands.

Corolla opening deep pink, fading to yellowish pink, with purple spots.

R. nigroglandulosum at Deer Dell

Leaves oblong-elliptic, oblong-ovate or oblong-lanceolate.

Thin rust-coloured indumentum with a sheen on old leaves (whitish to buff on newer leaves).

Corolla generally larger than in its varieties.

Leaves oblong-elliptic, oblong-ovate or oblong-lanceolate.

Petiole often glandular.

Calyx eglandular.

Corolla white to pinkish, or rarely yellow.

R. phaeochrysum var. *phaeochrysum* at Deer Dell

Ponticum
(Taliensia Group C)

R. phaeochrysum var. agglutinatum

D.F.Chamb.   **Taliensia Gp C**
R. agglutinatum Balf.f. & Forrest

Shrub or **tree** 0.6–4.5 m (2–15 ft) rarely 7.6 m (25 ft) high; **branchlets** glabrous or densely tomentose with a thin tomentum.

Leaves elliptic, oblong or oblong-lanceolate; **lamina** 4–9 cm long, 1.4–5 cm broad; **upper surface** dark green, glabrous; **underside** covered with a thin, continuous, agglutinate (plastered) or sometimes suede-like, brown, cinnamon or fawn, unistrate indumentum of long-rayed hairs.

Inflorescence in trusses of 10–15 flowers.

Corolla funnel-campanulate, 2–3.5 cm long, white, creamy-white, white suffused rose or pink, with or without crimson spots.

Ovary glabrous.

SW Sichuan, NW Yunnan, E and SE Tibet, Bhutan. 3,700–4,900 m (12,000–16,000 ft).

Epithet: Stuck together, referring to the indumentum on the lower surface of the leaves.   Hardiness 3.   April–May.

R. phaeochrysum var. levistratum D.F.Chamb.

R. dryophyllum Balf.f. & Forrest   **Taliensia Gp C**
(R. dumulosum Balf.f. & Forrest, R. sigillatum Balf.f. & Forrest, R. theiophyllum Balf.f. & Forrest)

Shrub or **tree** 0.6–7.6 m (2–25 ft) high; **branchlets** densely tomentose with a thin tomentum, eglandular.

Leaves oblong, oblong-lanceolate or lanceolate; **lamina** 5–14 cm long, 1.4–5.2 cm broad; **upper surface** olive-green or dark green; **underside** covered with a thin suede-like, continuous, fawn or brown, unistrate indumentum of hairs.

Petiole densely tomentose with a thin tomentum, eglandular.

Inflorescence in trusses of 8–16 flowers.

Calyx eglandular.

Corolla funnel-campanulate or campanulate, 2–4 cm long, white, creamy-white, white suffused rose, pink or pinkish-purple.

NW Yunnan, Upper Burma, SW Sichuan, SE Tibet, Bhutan Frontier. 3,300–4,600 m (10,800–15,000 ft).

Epithet: With a felted layer, referring to the indumentum.   Hardiness 3. April–May.

R. dumosulum Balf.f. & Forrest *is a more dwarf, small-leaved, form of* R. phaeochrysum *var.* levistratum.

Indumentum whitish, **plastered** and somewhat shiny.

Plants vary, even in cultivation, and there is a continuous sequence of variation between var. *phaeochrysum* and var. *agglutinatum*.

Leaves oblong-elliptic, oblong-ovate or oblong-lanceolate.

Petiole glandular or eglandular.

Calyx eglandular.

Corolla white, sometimes flushed cream or rose, or pink.

R. phaeochrysum var. *agglutinatum* R11335 in the Valley Gardens

New leaves have a yellowish indumentum, but otherwise very similar to *R. phaeochrysum*.

Leaves oblong, oblong-lanceolate or lanceolate.

Indumentum thin suede-like, fawn or brown.

Petiole eglandular.

Calyx eglandular.

Corolla white, sometimes flushed cream or rose, or pink, or pinkish-purple.

R. phaeochrysum var. *levistratum* (syn. *R. dryophyllum*) F20442 at Deer Dell

R. pronum Tagg & Forrest   Taliensia Gp C

Creeping or prostrate or matted **shrub**, 8–60 cm (0.25–2 ft) high; annual growths and internodes very short; **branchlets** glabrous; **leaf-bud scales** persistent, numerous, closely set.

Leaves oblong, oblanceolate or oblong-obovate; **lamina** 3–8.8 cm long, 1–2.8 cm broad; **upper surface** bluish-green or dark green (in young leaves, bluish-green), matt, glabrous; margin recurved; **underside** covered with a thick, spongy, fawn or brown continuous indumentum of bistrate hairs, with a surface pellicle.

Inflorescence in trusses of 6–12 flowers, **flower-bud scales** persistent.

Corolla campanulate, 3–4.5 cm long, creamy-yellow or white or pink, with or without crimson spots.

NW and mid-W Yunnan. 3,700–4,600 m (12,000–15,000 ft).
Epithet: Prostrate.   Hardiness 3.   May.

Distinctive growth habit. Reminiscent of a dwarf R. clementinae *but with narrower foliage.*

R. taliense Franch.   Taliensia Gp C

Shrub 1.2–3.7 m (4–12 ft) or rarely 60 cm (2 ft) high; **branchlets** moderately or densely tomentose with a thin brown tomentum, eglandular.

Leaves oblong-lanceolate, lanceolate or ovate-lanceolate; **lamina** 5–11 cm long, 1.8–4 cm broad; **upper surface** dark green, slightly rugulose, glabrous; **underside** covered with a thick, woolly, brown, continuous, bistrate indumentum of hairs, without pellicle.

Petiole densely tomentose with a thin brown tomentum, eglandular.

Inflorescence in trusses of 8–19 flowers.

Corolla campanulate, 2.5–3.5 cm long, creamy-yellow, pale yellow, white or white suffused with rose, with crimson spots.

Ovary glabrous, eglandular.

W and NW Yunnan. 3,000–3,700 m (10,000–12,000 ft).
Epithet: From Tali Range, Yunnan.   Hardiness 3.   April–May.

This species can be detected from the scent exuded by the plant on a warm day.

Creeping, or prostrate shrub.

Bluish foliage in young leaves.

Annual growths and internodes very short.

Leaf-bud scales persistent.

Flower-bud scales persistent.

Indumentum with a surface pellicle.

Corolla white, creamy-yellow, or pink.

R. pronum (J.C. Birck)

Growth habit of *R. pronum* at Glendoick

Thick, woolly brown indumentum (white in young leaves), often with curious dark spotting on top of the hairs in some clones; this is reported to be caused by red spider mite.

Branchlets moderately or densely tomentose.

Indumentum continuous, brown, without pellicle.

Petiole densely tomentose.

Corolla white, rose flushed, creamy or pale yellow with crimson spots.

R. taliense F6772 at RBG Edinburgh

Ponticum
(Taliensia Group C)

R. traillianum var. traillianum · Taliensia Gp C

R. traillianum Forrest & W.W.Sm.
(*R. aberrans* Tagg & Forrest)

Shrub or **tree**, 1–9 m (3–30 ft) high; **branchlets** with a thin tomentum.
Leaves oblong, oblong-obovate or oblong-lanceolate; **lamina** 5.2–17 cm long,
2–6.8 cm broad; **upper surface** pale green or dark green, glabrous; **underside**
covered with a thin, continuous, powdery or sometimes suede-like, brown, rust-
or sometimes cinnamon-coloured, or yellowish, unistrate indumentum of
radiate hairs with somewhat pear-shaped arms (pyriform).
Petiole with a thin tomentum.
Inflorescence in trusses of 9–15 flowers.
Corolla funnel-campanulate, 2.6–4 cm long, white, white suffused with rose, pink
or rose.

NW, W and mid-W Yunnan, SW Sichuan, SE Tibet. 3,000–4,600 m
(10,000–15,000 ft).
Epithet: After G.W. Traill (1836–1897), botanist and father-in-law of George Forrest.
Hardiness 3.   April–May.

This species can be detected from the scent exuded by the plant on a warm day.

R. traillianum var. dictyotum · D.F.Chamb.

R. dictyotum Balf.f. ex Tagg   **Taliensia Gp C**

Shrub 1.2–3.7 m (4–12 ft) high, branchlets glabrous or tomentose.
Leaves oblong, oblong-elliptic or oblanceolate; **lamina** 5–13.5 cm long, 2–5.5 cm
broad, **upper surface** dark green; **underside** covered with a thin felty,
continuous, brown, rust- or cinnamon-coloured, unistrate indumentum of
ribbon-like hairs.
Inflorescence in trusses of 8–15 flowers.
Calyx 1–2 mm long.
Corolla campanulate, 3.2–5 cm long, white or white suffused with rose, with or
without a crimson blotch, often with crimson spots.

SE Tibet, NW Yunnan, SW Sichuan. 3,400–4,300 m (11,000–14,000 ft).
AM 1965 (Exbury) 'Kathmandu', flowers white with crimson blotch and
crimson spots.
Epithet: With net veins.   Hardiness 3.   April–May.

The placing of this species under R. traillianum *is open to question; it may be
a hybrid of* R. phaeochrysum.

This has a powdery indumentum that can be easily rubbed-off.

Corolla white, white flushed rose, pink, or rose.

R. traillianum var. *traillianum* at Deer Dell

Thin felty indumentum.

Hair type differs from that of *R. traillianum* var. *traillianum*.

Indumentum continuous, brown, rust- or cinnamon-coloured.

Corolla white or white flushed rose.

R. traillianum var. *dictyotum* 'Kathmandu' AM at Exbury

Ponticum
(Taliensia Group C)

R. vellereum Hutch. ex Tagg

Taliensia Gp C

Shrub or **tree** 1.2–6 m (4–20 ft) high; **branchlets** glabrous, eglandular.

Leaves oblong-lanceolate or oblong-elliptic; **upper surface** dark green, shiny, glabrous, often with patches of white sheen; **underside** covered with a thick, spongy, silvery-white, yellowish or fawn, continuous, bistrate indumentum of hairs, with a surface pellicle not splitting.

Petiole eglandular.

Inflorescence in trusses of 12–20 flowers.

Pedicel eglandular.

Calyx glabrous, eglandular.

Corolla funnel-campanulate, 2.6–3.8 cm long, white, white suffused with pale pink, pale rose or pink, with or without carmine or purple spots.

Ovary glabrous, eglandular.

SE and S Tibet. 2,900–4,600 m (9,500–15,000 ft).

AM 1976 (Borde Hill) 'Lost Horizon', flowers white suffused pale rose, with red spots (KW5656).

AM 1979 (Borde Hill) 'Far Horizon' (KW5656).

Epithet: Fleecy.   Hardiness 3.   March–May.

Herbarium specimens of the closely related R. principis *have no pellicle on the indumentum, and* R. principis *appears not to be in cultivation*.

R. wiltonii Hemsl. & Wils.

Talensia Gp C

Shrub 1–5 m (3–16 ft) high; **branchlets** densely brown or whitish tomentose, eglandular.

Leaves oblanceolate or oblong-obovate; **lamina** 4.5–12.1 cm long, 1.6–4 cm broad; **upper surface** olive-green, shiny, markedly bullate, glabrous; margin recurved; **underside** covered with a thick woolly, brown or cinnamon-coloured, continuous, unistrate indumentum of hairs.

Petiole densely tomentose.

Inflorescence in trusses of 6–10 flowers.

Pedicel densely tomentose.

Corolla campanulate, 3–4 cm long, pink or white, with or without a crimson blotch, and with or without red spots.

W Sichuan. 2,300–3,300 m (7,500–11,000 ft).

AM 1957 (Exbury), flowers white suffused with pink, with crimson blotch.

Epithet: After Sir Colville E. Wilton, Chinese Consular Service, Ichang.   Hardiness 3.   April–May.

Thick, spongy, silvery-white indumentum with a pellicle.

Corolla white, white flushed pink, pink, or rose.

R. vellereum L&S2738 in the Valley Gardens

Markedly bullate, oblanceolate leaves. Upper surface olive-green, shiny.

Indumentum thick woolly, brown or cinnamon-coloured. Corolla usually pink.

R. wiltonii at Deer Dell

Ponticum
(Taliensia Group C)

Subsection *Taliensia* Sleumer
Group D. (Lacteum Group)
Lacteum Series

These species have a thin greyish or brown 'wash' of hairs that is distinctive to the naked eye, and that gives members of this group a different appearance when compared with other species in Subsection *Taliensia*.

R. beesianum (including the submerged R. colletum), R. dignabile,
R. lacteum, R. nakotiltum and R..wightii.

The unistrate indumentum of radiate or long-rayed hairs led Davidian to maintain this as a separate series. There is reason to keep R. beesianum, R. colletum, R. dignabile *and* R. wightii *(the new introduction only — the old is a natural hybrid) separate, as, although the radiate hairs of these species can be distinguished only under a microscope, the species have a thin greyish or brown 'wash' of hairs that is distinctive to the naked eye and differs from that of the other species in Subsection* Taliensia.

R. lacteum **F28248 at Deer Dell**

Leaves of Group D of Subsect. *Taliensia* **at Deer Dell.** Top row (left to right): *R. beesianum* (Colletum Form), *R. beesianum* (with a strongly winged petiole) and *R. wightii*. Bottom row (left to right): *R. lacteum* and *R. nakotiltum* (an old leaf).
Bar = 5 cm

R. beesianum Diels.   Taliensia Gp D

Shrub or **tree** 1.8–9 m (6–30 ft) high; **branchlets** eglandular.

Leaves oblong-lanceolate or oblanceolate; **lamina** 10–30 cm long, 3–8.3 cm broad; **upper surface** dark or pale green, glabrous; **underside** covered with a thin, continuous, brown, dark brown or grey, unistrate indumentum of radiate hairs.

Petiole thick, winged or ridged on each side, upper surface flat.

Inflorescence in trusses of 10–25 flowers.

Corolla broadly campanulate, 3.5–5.3 cm long, white, pink, rose or deep rose, with or without a crimson blotch.

NW Yunnan, NE Upper Burma, SE Tibet, SW Sichuan. 3,000–4,300 m (10,000–14,000 ft).

Epithet: After Messrs. Bees, nurserymen in Cheshire.   Hardiness 3.   April–May.

R. beesianum Colletum Form   Taliensia Gp D
R. colletum **Balf.f. & Forrest**

Leaves shorter and narrower than *R. beesianum*. Leaf and flower buds sticky. Leaf base tapered.

Petiole narrowly winged or ridged above: surface grooved.

Corolla white to pink to deep rose or red.

Similar to R. beesianum, *and submerged into that species*.

Leaf and flower buds sticky.

Petiole thick, winged or ridged, upper surface flat.

Leaf base rounded or obtuse.

Indumentum thin, continuous, brown, dark brown or grey.

Corolla white to pink to deep rose.

R. beesianum F16375 at Deer Dell

R. beesianum F30526 at Deer Dell

Ponticum
(Taliensia Group D)

R. dignabile Cowan

Taliensia Gp D

Shrub or small **tree**, 0.6–6 m (2–20 ft) high; **branchlets** moderately or sparsely floccose, eglandular or glandular.

Leaves oblong-elliptic, elliptic, oblong-obovate or oblanceolate; **lamina** coriaceous, 4–18 cm long, 2–6.5 cm broad; **upper surface** with vestiges of a juvenile tomentum; **underside** with closely scattered brown, unistrate indumentum of somewhat radiate hairs or sometimes with a thin veil of hairs.

Petiole 0.6–2.2 cm long, sparsely or moderately floccose, eglandular or glandular.

Inflorescence 8–15 flowers.

Pedicel 1.2–3 cm long, floccose, eglandular or rarely sparsely glandular.

Corolla campanulate, 2.5–4.3 cm long, pink, cream or lemon-yellow or white, with or without a purplish or dark crimson blotch at the base, and with or without crimson spots.

S and SE Tibet. 3,400–4,600 m (11,000–15,000 ft).

Epithet: Deemed worthy.   Hardiness 4?   April–May.

R. lacteum Franch.

Taliensia Gp D

Shrub or **tree** 1.2–9 m (4–30 ft) high; **branchlets** with a thin fawn tomentum, eglandular.

Leaves oval, ovate-elliptic, obovate or elliptic; **lamina** 6.8–18.5 cm long, 3.7–10 cm broad; **upper surface** pale green or dark green; **underside** covered with a thin, suede-like, continuous, fawn or brown, unistrate indumentum of radiate hairs.

Petiole eglandular.

Inflorescence in compact trusses of 12–30 flowers.

Corolla broadly campanulate, 3.5–5 cm long, yellow, pale yellow or clear canary yellow, sometimes tinged pink, rarely pure white, with or without crimson blotch.

W and NW Yunnan, NE Upper Burma. 3,000 m (10,000 ft).

FCC 1926 (Werrington), flowers sulphur-white with a dark crimson blotch.

FCC 1965 (Blackhills), flowers yellow without blotch or spots.

Epithet: Milky.   Hardiness 3.   April–May.

Close to
R. beesianum,
save for the very
thin indumentum:
closely scattered
hairs or a thin veil
of hairs.
Corolla usually
pinkish to white.

R. dignabile on the Doshong La, Tibet (P. Evans)

Leaf base rounded
or cordate.
Indumentum thin,
suede-like,
continuous, fawn or
brown.
Corolla almost
always yellow.

R. lacteum (Werrington form) at Deer Dell

Ponticum
(Taliensia Group D)

R. nakotiltum Balf.f. & Forrest   Taliensia Gp D

Shrub 1–3.7 m (3–12 ft) high; **branchlets** floccose, eglandular.

Leaves oblong or oblong-elliptic; **lamina** 6.5–12 cm long, 2.5–4.2 cm broad, apex acuminate, **upper surface** dark green; **underside** covered with a thin continuous, felty, grey or fawn, bistrate indumentum of hairs.

Petiole eglandular.

Inflorescence in trusses of 12–15 flowers.

Calyx minute, 1–1.5 cm long, glabrous, eglandular.

Corolla campanulate, 2.8–3.5 cm long, pale rose or white flushed with rose, with or without a crimson blotch, and with or without crimson spots.

NW Yunnan. 3,300–4,000 m (11,000–13,000 ft).

Epithet: Having the wool plucked off.   Hardiness 3.   April–May.

The plants in cultivation in the UK are probably natural hybrids between R. lacteum *and possibly* R. taliense *and do not match Davidian's herbarium description. They have pale yellow flowers. The only correct plant known to the authors is in W. Berg's garden near Seattle, USA, probably from F14060 or F14068, both of which were once at Werrington, Cornwall.*

R. wightii Hook.f.   Taliensia Gp D

Shrub or small **tree** 2–6 m; **branchlets** greenish-brown to brown; leaf and flower buds sticky.

Leaves 5–18 cm long, 5–7 cm broad, broadly elliptic to obovate; **upper surface** dark green and glabrous, **underside** with a compacted pale fawn to brown indumentum of radiate hairs.

Inflorescence 12–20 flowers.

Corolla campanulate, 3–4 cm long, white, cream or pale yellow.

Ovary with a dense red-brown tomentum.

Bhutan, Arunachal Pradesh, S Tibet. 3,400–4,500 m (11,000–14,700 ft).

Epithet: After R. Wight (1796–1872), a former superintendent, Madras Botanic Gardens.   Hardiness 3.   March–April.

Distinguished from R. lacteum *and* R. beesianum *by the elliptic leaf-shape. It has sticky buds like* R. beesianum, *but the flowers are of a different colour. All plants labelled* R. wightii *that were introduced before 1971 appear to be hybrids with a species from Subsection* Falconera. *These are have a distinctly one-sided truss of yellow flowers. See also* R. wightii *cultivated hybrid, p. 360.*

Leaf base obtuse.
**Indumentum thin
continuous, felty,
greyish to fawn.**
Corolla white, flushed
rose, or pale rose.

The supposed *R. nakotiltum* as in cultivation at Exbury

Foliage of *R. nakotiltum* in W. Berg's garden
(M.L.A. Robinson)

Leaf and flower buds
sticky.
Leaves usually
elliptic, indumentum
pale.
Corolla white, cream
or pale yellow, not
one-sided.

R. wightii on the Rhudong La (P. Evans)

R. × **bathyphyllum** Balf.f. & Forrest   Taliensia hybrid

Shrub 1–1.5 m (3–5 ft) high; **branchlets** with short internodes, densely woolly with rust- or cinnamon-coloured wool.

Leaves oblong; **lamina** 4–7.5 cm long, 1.3–2.5 cm broad; **upper surface** green, matt, glabrous; **underside** covered with a thick, woolly, rust- or cinnamon-coloured, continuous, unistrate indumentum of hairs.

Petiole densely woolly.

Inflorescence compact truss of 10–15 flowers.

Calyx minute, 0.5–1 mm long.

Corolla campanulate, 3–4 cm long, white or white suffused with rose, with crimson spots.

SE Tibet. 3,400–4,300 m (11,000–14,000 ft).

Epithet: Densely leafy.   Hardiness 3.   April–May.

This is a R. proteoides *hybrid: specimens attributable to* R. bathyphyllum *can be found in* R. proteoides *seedlings grown from wild seed.*

R. × **comisteum** Balf.f. & Forrest   Taliensia hybrid

Shrub broadly upright, 0.6–1.2 m (1–4 ft) high; annual growths and internodes short; **branchlets** stout, densely woolly or tomentose with fawn, brown or rust-coloured tomentum, eglandular or sparsely glandular; **leaf-bud scales** persistent.

Leaves oblanceolate or lanceolate; **lamina** coriaceous, small, 2.5–5 cm long, 0.6–1.7 cm broad; **upper surface** dark green, matt, not rugulose, glabrous or with vestiges of hairs; margin recurved; **underside** covered with a thick or somewhat thick, woolly, rust-coloured or dark brown or brown, continuous, bistrate indumentum of hairs, upper layer ramiform with short or long stem, and narrow curled or uncurled branches, lower layer rosulate.

Petiole densely woolly or tomentose with fawn, brown or rust-coloured tomentum, eglandular or glandular with short-stalked glands.

Inflorescence lax, 4–8 flowers.

Pedicel 1.5–2.6 cm long, densely woolly, eglandular or sparsely glandular with short-stalked glands.

Calyx floccose, eglandular.

Corolla campanulate or tubular-campanulate, 3–3.7 cm long, soft rose or deep soft rose or pale rose, with or without a few crimson spots.

Stamens 10.

Ovary densely tomentose with long brown hairs.

SE Tibet. 3,400–4,300 m (11,000–14,000 ft).

Epithet: To be taken care of.   Hardiness 3.   April–May.

Possibly a natural hybrid between R. proteoides *and* R. sanguineum.

R. × *bathyphyllum* at Deer Dell

R. × *comisteum* (J.C. Birck)

R. × detonsum Balf f. & Forrest Taliensia hybrid

Shrub 2.4–3.7 m (8–12 ft) high; **branchlets** floccose, glandular.

Leaves oblong-lanceolate or oblong; **lamina** 6.2–15.5 cm long, 2–5 cm broad; **upper surface** dark green, glabrous; **underside** with brown or fawn, closely or widely scattered tufts of hairs, or with a thin veil of hairs, or with a very thin, continuous or discontinuous, unistrate indumentum of long-rayed hairs.

Petiole floccose, glandular.

Inflorescence in trusses of 6–11 flowers.

Pedicel floccose, glandular.

Calyx glandular.

Corolla funnel-campanulate, 4.5–5.2 cm long, rose-pink or pink.

Stamens 10–14.

Ovary densely glandular.

NW Yunnan. 3,000–3,400 m (10,000–11,000 ft).

Epithet: Shorn.   Hardiness 3.   April–May.

R. × inopinum Taliensia hybrid
R. inopinum Balf.f.

Rounded **shrub** 1.5–3 m (5–10 ft) high; **branchlets** moderately or densely floccose, eglandular.

Leaves lanceolate or oblong-lanceolate; **lamina** 5.3–10.5 cm long, 1.6–3.6 cm broad, apex acuminate or acute; **upper surface** olive-green, shiny, glabrous; **underside** with small patches of closely or widely separated woolly, brown, discontinuous, detersile unistrate indumentum of hairs.

Petiole moderately or densely floccose.

Inflorescence in trusses of 6–10 flowers.

Calyx minute, 1 mm long.

Corolla campanulate 2.5–3.6 cm long, creamy-white, white suffused with pink or pale yellow, with deep crimson blotch and crimson spots.

W Sichuan.

Epithet: Unexpected.   Hardiness 3.   April–May.

R. × *detonsum* at Deer Dell

R. × *inopinum* at RBG Edinburgh

R. × paradoxum

Taliensia hybrid

R. paradoxum Balf.f.

Shrub 1.2–2.2 m (4–7 ft) high; **branchlets** rather densely or moderately hairy with long whitish hairs, eglandular.
Leaves oblong, oblong-elliptic or oblong-obovate; **lamina** 5–11.5 cm long, 2.3–4.6 cm broad; **upper surface** dark green, somewhat matt, glabrous; **underside** pale green, with large or small patches of closely or widely separated, woolly, brown, discontinuous, detersile, unistrate indumentum of hairs.
Petiole densely or moderately hairy.
Inflorescence in trusses of 8–10 flowers.
Pedicel densely tomentose.
Calyx moderately or densely tomentose.
Corolla campanulate, 3.4–5 cm long, white with a crimson blotch breaking into spots.
W Sichuan.
Epithet: Paradoxical.   Hardiness 3.   April–May.

R. purdomii Redh. & Wils.

Taliensia hybrid

Plants in cultivation under this name appear to be *R. maculiferum* itself or its hybrids.

R. roxieanum Forrest var. parvum Davidian

Taliensia hybrid

The hybrid variety differs from the species type in having smaller leaves (laminae 2.5–6 cm long, 0.6–1.3 cm broad) and in having a dwarf growth habit (1–2 or sometimes 3 ft).

NW Yunnan, SW Sichuan. 3,700–4,600 m (12,000–15,000 ft).
Epithet: Small.   Hardiness 3.   April–May.

Probably R. proteoides × R. roxieanum.

R. × *paradoxum* in the Valley Gardens

R. × wightii (natural hybrid in cultivation) Taliensia hybrid

Shrub or small **tree** 0.6–5 m (2–15 ft) high; **branchlets** with a thin fawn tomentum, leaf scars large, conspicuous.

Leaves oblong-elliptic, oblong-obovate or oblanceolate; **lamina** 6–10 cm long, 2–7.8 cm broad; **upper surface** dark green, glabrous; **underside** covered with a thin suede-like, continuous fawn, brown, rust or rarely cinnamon-coloured, unistrate indumentum of hairs.

Inflorescence in trusses of 12–20 flowers, large and lax one-sided or sometimes very small somewhat compact, bud-scales very viscid.

Pedicel tomentose.

Calyx minute, 0.5–1 mm long.

Corolla campanulate, 2.5–5 cm long, yellow, pale sulphur-yellow or lemon-yellow, with or without a crimson blotch, with crimson spots.

Sikkim, Nepal, Bhutan. 3,400–4,300 m (11,000–14,000 ft).

AM 1913 (Littleworth), flowers pale sulphur-yellow, crimson spots.

Epithet: See *R. wightii* Hook.f. (p. 352). Hardiness 3. April–May.

Subgenus *Hymenanthes*
Section *Ponticum* G. Don
Subsection *Thomsonia* Sleumer
Thomsonii Series

Shrubs or trees.

Bark smooth and peeling.

Leaves rounded-orbicular to elliptic or oval, usually glabrous at maturity or occasionally with a very thin indumentum.

Inflorescence lax.

Calyx often large.

Corolla 5-lobed with nectar pouches.

Stamens 10.

R. cerasinum, *R. cyanocarpum*, *R. eclecteum* (var. *eclecteum* and var. *bellatulum* [which includes the submerged *R. eclecteum* var. *brachyandrum*]), *R. hookeri*, *R. hylaeum*, *R. meddianum* (var. *meddianum* and var. *atrokermesinum*), *R. stewartianum*, *R. subansiriense*, *R. thomsonii* and *R. viscidifolium*.

Indumentum brown.
Inflorescence large and
lax, one-sided.
Corolla yellow.

R. × *wightii* – the one-sided truss of the natural hybrid in
cultivation at Deer Dell

Leaves of Subsect. *Thomsonia* at Deer Dell. Left side (top to
bottom): *R. meddianum* var. *meddianum* (2 leaves), *R. thomsonii*
and *R. viscidifolium*. Right side (top to bottom): *R. meddianum*
var. *atrokermesium*, *R. hylaeum* and *R. hookeri* (one leaf each).
Bar = 5 cm

R. cerasinum Tagg

Shrub 1.2–3.7 m (4–12 ft) high; **branchlets** glandular or eglandular.

Leaves oblong or oblanceolate; **lamina** 5–10 cm long, 1.8–4 cm broad; **upper surface** dark green, glabrous, glossy; **underside** glabrous.

Inflorescence in trusses of 3–7 pendulous flowers.

Pedicel glandular.

Corolla campanulate, 3.5–4–5 cm long, brilliant scarlet, cherry-red, deep crimson, or white with a broad cherry-red band around the summit; 5 deep-purple nectar pouches.

Style glandular throughout to the tip.

SE Tibet, Upper Burma, Assam. 2,800–3,800 m (9,000–12,500 ft).

AM 1938 (Nymans), flowers creamy-white with a cherry-red band around the margin (KW6923).

AM 1973 (Nymans) 'Cherry Brandy', cherry-red flowers (KW6923).

Epithet: Cherry-coloured.   Hardiness 3.   May–June.

R. cyanocarpum (Franch.) W.W.Sm.

Shrub or tree 1.2–7.6 m (4–25 ft) high; **branchlets** glabrous, eglandular, yellowish-green, glaucous.

Leaves orbicular, oval or broadly elliptic; **lamina** 5–12.6 cm long, 4–9 cm broad; **upper surface** bluish-green, glabrous; **underside** glaucous pale green, glabrous or minutely hairy.

Petiole stout, glaucous.

Inflorescence in trusses of 6–10 flowers.

Calyx cup-shaped, 0.2–1.1 cm long.

Corolla campanulate, 4–6 cm long, white suffused pale rose, pale pink, pale rose or white.

W and NW Yunnan. 3,000–4,000 m (10,000–13,500 ft).

AM (Leonardslee), flowers white, flushed rose.

Epithet: With blue fruits.   Hardiness 3.   March–April.

Oblong glossy leaves.

Flowers pendulous, brilliant scarlet, cherry-red or deep crimson, or white with a broad cherry-red band.

Style glandular throughout to the tip.

R. cerasinum KW6923 AM at Deer Dell

R. cerasinum KW5830 'Coals of fire' at Deer Dell

R. cerasinum KW8258 at Deer Dell

Rounded leaves, upper surface bluish, lower glaucous.

Branchlets glaucous. Calyx cup-shaped. Corolla usually white flushed pale rose, pale pink, or pale rose.

R. cyanocarpum at Deer Dell

Ponticum
(Thomsonia)

R. eclecteum

R. eclecteum Balf.f. & Forrest

Shrub 0.6–3 m (2–10 ft) high, or sometimes a **tree** up to 4.5 m (15 ft) high;
  branchlets glandular.
Leaves obovate, oblong-obovate or oblong; **lamina** almost sessile, 5–14.5 cm
  long, 2–6 cm broad; **upper surface** bright green, glabrous, glaucous;
  underside glabrous.
Petiole short, often broad, 0.3–1 cm long.
Inflorescence in trusses of 6–12 flowers.
Pedicel eglandular.
Calyx 0.2–2 cm long.
Corolla tubular-campanulate, 3–5.3 cm long, white, pink, rose, yellow, purple or
  red, with or without crimson spots.
Ovary densely or moderately glandular.
NW Yunnan, NE Upper Burma, SE and E Tibet, SW Sichuan. 3,000–4,400 m
  (10,000–14,500 ft).
AM 1949 (Exbury), flowers primrose yellow.
AM 1978 (Borde Hill) 'Kingdom Come', flowers white flushed yellow-green.
Epithet: Picked out.   Hardiness 3.   February–April.

Since both varieties of R. electeum *are now regarded as one: one being a hybrid
and the other submerged,* R. electeum *stands alone as a designation.*

R. eclecteum
R. eclecteum var. brachyandrum Tagg

Corolla crimson, deep rose-crimson or deep rose.

SE Tibet, NE Upper Burma, NW Yunnan. 3,700–4,300 m (12,000–14,000 ft).
Epithet: With short stamens.   Hardiness 3.   March–April.

Submerged into R. eclecteum: *as it differs only in flower colour.*

R. eclecteum var. bellatulum Balf.f. ex Tagg

NW Yunnan, SE Tibet. 3,000–4,300 m (10,000–14,000 ft).
Epithet: Neat.   Hardiness 3.   March–April.

Recent field studies suggest that R. electeum *var.* bellulatum *is a natural hybrid.*

R. faucium — see R. hylaeum

Leaves almost sessile: petiole short, often broad.

Leaf underside glabrous. Ovary densely or moderately glandular.
Close to *R. stewartianum*.
Corolla white, pink, rose, yellow, purple or red.

R. eclecteum KW6869 'Kingdom Come' AM at Deer Dell

R. eclecteum at Borde Hill

Petioles and pedicels long.
Leaves +/– oblong, not subsessile.

R. eclecteum var. *bellatulum* F21839 in the Valley Gardens

R. hookeri Nutt.

Shrub or small **tree** 2.4–6 m (8–20 ft) high; **stem** and **branches** with smooth, brown, flaking bark.

Leaves oblong, oblong-obovate or oblong-oval; **lamina** 6.3–17 cm long, 3–7.5 cm broad; **upper surface** dark green, glabrous; **underside** pale glaucous green, glabrous except the lateral veins studded with isolated bead-like hair tufts.

Inflorescence in trusses of 8–15 flowers.

Calyx 0.5–2 cm long, green, yellowish or reddish.

Corolla tubular-campanulate, 3.5–4.4 cm long, deep crimson, cherry-red or pink; 5 dark crimson nectar pouches.

Bhutan, Assam. 2,500–3,700 m (8,000–12,000 ft).

FCC 1933 (Bodnant), dark red flowers.

Epithet: After Sir J.D. Hooker (1817–1911).   Hardiness 3.   March–April.

R. hylaeum Balf.f. & Farrer

Shrub or **tree** 1.2–14 m (4–45 ft) high; **stem** and **branches** with smooth, brown flaking bark; **branchlets** glandular or eglandular.

Leaves oblong or oblanceolate; **lamina** 6–17.5 cm long, 2–5.8 cm broad; **upper surface** green, glabrous; **underside** pale green, glabrous, minutely punctulate, with scattered red hair bases on the veins.

Inflorescence in trusses of 10–12 or sometimes 18–24 flowers.

Pedicel eglandular.

Calyx cup-like, 4–8 mm long, eglandular.

Corolla tubular-campanulate, 3.6–4.5 cm long, pink, white tinged pink, rose or scarlet, with darker spots; 5 nectar pouches.

NE Upper Burma, E and SE Tibet, NW Yunnan, Assam. 2,100–3,800 m (7,000–12,000 ft).

Epithet: Belonging to forests.   Hardiness 2–3.   March–May.

The ovary is glabrous, *but in the closely related* **R. faucium,** *the ovary is densely glandular*. *If these two species are to be merged, as has been suggested, then as* R. hylaeum *was named in the 1920s it must take precedence, so* R. faucium *is included here, with a suggested taxonomic designation of* R. hylaeum *var.* faucium.

Hook-like hairs on the veins of the leaf underside.

Stems and branches with smooth, brown, flaking bark.

Corolla deep crimson, cherry-red or pink.

R. hookeri KW8238 at Deer Dell

Smooth, brownish flaking bark.

Leaf oblong or oblanceolate with a rounded apex; underside with scattered hair bases on the veins.

Ovary glabrous.

Calyx cup-like, eglandular.

Corolla pink, white tinged pink, rose or scarlet.

R. hylaeum KW6401 at Deer Dell

Ponticum
(Thomsonia)

R. meddianum var. meddianum

R. meddianum Forrest

Shrub 1–2.4 m (3–8 ft) high; **branchlets** eglandular, thinly glaucous.

Leaves oval, obovate or oblong-oval; **lamina** 5.5–11 cm long, 3.5–6.8 cm broad; **upper surface** pale green or dark green, glabrous, glaucous; **underside** glabrous.

Petiole broad.

Inflorescence 5–10 flowers.

Pedicel eglandular.

Calyx cup-shaped, 0.4–1 cm long, red, eglandular.

Corolla tubular-campanulate, 4–5.8 cm long, crimson, deep crimson, or almost black-crimson or scarlet without or sometimes with a dark blotch.

Ovary eglandular.

W Yunnan, NE Upper Burma. 2,700–3,700 m (9,000–12,000 ft).

Epithet: After G. Medd, Agent I.F. Company, Bhamo, Upper Burma. Hardiness 3.   April–May.

R. meddianum var. atrokermesinum Tagg

NE Upper Burma. 3,355 m (11,000 ft).

AM 1954 (Logan), flowers light red.

AM 1977 (Brodick) 'Bennan', flowers red with darker markings.

Epithet: Black-crimson.   Hardiness 3.   April–May.

A glandular expression of R. meddianum *with larger leaves and flowers. The epithet is misleading.*

R. meddianum var. *atrokermesinum* **F26495 (Borde Hill form) at Deer Dell**

Rounded leaf with the underside having a pale yellowish colouration.

Ovary eglandular.

Branchlets eglandular.

Leaves oval, obovate or oblong-oval, underside glabrous.

Petiole broad.

Calyx cup-shaped, red, eglandular.

Corolla crimson, deep crimson or scarlet.

R. meddianum var. *meddianum* **F24219 at Deer Dell**

Ovary densely glandular.

Branchlets and pedicels moderately glandular.

Obovate leaves.

Leaves and flowers larger.

R. meddianum var. *atrokermesinum* **F26495 – calyces at Deer Dell**

Ponticum
(Thomsonia)

R. stewartianum Diels.

(*R. aiolosalpinx* Balf.f. & Farrer, *R. stewartianum* var. *tantulum* Cowan & Davidian)

Shrub 0.6–3 m (2–10 ft) high; **branchlets** eglandular or glandular, glabrous.
Leaves obovate, oblong-obovate, elliptic or oval; **lamina** 5–12 cm long, 2.5–6.5 cm broad; **upper surface** bright green, glabrous; **underside** with a thin veil of hairs or with farinose indumentum or glabrous.
Inflorescence 3–7 flowers.
Pedicel glandular or eglandular.
Corolla tubular-campanulate, 3.6–5.4 cm long, pure white, white suffused with rose, yellow, creamy-white, pink, pale rose, purplish, or sometimes brilliant red or crimson.
Ovary moderately or densely glandular.
SE Tibet, NW Yunnan, NE and E Upper Burma, Assam. 3,000–4,500 m (10,000–14,500 ft).
AM 1934 (Exbury).
Epithet: After L.B. Stewart (1876–1934), a former Curator of the Royal Botanic Garden, Edinburgh.   Hardiness 3.   February–April.

R. subansiriense D.F.Chamb.

Shrub or **tree**, up to 14 m (45 ft); **stem** and **branches** smooth, peeling grey and red; young shoots tomentose.
Leaves oblong or oblong elliptic, 7–13 cm long, 2–4 cm broad, apex rounded, apiculate, base rounded to obtuse; **upper surface** glabrous; **underside** with veins vestigially pubescent and glandular with unstalked glands which are detersile and leave red punctulate bases, otherwise glabrous, epapillate.
Petiole glabrous.
Inflorescence dense, 10–15 flowers.
Pedicel glabrous, eglandular.
Calyx 4–5 mm, cup-shaped with ciliate margins.
Corolla tubular-campanulate, fleshy, scarlet, with a few flecks.
Ovary densely tomentose, eglandular.
Style glabrous.
NE India (Arunachal Pradesh). 2550–2750 m (8,400–9,200 ft).
Epithet: From the Subansiri division (Arunachal Pradesh).   Hardiness 3?   March.

Introduced from a single collection (C&H418) in 1995. Closely related to
R. hylaeum, *differing mainly in its scarlet corolla. Probably worthy of subspecies status only. Very early into growth and frequently frosted.*

Very similar to
R. eclecteum, but
the petiole is longer
and indumentum is
sometimes present.

Flower colour very
variable, from white,
yellow or shades of
pink, red or purple.

Leaf underside with
a thin veil of hairs
or with farinose
indumentum or
glabrous.

Ovary moderately or
densely glandular.

R. stewartianum Far926 (as *R. aiolosalpinx*) at Deer Dell

Most easily
identified from the
glandular or
punctulate leaf
veins, but
magnification may
be required.
Otherwise glabrous.

Ovary densely
tomentose,
eglandular.

Bark smooth, peeling
grey and red.

Corolla scarlet.

R. subansiriense at Baravalla
(M.L.A. Robinson)

New Foliage of
R. subansiriense C&H418
at Deer Dell

R. thomsonii Hook.f.

Shrub or **tree** 1–6 m (3.5–20 ft) high; **stem** and **branches** with smooth, brown, fawn or pink flaking bark; **branchlets** thinly glaucous.

Leaves orbicular, ovate or broadly elliptic; **lamina** 4–10 cm long, 3–7.5 cm broad; **upper surface** dark green or olive-green, glabrous; **underside** glaucous or pale glaucous green, glabrous.

Inflorescence in trusses of 6–10 or rarely 12–13 flowers.

Pedicel eglandular.

Calyx cup-shaped, 0.6–2 cm long.

Corolla campanulate, 3.5–6 cm long, deep blood red, deep crimson, crimson or very deep rose, often with a bloom; 5 nectar pouches.

Nepal, Sikkim, Bhutan, Assam, S Tibet. 2,400–4,300 m (8,000–14,000 ft).

AGM 1925.

AM 1973 (Windsor), flowers deep red.

Epithet: After Thomas Thomson (1817–1878), a former superintendent of Calcutta Botanic Garden.   Hardiness 3.   April–May.

R. viscidifolium Davidian

Shrub 0.6–2.4 m (2–8 ft) high; **branchlets** glabrous, glandular or eglandular.

Leaves oval or rounded; **lamina** 4–9.7 cm long, 2.8–6.6 cm broad, apex rounded; **upper surface** green, glabrous; **underside** glabrous, densely papillate, white, glandular, sticky to the touch.

Petiole glabrous, glandular or eglandular.

Inflorescence 1–2 flowers.

Pedicel glandular or eglandular.

Calyx cup-shaped, 4–9 mm long.

Corolla tubular-campanulate, 3.6–4.6 cm long, copper-red, spotted crimson; 5 crimson nectar pouches.

SE Tibet. 2,700–3,400 m (9,000–11,000 ft).

Epithet: With sticky leaves.   Hardiness 3.   April–May.

Leaves oval and glaucous below.

Calyx cup-shaped, yellow, whitish-green, green through flesh-coloured to crimson.

Bark smooth, peeling, brown fawn or pink.

Corolla deep red, deep crimson or deep rose, often with a bloom.

R. thomsonii at Deer Dell

Bark of *R. thomsonii* at Stonefield Castle, Scotland

The leaf underside, glabrous, densely papillate, white and is sticky to the touch (especially in young leaves).

Inflorescence 1–2 flowers.

Corolla copper-red to coppery-orange, spotted crimson.

R. viscidifolium LS&T3750 at Deer Dell

Ponticum
(Thomsonia)

Subgenus *Hymenanthes*
Section *Ponticum* G.Don
Subsection *Williamsiana* Sleumer
Thomsonii Series, Williamsianum Subseries

R. williamsianum Rehd. & Wils.

Compact, rounded or dome-shaped or spreading **shrub**, 0.6–1.5 m (2–5 ft) or
   rarely up to 2.4 m (8 ft) high or more; **branchlets** slender, setulose-glandular.
Leaves ovate or orbicular; **lamina** 1.5–4.2 cm long, 1.3–4 cm broad, base
   cordate or truncate; **upper surface** bright green (in young leaves, bronzy),
   glabrous; **underside** glaucous, papillate, glabrous or punctulate with vestiges
   of hairs.
Petiole often setulose-glandular.
Inflorescence in trusses of 2–3 (rarely up to 5) flowers.
Pedicel glandular.
Corolla campanulate, 3–4 cm long, rose or pink, with or without spots.
Style glandular throughout to the tip.
W Sichuan. 2,800 m (9,000 ft).
AM 1938 (Bodnant), flowers pink.
Epithet: After J.C. Williams (1861–1939) of Caerhays, Cornwall.   Hardiness 3.
   April–May.

*Can only be confused with its hybrids, which almost always have larger leaves and
more pointed leaf apices, as does the cultivar* R. williamsianum *'White'.*

R. williamsianum at Deer Dell

Compact, rounded growth habit and small rounded leathery leaves with cordate or truncate bases.

Compact, rounded or dome-shaped or spreading shrub.

Corolla rose or pink.

R. williamsianum at Deer Dell

Ponticum
(Williamsiana)

Subgenus Azaleastrum

Inflorescence axillary below the terminal buds.
Elepidote evergreen shrubs or trees.
Indumentum, if present, of simple hairs or glands.
Corolla 5-lobed; rotate or tubular-campanulate.
Stamens 5 or 10.

Subgenus Azaleastrum
Section Azaleastrum

Ovatum Series

In cultivation, usually small shrubs to 2 m but may be much larger
(to 8 m) in the wild.
Branchlets usually slender.
Young growths brightly coloured.
Leaf underside elepidote, glabrous.
Inflorescence axillary in the uppermost 1–4 leaves; normally single
flowers.
Ovary bristly.
Capsule short with the calyx persistent.
Stamens 5.

All species are rare in cultivation, *R. ovatum* being the most often found.

R. hongkongense Hutch.

Shrub 1.2–1.5 m (4–5 ft) high; **branchlets** glandular.
Leaves ovate, ovate-lanceolate or oval; **lamina** 3–8.4 cm long, 2–3 cm broad;
  upper surface dark green (in young leaves, crimson-purple), glossy, glabrous;
  underside pale green, glabrous.
Petiole rather densely minutely puberulous (young petioles crimson-purple).
Inflorescence axillary in the uppermost 1–4 leaves, 1(–2) flowers.
Pedicel crimson-purple, glandular.
Corolla rotate widely funnel-campanulate or almost bowl-shaped, 2–2.6 cm long,
  white with crimson spots.
Stamens 5.
Ovary setulose-glandular.

Hong Kong and Guangdong in adjacent mainland China. 750–1,200 m
  (2,400–4,000 ft).

Epithet: From Hong Kong.   Hardiness 1.   March–April.

Similar to the
closely related
R. ovatum, but:
young leaves
crimson-purple;
flowers may be
fragrant, and are
always white.

Leaves usually ovate-
lanceolate.

Young petioles crimson-
purple.

Inflorescence axillary,
1(–2) flowers.

Pedicel crimson-purple,
glandular.

R. hongkongense (Davidian)

Azaleastrum
(Azaleastrum)

R. leptothrium Balf.f. & Forrest

Shrub or **tree** 60 cm–7.5 m (2–25 ft) high; **branchlets** slender, eglandular or sometimes glandular, minutely puberulous; **young growth** reddish.

Leaves evergreen, lanceolate or oblong-lanceolate; **lamina** chartaceous, 2.5–10 cm long, 1–3 cm broad; **upper surface** dark green, glossy, glabrous, midrib minutely puberulous, primary veins raised; **underside** pale green, glabrous.

Petiole rather densely minutely puberulous.

Inflorescence axillary in the uppermost 1–4 leaves, 1 flower.

Pedicel glandular, minutely puberulous.

Calyx 3–9 mm long.

Corolla rotate somewhat flat, or almost bowl-shaped, 2–3.2 cm long, 2.5–4 cm broad, rose, pale pink, purple, purplish-red or deep magenta-rose, with or without crimson spots.

Ovary setulose.

W, mid-W and NW Yunnan, NE Upper Burma. 2,150–3,550 m (7,000–11,803 ft).

Epithet: With thin leaves.   Hardiness 1–3.   April–May.

R. ovatum (Lindl.) Maxim.

Shrub or **tree** 1–5 m (3–16 ft) high; **branchlets** slender, glandular or eglandular, rather densely minutely puberulous; **young growths** reddish-brown.

Leaves ovate, ovate-elliptic or ovate-lanceolate; **lamina** 2–6 cm long, 1–2.4 cm broad; **upper surface** dark green, glossy, glabrous, midrib minutely puberulous, primary veins raised; **underside** pale green, glabrous, midrib prominent, primary veins impressed.

Petiole rather densely minutely puberulous.

Inflorescence axillary in the uppermost 1–3 leaves, 1 flower.

Corolla rotate somewhat flat or almost bowl-shaped, 1.6–2.6 cm long, 2.5–3 cm broad, white, pink, rose, pale purple or very pale lilac, with or without pink or darker spots.

Stamens 5.

Zhejiang, Anhui, Hubei, Jiangsu, Fujian, Taiwan to Guangdong, Guizhou, mid-W Yunnan, NE Upper Burma. 175–2,750 m (575–9,000 ft).

Epithet: Egg-shaped.   Hardiness 1–3.   May–June.

The leaf is a similar shape (lanceolate or oblong-lanceolate) to that of *R. vialli*, but is chartaceous.

Young growth deep bronze.

Flowers not fragrant.

Branchlets minutely puberulous.

Inflorescence axillary, 1 flower.

Corolla pale pink to rose, purplish-red, magenta-rose, or purple.

R. leptothrium Cherry194 at Deer Dell

New foliage of *R. leptothrium* Cherry194 at Deer Dell

Small shiny ovate, ovate-elliptic or ovate-lanceolate leaves.

Young growth reddish-brown.

Flowers usually not fragrant.

Branchlets slender, rather densely minutely puberulous.

Petiole rather densely minutely puberulous.

Inflorescence axillary, 1 flower.

Corolla white flushed pink, purple or lilac.

R. ovatum at Deer Dell

Azaleastrum
(Azaleastrum)

R. vialii Delavay & Franch.

Shrub 1.2–4.5 m (4–15 ft) high; **branchlets** slender, rather densely minutely puberulous.

Leaves evergreen, obovate, oblanceolate, lanceolate, oblong-lanceolate or oblong-elliptic; **lamina** 2.6–8 cm long, 1.3–3.1 cm broad, apex rounded or acute, conspicuously mucronate; **upper surface** bright or dark green, glossy, glabrous, rather densely minutely puberulous; margin entire; **underside** pale green, glabrous.

Petiole 0.5–1.5 cm long, slightly grooved above, rather densely minutely leaves, 1–2 flowers.

Pedicel 3–6 mm long, densely setulose-glandular, glabrous.

Calyx crimson or pink, margin glandular.

Corolla tubular, slightly widened towards the top, 2.5–3.2 cm long, corolla tube 1.6–2.3 cm long, much longer than the lobes, crimson or pink.

Stamens 5; filaments glabrous.

S and W Yunnan. 1,200–1,850 m (4,000–6,000 ft).

Epithet: After Pere Paul Vial, of the French missions in Yunnan.   Hardiness 2–3. January–March.

Subgenus Azaleastrum
Section Choniastrum
Stamineum Series

Inflorescence axillary in the top 1–3 leaves, 1–8 flowers.
Small shrubs to large trees.
Bark often smooth.
Leaves usually glossy, apex acutely acuminate.
Young growth often tinged red.
Calyx minute.
Corolla tubular-funnel-shaped, often fragrant.
Ovary usually slender.
Stamens 10.

Corolla tubular, crimson or pink, slightly widened towards the top. Young growth pinkish-purple.

Branchlets slender, rather densely minutely puberulous.

Young growth pinkish. Inflorescence axillary, 1–2 flowers.

Pedicel densely setulose-glandular.

Corolla tube much longer than the lobes.

R. vialii at Deer Dell

New foliage of *R. vialii* at Deer Dell

R. championae Hook.f.

Shrub or **tree** 2.4–6 m (8–20 ft) high; **branchlets** bristly, bristly-glandular.

Leaves oblong, elliptic-lanceolate or lanceolate; **lamina** 7.8–18.5 cm long, 2.3–6.2 cm broad, apex acuminate or acutely acuminate; **upper surface** dark green, bristly or not bristly, margin rather densely bristly; **underside** bristly, often bristly-glandular.

Petiole bristly-glandular.

Inflorescence axillary in the uppermost 1–3 leaves, 2–6 flowers.

Calyx densely bristly-glandular.

Corolla tubular-funnel-shaped, 4.3–5.6 cm long, pink, rose or white, with or without a yellow blotch.

Hong Kong, Guangdong, Guangxi, Fujian, Zhejiang, Jiangsu. Up to 185 m (600 ft).

Epithet: After Mrs. Champion, the wife of its discoverer, J.G. Champion (1815–1854).   Hardiness 1–2.   April–May.

R. hancockii Hemsl.

Shrub or **tree** 1–4.5 m (3–15 ft) high; **branchlets** glabrous, eglandular.

Leaves evergreen, oblanceolate, elliptic-lanceolate, oblong-obovate, obovate, elliptic or oblong-lanceolate; **lamina** coriaceous, 8.5–13.5 cm long, 2.8–5 cm broad, apex acuminate, base tapered or obtuse, decurrent on the petiole; **upper surface** green, glabrous, eglandular, margin slightly recurved or flat; **underside** pale green, glabrous, eglandular.

Petiole narrowly winged on both sides or not winged, glabrous or rarely sparsely setulose, eglandular.

Inflorescence axillary in the uppermost 1–3 leaves, umbellate, 1 flower.

Pedicel 1.2–2.6 cm long, glabrous or hairy, eglandular.

Calyx minute glabrous or hairy, eglandular.

Corolla tubular-funnel-shaped, 5–7.8 cm long, white, with a yellow blotch at the base, or sometimes pink.

Stamens 10, shorter than the corolla; filaments densely pubescent towards the lower half or one-third of their length.

Ovary slender, tapering into the style, densely tomentose, eglandular.

Style glabrous or hairy at the base.

SE and S Yunnan. 1,500–2,000 m (4,900–6,550 ft).

Epithet: After W. Hancock (1847–1914), Chinese Imperial Customs.   Hardiness 2?   May.

Bristly branchlets.
Leaf margin rather
densely bristly,
underside bristly.
Petiole bristly
glandular.
Calyx densely bristly
glandular.

Inflorescence axillary,
2–6 flowers.

Corolla pink, rose or
white.

R. championae courtesy of Hong Kong Herbarium

Closely related to
R. ellipticum but
leaves obovate,
oblong-obovate,
elliptic to oblong-
lanceolate, apex
acuminate.
Inflorescence with
1 flower per stem.
Corolla large
5–7.8 cm long,
white, occasionally
pink.
Ovary densely
tomentose.

R. hancockii (S. Hootman)

Azaleastrum
(Choniastrum)

R. latoucheae Franch.

(*R. wilsonae* Hemsl. & E.H.Wilson)

Shrub or **tree**, 1–3 m (3–10 ft) high; **branchlets** slender, glabrous.

Leaves evergreen, oblanceolate or lanceolate; **lamina** 5–10 cm long, 2–2.9 cm broad, apex acutely acuminate; **upper surface** green, glabrous; **underside** pale green, glabrous.

Petiole thinly tomentose or glabrous.

Inflorescence axillary in the uppermost 1–3 leaves, 1 flower, flower-bud elongate, apex acuminate, flower bud scales persistent.

Pedicel glabrous.

Corolla widely tubular-funnel-shaped with a short tube, 3.5–4.3 cm long, pink or purplish.

Guangdong, Jiangxi, Zhejiang. 150–600 m (500–2,000 ft).

Epithet: After Madame de la Touche, collector in Fujian.   Hardiness 1–2. April–May.

Rare in cultivation. May be found labelled R. wilsonae.

R. moulmainense Hook.f.

(*R. ellipticum* Maxim, *R. mackenzieanum* Forrest, *R. oxyphyllum* Franch., *R. pectinatum* Hutch., *R. stenaulum* Balf.f. & W.W.Sm., *R. westlandii* Hemsl.)

Shrub or **tree** 1–12 m (3–40 ft) high; **branchlets** slender, glabrous, eglandular.

Leaves evergreen, elliptic-lanceolate, oblong-lanceolate or lanceolate, apex acutely acuminate; **upper surface** green or olive-green, glabrous; **underside** pale green, glabrous, eglandular.

Petiole glabrous, eglandular.

Inflorescence axillary, 2–8 flowers.

Pedicel glabrous, eglandular.

Calyx minute, 0.5–1 mm long, glabrous, eglandular.

Corolla tubular-funnel-shaped, 3.5–6 cm long, fragrant, white, white flushed yellow, pink, rose, rose-red or lilac, with or without a yellow or pale green blotch.

Stamens 10.

Burma, SE and E Tibet, NW, W and NE Yunnan, Guizhou to Guangxi and Thailand. 400–3,650 m (1,300–12,000 ft).

AM 1937 (Exbury, and the Earl of Stair, Stranraer).

Epithet: From Moulmein, Burma.   Hardiness 1–2.   April–May.

A very variable species. Note the altitudinal and geographical range. Hard to identify, and more field work is needed. Differs from R. stamineum *in the flowers.*

Inflorescence axillary, 1 flower.

Corolla widely tubular-funnel-shaped, with a short tube, pink or purplish.

R. latoucheae (*R. wilsonae*): note the two single-flowered inflorescences and the terminal growth bud (S. Hootman)

Leaves apex acutely acuminate.

Corolla tubular-funnel-shaped, fragrant.

Stamens and style usually included in the corolla.

Branchlets slender, glabrous, eglandular.

Inflorescence axillary, 2–8 flowers.

Corolla white, white flushed yellow, pink, rose, rose-red or lilac.

R. moulmainense (as *R. stenaulum*) at Caerhays Castle (M.L.A. Robinson)

Azaleastrum (Choniastrum)

At present, the following species has been submerged into
R. moulmainense:

R. ellipticum Maxim

Shrub up to 4.5 m (15 ft) high; **branchlets** eglandular.
Leaves oblong-lanceolate or lanceolate; **lamina** 7.5–10.5 cm long, 1.7–3.8 cm
  broad, apex acutely acuminate, base decurrent on the petiole; **upper
  surface** green, glabrous; **underside** pale green, glabrous, eglandular.
Inflorescence axillary in the uppermost 1–3 leaves, 1–3 flowers, flower-bud
  elongate, apex acuminate.
Pedicel glabrous, eglandular.
Calyx glabrous, eglandular.
Corolla tubular-funnel-shaped, 5–6 cm long, pink or white.
Stamens 10 or rarely 8.
Taiwan, South China to Japan. 30–2,500 m (100–8,200 ft).
Epithet: Elliptic.   Hardiness 1–2.   April–May.

R. stamineum Franch.

Shrub or **tree**, 1–10 m (3–33 ft) high; **branchlets** slender, glabrous.
Leaves lanceolate, oblong-lanceolate, oblanceolate or elliptic-lanceolate; **lamina**
  6–11 cm long, 2–4.5 cm broad, apex acutely acuminate; **upper surface** green,
  glabrous; **underside** pale green, glabrous.
Petiole glabrous.
Inflorescence axillary in the uppermost 1–3 leaves, 2–8 flowers.
Pedicel glabrous.
Calyx glabrous.
Corolla narrowly tubular-funnel-shaped, 2.5–3.7 cm long, fragrant, white or rose,
  with a yellow blotch.
Stamens 10, much longer than the corolla.
NE Yunnan, W and E Sichuan, Hubei, Guizhou, Hunan, Anhui, Guangxi. 350 2,150 m
  (1,200–7,000 ft).
AM 1971 (Windsor), flowers white with a yellow blotch.
Epithet: With prominent stamens.   Hardiness 1–2.   April–May.

Inflorescence axillary, 1–3-flowers.

Corolla tubular-funnel-shaped, white or pink, larger than *R. moulmainense*.

R. moulmainense (R. ellipticum): note the position of the growth bud at Deer Dell

R. moulmainense (R. ellipticum) in the Savill Gardens

Corolla white or rose, often scented.

Stamens 10, much longer than the corolla.

The corolla lobes are long and narrow, and are usually recurved, making the stamens even more noticeable.

Leaf apex acutely acuminate.

Inflorescence axillary, 2–8-flowers.

R. stamineum at Glendoick (M.L.A. Robinson)

Azaleastrum
(Choniastrum)

Subgenus Pentanthera (G.Don) Pojark

Deciduous elepidote shrubs, or rarely trees.
Indumentum, if present, of simple hairs.
Flowers from terminal buds (usually several), but leafy shoots from separate lateral buds below.
Stamens 5–10.
Seeds with or without wings.

Section Pentanthera G.Don

Azalea Series

Leaves, though close together, always alternate.
Corolla zygomorphic, outer surface covered with hairs.
Stamens 5.
Seed lacking wings.

Differences between Subsections Pentanthera *and* Sinensia *are based **entirely** on flower shapes.*

The principal taxonomic differences between species in Section Pentanthera *are based on flower characteristics. It is therefore not surprising that it is difficult to identify the species within this Section out of flower. Virtually nothing can be gleaned from leaf shapes, though the presence or absence of hairs on the leaf surfaces helps a little. Some information can be gathered from the habit of growth and the detail of the flower bud, but even so, exact identification usually needs the flowers.*

Subsection Pentanthera

Luteum Subseries

Corolla narrowly tubular-funnel-shaped, outside usually pubescent or glandular.
Stamens as long as or longer than the corolla.

From N America except for R. luteum.

Flower groups in Subsection Pentanthera

Group A. Flowering after the leaves emerge; flower buds **glabrous**:

R. arborescens
Young shoots glabrous; leaves lustrous, not matt; flower buds chestnut brown; flowers white to pink; very fragrant.

R. cumberlandense
Often stoloniferous; young shoots pubescent; flower buds with ciliate margins; flowers orange through to red; not fragrant.

R. prunifolium
Young shoots glabrous; flower bud scales with ciliate margins; flowers red or scarlet, occasionally orange, very late; not fragrant.

Group B. Flowering **before or with** leaf emergence; flower buds **glabrous**:

R. alabamense
Flowers white with a yellow blotch; lemon scented.

R. atlanticum
Flower buds glabrous or with small hairs; leaf underside glaucous; flowers white or pink tinged; highly scented.

R. calendulaceum
Flowers orange-red; little or no scent.

R. luteum
Flowers yellow with a deeper yellow blotch, fragrant.

R. periclymenoides
Flower buds with ciliate margins; flowers usually pink or white; varies from no fragrance to slightly fragrant.

Group C. Flowers **before or with** leaf emergence; flower buds **pubescent**:

R. austrinum
Flower buds not densely pubescent, greyish with glandular margins; flowers yellow-orange; fragrance sweet and musky.

R. canescens
Flower buds densely covered with greyish hairs, also on the bud scale margins; flowers white to deep reddish pink; not fragrant.

R. flammeum
Flower buds usually pubescent but always with ciliate bud scale margins of longish hairs; flowers orange-scarlet; not fragrant.

R. prinophyllum
Occasionally stoloniferous; flower buds minutely puberulous, bud scale margins strongly ciliate; flowers usually pink in cultivation, rarely white; fragrant of cloves.

Group D. Flowering **after** the leaves emerge; flower buds **pubescent**:

R. eastmanii
Flower bud scales glabrous, but margins pubescent near the apex, glandular along the lower two-thirds; flowers white usually with a yellow blotch. Very fragrant with a far-reaching fragrance.

R. occidentale
Flower buds only slightly pubescent but with ciliate bud scale margins; strong growth of (for this Subsection) comparatively thick shoots; flowers white, pink, orange, red, or yellow; fragrant. The flowers may, rarely, open with or before the leaves.

R. viscosum
Usually stoloniferous; young shoots pubescent; leaf surface matt or only slightly shining; flower buds usually finely pubescent; flowers white or pink; very fragrant.

R. arborescens (Pursh) Torr. Pentanthera Gp A

Shrub 1–3 m (3–10 ft) or rarely small tree up to 6 m (18 ft) high; young shoots glabrous, not setulose.

Leaves deciduous, obovate, oblanceolate or oblong-lanceolate; **lamina** 3–8 cm long, 1.1–3 cm broad; **upper surface** bright green, glabrous, not strigose; **underside** pale green or glaucescent, glabrous.

Inflorescence 3–8 flowers, flowers opening after the leaves are developed, very fragrant.

Corolla tubular-funnel-shaped, 3.5–5.2 cm long, white or pink, with or without a yellow blotch.

Pedicel glandular.

Stamens 4.5–6 cm long, about twice as long as the corolla-tube.

Extends from New York and Pennsylvania to Georgia, Alabama, N and S Carolina, Kentucky and Tennessee. Up to and over 1,600 m (5,200 ft).

AM 1952 (M. Adams-Acton, London) 'Ailsa', flowers white with yellow blotch.

Epithet: Becoming tree-like. Hardiness 3. June–July.

R. arborescens var. richardsonii Rehder

Low compact or wide-spreading shrub, usually about 1 m (3 ft) high, with smaller and more glaucous leaves. 1,150–1,600 m (3,800–5,200 ft).

Epithet: After Mr. H. Richardson. Hardiness 3. June–July.

No longer recognised as separate: R. arborescens *is very variable within its wide geographical range.*

R. cumberlandense E.L.Braun Pentanthera Gp A

R. bakeri (Lemmon & McKay) Hume

Shrub 0.6–2 m (2–10 ft) high, or low forms 15 cm (6 inches) high, stoloniferous.

Leaves deciduous, obovate; **lamina** 3–7 cm long, 1.5–2.5 cm broad; **upper surface** dark green, glabrous or pubescent; **underside** more or less glaucous, glabrous to sparsely pubescent, eglandular.

Inflorescence 3–7 flowers, flowers opening after the leaves.

Corolla tubular-funnel-shaped, 2.5–4 cm long, 3.5–5 cm across, yellow, orange, reddish-orange or red, with orange-yellow blotch, outside densely puberulous, glandular.

From Kentucky to Georgia, Alabama and N Carolina.

Epithet: After Dr. W.F. Baker, Amory University, USA. Hardiness 3. May–June.

Rare in cultivation. The type form of R. bakeri *is a probable hybrid, so the name* R. cumberlandense *has to be used for the true species.*

Similar to
R. viscosum, but has
glabrous light brown
branchlets.

Leaves lustrous
rather than glossy,
but not matt, and
totally smooth not
bristly.

Crushed leaves smell
of newly mown grass.

Flower buds
glabrous, eglandular.

Flowers sweetly
scented with very
long stamens: about
twice as long as the
corolla-tube.

Leaf underside glabrous.
Flowers opening after
the leaves, white or
pink, very fragrant.

R. arborescens in NE USA (M. Creel)

Branchlets usually
moderately bristly.
Very similar to
R. calendulaceum,
but in
R. cumberlandense
the growth is
stoloniferous, more
compact.

The flowers are
produced 4–5 weeks
after the leaves, are
often deeper shades
of red, and are
slightly fragrant.

Leaf underside glabrous
to sparsely pubescent.

R. cumberlandense at Deer Dell

Pentanthera
(Pentanthera Group A)

R. prunifolium (Small) Millais

Pentanthera Gp A

Medium or large **shrub** up to 4.5 m (15 ft) high, often stoloniferous; **young shoots** dark purplish-red, glabrous, not setulose, eglandular.

Leaves deciduous, oblong-obovate, obovate, elliptic or oblong; **lamina** 3–12.5 cm long, 1.5–3.8 cm broad; **upper surface** bright green or dark green, glabrous, not strigose, eglandular; margin strigose; **underside** pale green, glabrous, not strigose, eglandular, midrib strigose or not strigose.

Inflorescence 4–5 flowers, flowers appearing after the leaves are formed.

Pedicel setulose.

Corolla tubular-funnel-shaped, 3–3.5 cm long, crimson, vermilion, scarlet, red, reddish-orange or orange, with or without orange blotch, tube glandular outside.

Stamens 5.5–6.5 cm long.

SW Georgia, E Alabama.

AM 1950 (Windsor) 'Summer Sunset', flowers vermilion.

Epithet: With leaves like those of a plum.   Hardiness 3.   July–September.

R. alabamense Rehder

Pentanthera Gp B

Shrub 1–1.5 m (3–6 ft) high, stoloniferous; **young shoots** densely or sparsely strigose, eglandular.

Leaves deciduous, obovate, elliptic, oblong or oblong-obovate; **lamina** 3.1–6.3 cm long, 1.3–3.1 cm broad; **upper surface** dark green or bright green, often finely strigillose; margin strigose; **underside** often glaucous, rather densely pubescent.

Inflorescence 6–12 flowers, flowers opening with the leaves, lemon-scented.

Pedicel villous, glandular.

Corolla tubular-funnel-shaped, 2–3 cm long, 2.4–3.8 cm across, white, usually with yellow blotch, outside glandular with long-stalked glands.

Central Alabama, Georgia, S Carolina, USA.

Epithet: From Alabama, USA.   Hardiness 3.   May.

Pink forms are hybrids with R. canescens. *Rare in cultivation in the UK as it requires considerable summer heat to succeed.*

Branchlets smooth, dark purplish-red, glabrous.

Flower bud scales glabrous with ciliate margins, eglandular.

Very late flowering – the leaves open well before the flowers.

Leaf underside glabrous, not strigose, eglandular.

Corolla usually shades of red or orange.

Flower buds of *R. prunifolium* in the wild. Note the ciliate margins on the bud scales (M. Creel)

Leaf underside glaucous or sometimes pale green.

Out of flower almost identical to *R. canescens*, but the leaves of *R. alabamense* have a spicy odour if crushed.

Within the Subsection, only this species, *R. arborescens* and *R. eastmanii* have a white corolla with a yellow blotch.

Fragrant.

Stoloniferous.

Leaf underside rather densely pubescent.

Flowers opening with the leaves, lemon-scented.

R. alabamense in the wild (M. Creel)

Pentanthera
(Pentanthera Group A/B)

R. atlanticum (Ashe) Rehder   Pentanthera Gp B

Low shrub, 30–60 cm (1–2 ft) high or rarely taller, stoloniferous; **young shoots** setulose.

Leaves deciduous, obovate or oblong-obovate; **lamina** 2.2–6.3 cm long, 1.2–2.2 cm broad; **upper surface** bluish-green or bright green, glabrous, not strigose; **underside** bright green or glaucescent, glabrous, not strigose, often glandular.

Inflorescence 4–10 flowers, flowers opening before or with the leaves, very fragrant.

Corolla tubular-funnel-shaped, 3.5–4.8 cm long, white, usually flushed pink or purple, without blotch, tube outside and middle of all the lobes glandular with long-stalked glands.

S Pennsylvania and Delaware to Virginia, N and S Carolina, and Georgia.

AM 1964 (Windsor) 'Seaboard', flowers white, with pale pink corolla tube.

Epithet: From the Atlantic seaboard.   Hardiness 3.   May.

R. calendulaceum (Michx.) Torr.   Pentanthera Gp B

Shrub 1–3 m (4–10 ft) or rarely up to 5 m (16 ft) high; **young shoots** pubescent and strigose, older branchlets glabrous.

Leaves deciduous, elliptic, oblong-obovate, oblanceolate or lanceolate; **lamina** 3.5–8 cm long, 1.3–3.5 cm broad; **upper surface** pubescent or strigose; **underside** rather densely or moderately pubescent or strigose.

Petiole densely pubescent, and strigose.

Inflorescence 5–8 flowers, flowers opening with or shortly after the leaves.

Corolla tubular-funnel-shaped, 2.5–4 cm long, yellow, yellowish-orange, orange, red or scarlet, with orange blotch, tube glandular outside, lobes glandular or eglandular.

Pennsylvania, W Virginia, N Carolina, Georgia.

AM 1965 (Windsor) 'Burning Light', flowers coral red with orange blotch.

Epithet: Like a *Calendula*.   Hardiness 3.   Late May–June.

Similar to a number of Ghent azaleas, especially 'Coccinea speciosa'.

A low-growing shrub with thin young growth, and, usually, leaves with at least a glaucous 'wash' above; totally glabrous. Flower buds mid-brown, almost always glabrous, eglandular.

The long tube of the corolla is conspicuously glandular with long stalked glands.

Leaf upper surface bluish-green, glabrous, not strigose; underside glabrous, not strigose.

Flowers opening before or with the leaves, white, very fragrant.

R. atlanticum showing the narrowly tubular campanulate corolla (D. White)

R. atlanticum in NE USA (M. Creel)

R. atlanticum at Deer Dell

Growth not stoloniferous. Forms with glabrous leaves are found occasionally. Similar to *R. cumberlandense*, but the flower bud scales are glabrous with palely glandular margins.

Flowers appear before or with the new growth, and are always on the yellow side of red.

Leaf upper surface pubescent or strigose; underside rather densely or moderately pubescent or strigose.

Corolla yellowish-orange, orange, tube pubescent and glandular on the outside.

R. calendulaceum (Davidian)

R. luteum Sweet   Pentanthera Gp B
(*Azalea pontica* L.)

Shrub 0.6–3.6 m (2–12 ft) high, stoloniferous; **young shoots** setulose or not setulose, glandular with long-stalked glands.
Leaves deciduous, oblong, oblong-obovate or oblong-lanceolate; **lamina** chartaceous, 5–11 cm long, 1.3–4 cm broad; **upper surface** dark green or pale green, strigose-glandular; margin serrulate, strigose; **underside** strigose-glandular.
Inflorescence 7–12 flowers.
Corolla tubular-funnel-shaped, 3.2–4.5 cm long, yellow with a darker (yellow) blotch, flowers opening before the leaves, very fragrant.
Pedicel glandular, outside glandular, viscid.
Calyx glandular.
Stamens 5.
Caucasus, N Turkey, SW Russia, Georgia, Ukraine. Sea-level to 2,200 m (7,200 ft). AGM 1930.
Epithet: Yellow.   Hardiness 3.   May–June.

Very common in cultivation, sometimes still labelled 'Azalea pontica'*: distinctive in its yellow fragrant flowers and in its red or purple autumn leaf colour.*

R. periclymenoides (Michx.) Shinners   Pentanthera Gp B
(*R. nudiflorum* [L.] Torr.)

Shrub 0.6–3 m (2–10 ft) high, occasionally stoloniferous; **young shoots** glabrous or sparsely pubescent, strigose, eglandular.
Leaves deciduous, elliptic, obovate or oblong; **lamina** chartaceous, 3–9 cm long, 1.3–3 cm broad; **upper surface** bright green, strigose or not strigose, midrib densely pubescent; margin strigose; **underside** green, strigose or not strigose, midrib strigose or not strigose.
Inflorescence 6–12 flowers, flowers opening before or with the leaves, fragrant.
Corolla tubular-funnel-shaped, 2.6–3.5 cm long, white, pale pink, pink or pinkish-purple, without blotch, outside pubescent, eglandular.
On the Appalachian Mountains, and extends from Massachusetts to N Carolina, W to Central New York, Pennsylvania, S Ohio, E Kentucky, E Tennessee, N Georgia, N Alabama. Up to 1,150 m (3,800 ft).
Epithet: Like honeysuckle.   Hardiness 3.   May.

Also known as 'nudiflorum', *which may assist identification.*

R. periclymenoides var. *glandiferum* (Porter) Rehder
The variety differs from the species type in that the pedicels and the outside of the corolla (both tube and lobes) are glandular. The flowers are usually pink or carmine.
Epithet: Glandular.   Hardiness 3.   May.

Very rare in cultivation.

Both surfaces of the leaves are strigose glandular.

The young shoots, petioles, pedicels, corolla tube and lobes, and the calyx are all glandular.

Corolla yellow with a darker (yellow) blotch.

Stoloniferous.

Flowers opening before the leaves.

R. luteum (Davidian)

Leaves not glaucous; usually glabrous except the midrib.

Flower bud scales glabrous with ciliate margins.

Corolla tube eglandular but ciliate, appearing woolly (pubescent).

Leaves elliptic, obovate, or sometimes oblong.

Inflorescence 6–12 flowers, flowers opening before or with the leaves.

Corolla white, pink, or pinkish-purple.

R. periclymenoides in the Valley Gardens

R. austrinum (Small) Rehder   Pentanthera Gp C

Shrub up to 2.75 m (9 ft) high, or small tree 4.5 m (15 ft) high or more; **branchlets** pubescent, glandular.
Leaves deciduous, elliptic, oblong-obovate, obovate or ovate; **lamina** 3–9 cm long, 0.6–4 cm broad, **both surfaces** rather densely minutely pubescent, margin strigose.
Petiole pubescent, glandular.
Inflorescence 6–15 flowers, flowers opening before or with the leaves, fragrant.
Pedicel and **calyx** pubescent, glandular.
Corolla tubular-funnel-shaped, 3.2–3.4 cm long, yellow or orange, the tube usually purplish, without blotch.
Stamens 5–6 cm long, nearly twice as long as the corolla tube.

From N Florida, and the Georgia–Alabama coastal plain to SE Mississippi.
Epithet: Southern.   Hardiness 3.   May.

Rare in cultivation in the UK as it needs considerable summer heat to ripen its growth. Plants with red or reddish flowers are likely to be hybrids with R. canescens.

R. canescens (Michx.) Sweet   Pentanthera Gp C
(*R. bicolor* Pursh, *R. candidum* Small)

Shrub or **tree** up to 4.5 m (15 ft) high or more; **young shoots** pubescent, eglandular.
Leaves deciduous, oblong-obovate, oblanceolate or oblong; **lamina** 4.3–9 cm long, 1.3–4 cm broad; **upper surface** sparsely pubescent or glabrous, eglandular; margin strigose; **underside** moderately or densely pubescent or glabrous.
Petiole eglandular.
Inflorescence 6–15 flowers, flowers opening before or with the leaves, fragrant.
Calyx glandular with long-stalked glands.
Corolla tubular-funnel-shaped, 2.8–3.8 cm long, tube long, slender, nearly twice as long as the lobes, white, pink or white flushed pink, the tube pink or reddish, usually without blotch, outside pilose, moderately or rather densely glandular; lobes 1–1.5 cm long.
Stamens 5, 3.5–6.3 cm long, twice or nearly 3 times as long as the corolla tube.

Gulf Coastal Plains, extends from N Carolina to Georgia, Florida, west to southern Tennessee and SE Texas.
Epithet: Hoary.   Hardiness 3.   April–May.

Candidum Group has palely glaucous leaf undersides.

Leaf margins, petioles, flower bud scale margins and pedicels pubescent, glandular.

Flower bud scales somewhat pubescent with greyish hairs.

Corolla yellow or orange, the tube usually purplish.

Flowers slightly fragrant.

Both surfaces of the leaf rather densely minutely pubescent.

Flowers opening before or with the leaves.

Stamens nearly twice as long as the corolla tube.

R. austrinum in the wild (M. Creel)

Young shoots greyish pubescent but not glandular.

Flower buds initially densely greyish pubescent.

The flower has very long glandular tubes, nearly twice as long as the lobes.

Stamens twice or nearly 3 times as long as the corolla tube.

Young shoots pubescent, eglandular.

Petiole eglandular.

Flowers opening before or with the leaves.

Corolla white to pink, outside pilose, moderately or rather densely glandular.

A typical form of *R. canescens* in NE USA (M. Creel)

R. flammeum (Michx.) Sargent   Pentanthera Gp C

Shrub much-branched, 0.3–1.8 m (1–6 ft) high, stoloniferous; **young shoots** densely pubescent, strigillose.

Leaves deciduous, obovate, elliptic or oblong; **lamina** chartaceous, 3–6 cm long, 1.3–2.5 cm broad; **upper surface** bright green or dark green, ciliate, strigillose; **underside** often finely pubescent, midrib strigose, eglandular.

Inflorescence terminal or terminal and axillary, 6–15 flowers, flowers opening with or before the leaves.

Corolla tubular-funnel-shaped, 3–3.5 cm long, 3–5 cm across, scarlet, bright red or orange, with large orange blotch, outside pubescent, pilose.

Central Georgia to S Carolina, on dry sandy bluffs along the Savannah and Edisto river basins.

Epithet: Flame-coloured.   Hardiness 3.   May–June.

Earlier flowering than R. cumberlandense.

R. prinophyllum (Small) Millais   Pentanthera Gp C
(*R. roseum* [Lois.] Rehder & E.H.Wilson)

Shrub 0.6–4.5 m (2–15 ft) high, rarely stoloniferous; **young shoots** rather densely finely pubescent, strigose or not strigose.

Leaves deciduous, elliptic, obovate or oblong-obovate; **lamina** 3–7 cm long, 1.3–3 cm broad; **upper surface** bluish-green or green, villous; margin strigose; **underside** pale green, rather densely villous, strigose.

Petiole densely pubescent, strigose.

Inflorescence 5–9 flowers, flowers opening with the leaves, clove-scented.

Pedicel villous, glandular.

Corolla tubular-funnel-shaped, 2.8–3.5 cm long, pink or purplish-pink, rarely white, with or without brown-red blotch, flowers opening with the leaves, clove-scented; glandular.

Stamens about twice as long as the corolla-tube.

Extends from SW Quebec, through Massachusetts, Pennsylvania, SE Missouri and Tennessee to Virginia.

AM 1955 (Tower Court) as *R. roseum,* flowers pink.

Epithet: With leaves like prinos.   Hardiness 3.   May.

May be found labelled 'R. roseum'.

Leaf upper surface ciliate.

Flower bud scale margins pubescent with longish hairs but not glandular.

Corolla tube eglandular.

Leaf underside often finely pubescent.

Flowers opening with or before the leaves.

Inflorescence 6–15 flowers.

Corolla scarlet, bright red or orange, with large orange blotch, outside pubescent, pilose.

R. flammeum in NE USA (M. Creel)

Very similar to *R. periclymenoides*, but the leaves are somewhat glaucous, the shoots and flower buds are usually pubescent, and the corolla tube is glandular.

Stamens somewhat shorter than in most of the Subsection: about twice as long as the corolla tube.

Leaf underside rather densely villous.

Petiole densely pubescent.

Flowers opening with the leaves, pink or purplish-pink, clove-scented.

R. prinophyllum in the Valley Gardens

Pentanthera
(Pentanthera Group C)

R. eastmanii Kron & Creel   Pentanthera Gp D

Shrub or **small tree** to 5 m (16 ft) high, not stoloniferous; **branchlets** reddish
brown densely pubescent, eglandular.
Leaves deciduous, ovate or obovate to elliptic; **lamina** 4.3–7.1 cm long,
1.8–2.9 cm broad, eglandular; **upper surface** dark green, sparsely or
densely pubescent; **underside** moderately or densely pubescent, midrib
densely pubescent.
Flower bud scales chestnut brown, margins pubescent near apex, glandular
along the lower two-thirds.
Inflorescence 5–9 flowers, flowers opening after the leaves.
Corolla tubular-funnel-shaped, 2.0–4.3 cm long, 2.6–4.1 cm across, white with
a yellow blotch, fragrant, outside pubescent and glandular, the glands usually
weakly developed, stamens and style exserted.
Ovary densely pubescent, eglandular.
S Carolina.
Epithet: After C. Eastman of Columbia, S Carolina, USA, the discoverer.
Hardiness 3.   May.

*Described in 1999. Likely to be a new species, superficially resembling a late-
flowering* R. alabamense. *However* R. eastmanii *flowers after the leaves have
emerged and the corolla tube is less noticeably glandular. Probably not, as yet, in
cultivation in the UK.*

R. occidentale (Torr. & Gray) A.Gray   Pentanthera Gp D
(*Azalea californica* Torr. & A.Gray ex Durand)

Shrub 0.6–4.5 m (2–15 ft) high, **young shoots** glabrous or minutely puberulous.
Leaves deciduous, oblanceolate, lanceolate, oblong-lanceolate, obovate or
elliptic; **lamina** 4–9 cm long, 1.5–3 cm broad; **upper surface** bright green or
dark green, often minutely pubescent; margin rather densely strigose; .
underside often minutely pubescent, midrib strigose.
Inflorescence 5–12 flowers, flowers opening with or after the leaves, fragrant.
Calyx 1–6 mm long.
Corolla tubular-funnel-shaped, 4.2–5 cm long, white, pink, orange-pink, red or
yellow, with or without stripes and spots, usually with yellow or orange blotch,
outside villous, glandular.
From S Oregon to S California. Sea level up to 2,750 m (9,000 ft).
AM 1944 (Kew), flowers white, heavily flushed rose-pink, with a yellow blotch.
Epithet: Western.   Hardiness 3.   May–July.

*A variable species with many selected forms. In selected forms, the truss may
commonly contain up to 25 flowers, with up to 54 in the best-selected forms.*

Similar in appearance to *R. occidentale*, but the petioles and leaves are eglandular in *R. eastmanii*. Branchlets reddish brown densely pubescent, eglandular.

Flower bud scales chestnut brown, margins unicellular pubescent near apex, glandular along the lower two-thirds.

Flowers opening after the leaves.

Corolla white with a yellow blotch, glandular, the glands usually weakly developed.

Out of leaf and flower, the densely hairy twigs distinguish this species from *R. arborescens*. The glandular flower bud margins are unique among the white-flowered species of this Subsection, but occur in *R. calandulaceum* and *R. cumberlandense*. Hard, therefore, to identify out of flower.

R. eastmanii from Crane Creek, Richland County, S Carolina (M. Creel)

A strong grower with stout branches. Leaves often shining and usually oblanceolate.

Flower buds pubescent, occasionally glandular.

Corolla less narrowly tubular than other species in the Subsection; outside villous, glandular; very fragrant.

Stamens only a little longer than the corolla (unique in this Subsection).

Leaf underside midrib strigose.

Flowers white, pink, yellow, orange or red, opening with or after the leaves.

R. occidentale at Hindleap Lodge (M.L.A. Robinson)

R. viscosum (L.) Torr.   Pentanthera Gp D

(*R. oblongifolium* [Small] Millais, *R. serrulatum* [Small] Millais)

Shrub 1–4.5 m (3–15 ft) high, stoloniferous; **young shoots** setulose or strigose, eglandular.

Leaves deciduous, obovate, oblong-obovate or oblanceolate; **lamina** 1.5–7.3 cm long, 0.5–3.5 cm broad; **upper surface** dark green, usually glabrous; margin strigose; **underside** pale green or glaucescent, usually glabrous except midrib strigose.

Petiole setulose or strigose.

Inflorescence 3–9 flowers, flowers opening after the leaves are developed, spicy fragrance.

Pedicel setulose-glandular.

Calyx setulose-glandular or setulose.

Corolla slender, narrowly tubular-funnel-shaped, 3–4.5 cm, long, white or sometimes white suffused with pink, tube setulose-glandular outside, lobes glandular along the middle.

Maine, Massachusetts, Connecticut to N Carolina, south eastern S Carolina, west to Ohio, SE Tennessee, Louisiana.

AGM 1937.

Epithet: Sticky.   Hardiness 3.   June–July.

R. viscosum var. rhodanthum *has pink flowers but has no separate botanical status.*

R. viscosum var. glaucum (Ait.) Torr.

The variety differs from the species type in that the leaves are glaucous below, and sometimes on both surfaces.

AM 1921 (F.G. Strover, London), as *Azalea viscosa glauca*, flowers white.

Epithet: Glaucous.   Hardiness 3.   June–July.

This is one of a number of extreme named forms of R. viscosum.

R. viscosum Oblongifolium Group   Pentanthera Gp D

R. oblongifolium (Small) Millais

Shrub up to 1.8 m (6 ft) high, sometimes stoloniferous; **young shoots** pubescent or not pubescent, strigose or sometimes setulose.

Leaves deciduous, obovate or oblanceolate; **lamina** chartaceous, 3.8–10 cm long, 1.5–3.8 cm broad; **upper surface** dark green, usually glabrous; margin strigose; **underside** pale green, often glabrous except midrib pubescent, strigose.

Petiole puberulous.

Inflorescence 6–12 flowers, flowers opening after the leaves, faintly clove-scented.

Pedicel rather densely setulose-glandular.

Calyx glandular.

Corolla tubular-funnel-shaped, 2.8–4 cm long, white or sometimes pink, without blotch, tube setulose-glandular outside, lobes glandular.

Stamens 5.

Arkansas, Oklahoma, Texas.

Epithet: With oblong leaves.   Hardiness 3.   May–July.

Branchlets bristly (setulose or strigose), eglandular.

Leaf margins strigose. Flower buds sparsely pubescent.

Flowers sticky (glandular), white to pink, very fragrant.

Stamens and style always white.

Stonoliferous shrub.

Leaf margin strigose; underside midrib strigose.

Petiole setulose or strigose.

Pedicel and calyx setulose glandular or setulose.

Flowers opening after the leaves, spicy scented, white or flushed pink.

R. viscosum at Deer Dell

Larger leaves; earlier flowering.

Leaves obovate or oblanceolate; underside midrib pubescent strigose. Petiole puberulous. Pedicel densely setulose glandular. Calyx glandular.

R. viscosum pink form (E. Daniel)

R. viscosum Serrulatum Group

Pentanthera Gp D

R. serrulatum (Small) Millais

Shrub up to 6 m (20 ft) high or more; **young shoots** reddish-brown, strigose.
Leaves deciduous, obovate, oblong-obovate or elliptic; **lamina** 3.8–8.1 cm
    long, 1.5–3.8 cm broad; **upper surface** dark green, glabrous, not strigose;
    margin serrulate, strigose; **underside** pale green, glabrous, not strigose
    except midrib strigose.
Petiole strigose.
Inflorescence 6–10 flowers, flowers opening after the leaves, fragrant,
    clove-scented.
Pedicel setulose-glandular.
Calyx setulose-glandular.
Corolla tubular-funnel-shaped, 3.5–4.5 cm long, white or sometimes pale
    pink, tube setulose-glandular outside, lobes glandular along the middle.

Middle Georgia to central Florida, and westward along the coast to SE Louisiana.
Epithet: With small teeth.   Hardiness 3.   July–September.

Subgenus Pentanthera (G.Don) Pojark
Section Pentanthera G.Don
Subsection Sinensia (Nakai) Kron
Luteum Subseries

**Corolla broadly funnel-shaped, outer surface with unicellular hairs;
upper corolla lobe spotted.**
Stamens 5, usually shorter or as long as the corolla.

From China or Japan.

Low growing and stoloniferous.

Leaf margins slightly toothed (serrulate).

Late flowering.

Young shoots reddish-brown.

Petiole strigose.

Pedicel setulose glandular.

Calyx setulose-glandular.

Flower bud of *R. viscosum* (*R. serrulatum*) in the wild. Note the sparse pubescence (M. Creel)

R. canescens (Pentanthera Group C) in NE USA (M. Creel)

R. molle ssp. molle

R. molle (Bl.) G.Don.
(*R. sinensis* Sweet & Lodd, *Azalea* mollis Blume)

Shrub 0.3–1.2 m (1–4 ft) high; **young shoots** villous, eglandular.
Leaves deciduous, oblong, oblanceolate or oblong-lanceolate; **lamina** 6–15 cm long, 1.8–5.6 cm broad; **upper surface** often pubescent, eglandular; margin serrulate, strigose; **underside** often rather densely pubescent, eglandular.
Petiole often setulose, eglandular.
Inflorescence 6–12 flowers, flowers opening before or with the leaves.
Calyx minutely puberulous.
Corolla widely funnel-shaped, 4.3–5.6 cm long, golden-yellow, yellow or orange, with a large greenish blotch separated into dots.

E China, in Hubei, Zhejiang and Jiangsu.
Epithet: With soft hairs.   Hardiness 3.   May–June.

Note the differences between this species and R. luteum, which has glandular new growth, a moderately or sparsely glandular leaf underside, and a blotched flower. R. molle is rare in cultivation outside China.

R. molle ssp. japonicum Kron

R. japonicum (A.Gray) Suringar
(*Azalea japonica* A.Gray)

Shrub 0.6–1.8 m (2–6 ft) high, occasionally stoloniferous; **young shoots** setulose, eglandular.
Leaves deciduous, obovate, oblong-obovate or oblanceolate; **lamina** chartaceous, 5–11.5 cm long, 1.8–3.5 cm broad, **upper surface** dark green, strigose, eglandular, margin strigose; **underside** pale green, not pubescent, not setulose, strigose only on the lateral veins, midrib strigose.
Petiole setulose.
Inflorescence 6–12 flowers, flowers opening before the leaves.
Pedicel setulose.
Calyx margin densely setulose.
Corolla widely funnel-shaped, 4.8–6.2 cm long, orange, yellow, salmon-red or brick-red, with large orange blotch.
Stamens 5.

Japan. Up to 1,000 m (3,300 ft).
Epithet: From Japan.   Hardiness 3.   May.

The yellow-flowered form is rare in cultivation.

Leaf underside densely pubescent with white hairs, partly detersile.

Flowers always close to yellow.

Capsule sparsely pubescent.

Young shoots villous, eglandular.

Leaf upper surface often pubescent; underside often rather densely pubescent.

Corolla golden-yellow, yellow or orange, with a large greenish blotch separated into dots.

R. molle ssp. *molle* (Davidian)

Leaf underside lamina glabrous: not pubescent, setulose only on the lateral veins.

Capsule densely pubescent.

Leaf upper surface strigose, margin strigose, underside midrib strigose.

Petiole setulose.

Flowers opening before the leaves.

Corolla orange, yellow, salmon-red or brick-red, with large orange blotch.

R. molle ssp. *japonicum* at Beresford Gardens showing the wider corolla (Mrs. E. Wood)

R. molle ssp. *japonicum* (Davidian)

Subgenus Pentanthera (G.Don) Pojark
Section Rhodora (L.) G.Don
Canadense Subseries

Corolla divided to, or nearly to, the base, so 2-lipped. The
two species here share only this characteristic and are not
closely related.
Leaves scattered, not in whorls.
Flowers not fragrant.
N American.

Recent molecular studies (Goetsch *et al.* 2005) have led to the
suggestion that *R. canadense* should be assigned to Subsection
Pentanthera, and *R. vaseyi* to Subsection *Sciarhodion*, which is
more in line with Davidian's classification.

R. canadense (L.) Torr.
(*R. rhodora* S.F.Gmelin)

Shrub 0.3–1.2 m (1–4 ft) high; **young shoots** puberulous, later glabrous.
Leaves deciduous, scattered, oblong, elliptic or oblong-oval; **lamina** 2–5.5 cm
long, 0.6–2.3 cm broad; **upper surface** pale bluish-green, strigose; **underside**
rather densely minutely puberulous (downy).
Petiole rather densely puberulous.
Inflorescence 3–6 flowers, flowers opening before the leaves.
Corolla rotate-campanulate, 1.6–1.8 cm long, 2-lipped, rose-purple.
Stamens 10.
From Labrador, Newfoundland, Quebec to New England, central New York, NE
Pennsylvania and N New Jersey.
Epithet: From Canada.   Hardiness 3.   April–May.

R. canadense f. albiflorum (Rand & Redfield) Rehder
Corolla white.
Epithet: With white flowers.   Hardiness 3.   April–May.

A white form of *R. canadense* (B. Starling)

A low upright and stoloniferous shrub with matt bluish-green leaves, tomentose below. The flower has the three upper lobes of the corolla fused together to form the upper lip, the lower lip being divided into two lobes. The corolla has virtually no tube, usually rose-purple. Stamens 10.

Leaf underside minutely puberulous.

R. canadense (B. Starling)

Pentanthera
(Rhodora)

R. vaseyi A.Gray

Shrub 1.5–4.6 m (5–15 ft) high; **young shoots** at first sparsely pilose, afterwards glabrous or glabrescent.

Leaves deciduous, scattered, elliptic, elliptic-lanceolate or oblong-lanceolate; **lamina** 5–12.5 cm long, 1.3–4.3 cm broad, apex acuminate; **upper surface** dark green, glabrous; margin strigose; **underside** light green, not setulose except midrib setulose.

Inflorescence 4–8 flowers, flowers opening before the leaves.

Pedicel glandular.

Corolla rotate-campanulate, 2.5–3 cm long, 3.5–3.8 cm across, 2-lipped, upper lip of 3 lobes, lower lip of 2 lobes, light rose or pale pink, with orange or reddish-brown spots.

Stamens 7.

N and S Carolina. 900–1,830 m (3,000–5,700 ft).

AM 1969 (Exbury) 'Suva', flowers light rose with reddish-brown spots.

Epithet: After G.R. Vasey (1822–1893), who discovered it in N Carolina in 1878. Hardiness 3.   April–May.

R. vaseyi f. album Nicholson

Corolla white.

In cultivation, a clone has been known as 'White Find'.

Epithet: With white flowers.   Hardiness 3.   April–May.

Leaves the same shape as those of the species type, but somewhat broader in cultivation.

Subgenus Pentanthera (G.Don) Pojark
Section Sciadorhodion Rehder & E.H.Wilson
Schlippenbachii Subseries mainly

Leaves in clusters or whorls of five at the ends of the branches.
Leaves more or less obovate.
Corolla zygomorphic, not divided, outer surface glabrous.
Stamens 10.

While Davidian included them with the other species of Section Brachcalyx, modern taxonomy places R. quinquefolium and R. schlippenbachii in this Section, presumably because of the shape of the leaves and their grouping into fives.

However, in these two species, the leaves and flowers emerge from the same bud, a taxonomic characteristic used to define Subgenus Tsutsusi Section Brachcalyx, so, while the modern treatment will assist identification, it is unsatisfactory.

Japan, Korea and E Russia.

The leaf shape is willow-like, at least 3 times as long as broad, and with undulate margins.

Corolla light rose or pale pink, with orange or reddish-brown spots.

Stamens usually 7, occasionally 5 or 6. Bright red autumn colour.

Leaf apex acuminate.

R. vaseyi (white form) at Benmore

R. vaseyi at Muncaster, Cumbria

Pentanthera
(Rhodora/Sciadorhodion)

R. albrechtii Maxim.

Lax **shrub** 1–2.5 m (3–8 ft) high; **young shoots** at first rather densely setulose-glandular, later eglandular.

Leaves deciduous, in clusters of 5 at the end of branchlets, obovate or oblong-obovate; **lamina** 3.6–9 cm long, 1.3–4 cm broad, **both surfaces** rather densely strigose.

Petiole setulose-glandular.

Inflorescence 2–5 flowers, flowers opening before or with the leaves.

Corolla rotate-campanulate, 2.3–3.5 cm long, 3.5–5 cm across, red-purple, bright purplish-rose or deep rose, with olive-green spots.

Japan. 1,000–2,000 m (3,300–6,550 ft).

AM 1943 (Bodnant).

FCC 1962 (Bodnant) to 'Michael McLaren', flowers purple with yellowish-green spots.

Epithet: After Dr. M. Albrecht, Russian naval surgeon.   Hardiness 4.   April–May.

R. pentaphyllum Maxim.

Shrub or **tree** 1.8–7.5 m (6–25 ft) high, young shoots at first sparsely pilose, soon glabrous.

Leaves deciduous, in pseudowhorls of 5 at the end of the branchlets, broadly elliptic, elliptic, ovate or nearly oval; **lamina** chartaceous, 2.5–6.3 cm long, 1.4–3.1 cm broad; **upper surface** dark green, strigose or not strigose; margin finely serrulate, ciliate; **underside** paler, not strigose or very sparsely strigose.

Petiole setulose.

Inflorescence 1–2 flowers, flowers opening before the leaves.

Pedicel glandular.

Calyx 1–5 mm long, eglandular.

Corolla rotate-campanulate, not divided, 2.3–3 cm long, 4–5 cm across, bright rose-pink, sometimes with red-brown flecks on upper 3 lobes.

Stamens 10.

Ovary glabrous, eglandular.

Honshu, Skikoku and Kyushu (Japan).

AM 1942 (Bodnant), flowers rose.

Epithet: With five leaves.   Hardiness 3.   March–May.

Rare in cultivation. A white-flowered form has been recorded.

The leaves appear to be in a whorl of 5, but they are alternate – not appearing from the same node. This arrangement is sometimes called a pseudowhorl.

Stamens dimorphic, the longer ones being glabrous, and the shorter densely pubescent at the base.

Corolla red-purple, bright purplish-rose or deep rose.

Flowers open before or with the leaves.

Shrub of *R. albrechtii* and Carolina Rivas-McQuire at Deer Dell

R. albrechtii at Deer Dell

As with *R. quinquefolium*, the leaves often have a red-purple margin.

Bark not corky.

Flowers and leaves emerge from separate buds.

Corolla bright rose-pink.

Flowers open before the leaves.

R. pentaphyllum at Deer Dell

Pentanthera
(Sciadorhodion)

R. quinquefolium Bisset & S.Moore

Shrub or small **tree**, 1.2–7.6 m (4–25 ft) high; **branchlets** glabrous, eglandular.

Leaves deciduous, in pseudowhorls of 4 or 5 at the end of the branchlets, broadly obovate, oval or broadly elliptic; **lamina** 3–5 cm long, 1.8–3 cm broad; **upper surface** pale green, often with a red-purple margin, glabrous; midrib pubescent; margin ciliate; **underside** pubescent towards the base or glabrous.

Petiole pubescent.

Inflorescence 1–3 flowers, flowers produced with the leaves from the same terminal bud.

Corolla rotate-campanulate, 2.3–2.8 cm long, 3–4.3 cm across, pure white with green spots.

Honshu and Shikoku (Japan). 700–1,500 m (2,300–4,900 ft).

AM 1931 (Cawdor, Haslemere), flowers white, spotted pale green.

AM 1958 (Exbury) 'Five Arrows', flowers white with olive green spots.

Epithet: Leaves in fives.   Hardiness 3.   April–May.

R. schlippenbachii Maxim.

Shrub 1–5 m (3–16 ft) high; **young shoots** glandular with long-stalked glands, later eglandular.

Leaves deciduous in pseudowhorls of 5 at the end of the branchlets, obovate; **lamina** 5–8.8 cm long, 3–6.8 cm broad; **upper surface** dark green, glabrous; **underside** pale green, glabrous, midrib pubescent.

Petiole flattened, glandular, glands long-stalked.

Inflorescence 3–6 flowers, flowers opening with or immediately before the leaves.

Pedicel rather densely glandular.

Corolla broadly rotate-funnel-shaped, 3–4 cm long, 5.6–8 cm across, pale to rose-pink or white, spotted red-brown.

Korea and the Korean archipelago, bordering parts of Russia and Japan. 400–1,800 (1,300–5,500 ft).

AM 1896 (Veitch, Chelsea), flowers soft pink.

FCC 1944 (Bodnant), flowers pink.

FCC 1965 (Leonardslee) 'Prince Charming', flowers pink with darker tinges, spotted deep crimson.

Epithet: After Baron A. von Schlippenbach, Russian naval officer and traveller.   Hardiness 3.   April–May.

The species is larger in all its parts than the rest of this Subsection.

After flowering, R. schlippenbachii *might be confused with the equally large-leaved* R. nipponicum, *but the leaves are alternate in that species.*

Bark corky.

Leaves in pseudowhorls of 4 or 5, often with a red-purple margin.

Flowers and leaves emerge from the same bud.

Corolla white.

Flowers open with the leaves.

New foliage of *R. quinquefolium* at Deer Dell

R. quinquefolium at Hergest Croft

Large broadly obovate leaves in pseudowhorls of five.

Corolla large, pink or white, sometimes slightly fragrant.

Petiole, pedicel, calyx margin, ovary and capsule glandular.

Leaves and flowers emerge from the same bud.

Flowers open with, or immediately before, the leaves.

R. schlippenbachii at Deer Dell

Pentanthera
(Sciadorhodion)

Subgenus Pentanthera (G.Don) Pojark
Section Viscidula Matsum. & Nakai
Nipponicum Subseries
A monotypic section

Bark peeling.
Leaves alternate.
Petiole subsessile.
Flowers and leaves from separate buds.
Corolla tubular campanulate with short lobes.
Seeds with tails at either end.

Japan only.
Recent molecular studies (Goetsch *et al*. 2005) have led to the suggestion that this species should be included in Section *Tsutsusi*.

R. nipponicum Matsum.

Shrub 1–1.8 m (3–6 ft) high, **bark** shredding and exposing the polished reddish-brown stem and branches; **young shoots** glandular, later glandular or eglandular.
Leaves deciduous, obovate; **lamina** 5–17.5 cm long, 4–8.8 cm broad; **upper surface** bright green, adpressed-bristly; margin bristly; **underside** adpressed-bristly.
Petiole subsessile (scarcely stalked), leaf-base decurrent on the petiole.
Inflorescence 6–15 flowers, flowers from terminal buds, the leafy shoots from separate axillary (lateral) buds below, flowers opening with or after the leaves.
Pedicel pendulous.
Corolla tubular-campanulate, 1.5–2.4 cm long, 0.8–1 cm broad, white, or tube white and greenish, lobes flushed pink.
Stamens 10.
Central Japan. 900 m (3,000 ft).
Epithet: From Japan.   Hardiness 3.   May–June.

The tubular or tubular-campanulate corollas are unique among azaleas and are similar to those of the genus Menziesia.

Large rugose leaves, superficially similar to those of *R. schlippenbachii*, but alternate.

Petiole subsessile (scarcely stalked).

Good autumn colour.

Small inflorescence often hidden beneath the leaves.

Leaves obovate, underside adpressed-bristly.

Inflorescence 6–15 flowers.

Flowers white or greenish-white, from terminal buds, pedicel pendulous.

Leafy shoots from separate axillary (lateral) buds below the flowers.

R. nipponicum (S. Hootman)

R. nipponicum (B. Starling)

Subgenus Tsutsusi (Sweet) Pojark

Leaf and flower buds from within the same bud scales.
Indumentum of simple hairs or bristles, sometimes ribbon-like,
    sometimes stiff and glandular.
Inflorescence terminal.
Ovary strigose or glandular.
Seeds without tails.

Section Brachycalyx Sweet
Schlippenbachii Subseries

Leaves, usually rhombic, in pseudowhorls of 2 or almost always 3,
    always deciduous.
Flowers appear before or with the leaves.
Pedicel densely pilose.
Corolla funnel-shaped or funnel campanulate.
Calyx minute, often pilose.
Stamens almost always 10.

Japan and SE Asia.

*The taxonomy of this Section probably needs revision, with some species
relegated to subspecific or varietal status. In particular, plants labelled
R. reticulatum in cultivation may be one of a number of species, subspecies
or forms.*

R. amagianum Makino

Shrub or small **tree**, up to 5 m (16 ft) high; **young shoots** rather densely rufous
    or white-pubescent, later becoming glabrous.
Leaves deciduous, in pseudowhorls of 3 or 4 at the end of the branchlets,
    broadly ovate or rhombic; **lamina** 4–8.4 cm long, 3–9 cm broad; **upper
    surface** dark green, lustrous, hairy; margin hairy; **underside** pale green or pale
    glaucous-green, hairy or glabrous.
Inflorescence 3–4 flowers, flowers opening after or with the leaves.
Corolla rotate-funnel-shaped, 2.8–4 cm long, 3.5–5 cm across, soft shade of
    scarlet or brick-red, with darker spots.
Hondo, Japan. 100–1,000 m (300–3,300 ft).
AM 1948 (Bodnant), flowers pale scarlet, with red spots.
Epithet: From Mount Amagi, Japan.   Hardiness 3.   June–July.

Very closely allied to R. weyrichii.

Composite of leaves of *R. farrerae* – autumn colour in the centre (Rhododendron Species Foundation). Bar = 1 cm

Large lustrous rhombic leaves, usually in threes.

Pedicel relatively long.

Corolla more reddish than its relatives.

Shoots rather densely rufous or, usually, white pubescent.

Leaves in whorls of 3 or 4 at the end of the branchlets.

Flowers opening after or with the leaves.

Corolla soft shade of scarlet or brick-red.

***R. amagianum* at Deer Dell**

R. farrerae Tate

Shrub 0.6–3 m (2–10 ft) high; **branchlets** short, shining, brown.

Leaves deciduous or semi-evergreen, usually in whorls of 3 at the end of the branchlets, ovate; **lamina** 2.1–3.5 cm long, 1.3–2.2 cm broad; **upper surface** dark green, often bristly; **underside** not bristly or sparsely bristly.

Petiole often densely strigose-setulose, eglandular.

Inflorescence 1–2 flowers, flowers opening before the leaves.

Pedicel densely villous.

Calyx densely pubescent.

Corolla rotate-funnel-shaped, 2.3–3.1 cm long, pale rose or deep rose, with red-purple spots.

Stamens 8–10.

Ovary densely adpressed-pilose.

Hong Kong, Guangdong and Guangxi provinces. About 600 m (2,000 ft).

Epithet: After the wife of Capt. Farrer, East India Co., 1829.   Hardiness 1–2. June.

Tender, and extremely rare in cultivation.

R. mariesii Hemsl. & E.H.Wilson

Shrub 1–6 m (3–20 ft) high; **young shoots** at first adpressed-pilose, later glabrous, eglandular.

Leaves deciduous, in whorls of 2 or 3 at the end of the branchlets, ovate, ovate-lanceolate or elliptic; **lamina** chartaceous, 3–7.5 cm long, 1.8–4.4 cm broad; **upper surface** dark green, glabrous; **underside** glabrous.

Inflorescence 1–2 flowers, occasionally up to 5 flowers, flowers opening before the leaves.

Calyx minute, 0.5–1 mm long, densely villous.

Corolla rotate-funnel-shaped, deeply cleft, 2.5–3.2 cm long, rose-purple, rose or pink, spotted with red-purple, outside glabrous.

Zhejiang, Anhui, Guangdong, Hunan, Guizhou, Sichuan, Taiwan. 200–1,300 m (650–4,250 ft).

Epithet: After C. Maries (c.1851–1902), collector for Messrs. Veitch.   Hardiness 3?   April–May.

Rarely found in cultivation.

**Small ovate
leathery leaves,
often semi-
evergreen.
Petiole short.**

Leaves usually in whorls
of 3 at the end of the
branchlets, underside
often sparsely bristly.

Inflorescence 1–2
flowers.

Pedicel densely villous.

Calyx pale or deep rose,
densely pubescent.

Flowers opening before
or with the leaves.

R. farrerae in Japan (Y. Kurashige)

**Hardier than
R. farrerae, to which
it is related.**

**It has larger and
chartaceous leaves
(ovate, ovate-
lanceolate or
elliptic) and a longer
petiole.**

Leaves usually in whorls
of 3 at the end of the
branchlets, underside
glabrous.

Inflorescence 1–2
flowers: corolla deeply
cleft.

Calyx densely villous.

Flowers pink, rose or
rose-purple, opening
before the leaves.

R. mariesii (Davidian)

Tsutsusi
(Brachycalyx)

R. reticulatum D.Don ex G.Don
(*R. rhombicum* Miq.)

Shrub or **tree**, 1–3 m (3–10 ft) high; **young shoots** densely adpressed-pilose, later glabrous, eglandular.

Leaves deciduous in whorls 2 or 3 at the end of the branchlets, broadly ovate or ovate or rhombic; **lamina** generally 3–7 cm long, 2–5 cm broad; **upper surface** dark green, later glabrous or almost glabrous; **underside** somewhat reticulate, pubescent or nearly glabrous.

Petiole pilose.

Inflorescence 1–2 rarely 3–4 flowers, flowers opening before the leaves.

Calyx minute, 0.5–1 mm long, pilose.

Corolla rotate-funnel-shaped, 2–2.8 cm long, 3.5–5 cm across, purple, rose-purple or reddish-purple, unspotted or with dark purple blotches.

Stamens 10.

Honshu, Kyushu, Shikoku in Japan. 300–1,000 m (650–3,300 ft).

Epithet: Netted, referring to the venation.   Hardiness 3.   April–May.

Commonly cultivated.

Leaf apex acute; underside somewhat reticulate.

Flower buds ovate.

Flower bud scales pilose, scale margins densely ciliate.

Petiole pilose.

Ovary densely pilose.

Young shoots densely adpressed-pilose, later glabrous.

Leaves deciduous in whorls 2 or 3 at the end of the branchlets. Flowers opening before the leaves.

Corolla purple, rose-purple or reddish-purple, or magenta.

R. reticulatum (D. White)

The Japanese recognise numerous species that are related to R. reticulatum *from different Japanese regions. These appear to be somewhat confused, even in Japan. The enthusiast may wish to identify some of these, so brief notes of their differences from* R. reticulatum *are given below.* R. 'nudipes' *and* R. 'wadanum' *are certainly in cultivation.*

R. decandrum Makino

Petiole sparsely glandular; ovary glandular, corolla magenta, spotted.

S Honshu, Shikoku.

Epithet: With 10 stamens.

R. dilatatum Miq.

Stamens 5; pedicel and calyx densely of moderately glandular. Corolla purple or rose purple, spotted.

S Honshu.

Epithet: Spread out, referring to the flowers.

Because of its 5 stamens, **R. dilatatum** *may merit specific status.* **R. reticulatum f. pentandrum** *may belong here. The original plant of* **R. reticulatum var. albiflorum** *has 10 stamens, but it is unclear whether it usually has five. If 10, it is correctly named: if 5, it should be referred to as* **R. dilatatum f. leucanthum Sugimoto.**

R. hidakanum H.Hara

Young shoots glandular initially; greyish; leaf apex cuspidate. Petiole glandular; corolla magenta; calyx green; ovary glandular.

S Hokkaido.

R. kiyosumense Makino

Leaf underside glabrous except for the midrib; margins minutely denticulate; apex acuminate. Pedicel petiole and calyx eglandular. Flower bud glabrous. Corolla purple. Ovary densely bristly.

S Honshu.

R. mayebarae Nakai & H.Hara

Leaf underside glabrous except for the midrib; margins minutely denticulate. Pedicel, petiole and calyx eglandular. Pedicel with dense brown hairs. Flower bud densely pilose. Ovary bristly.

Kyushu.

R. nudipes Nakai

Leaf apex acute with a blunt tip; leaf surfaces with detersile long brown hairs; petiole glabrous. Stamens usually 8–10. Ovary densely pilose with long brown hairs.

S Honshu, Shikoku.

Epithet: Naked, referring to the petiole.

R. nudipes **in W. Berg's garden (M.L.A. Robinson)**

R. viscistylum Nakai

Leaf underside sometimes viscous; ovary glandular; style viscous; corolla reddish purple with spots.

Epithet: With a sticky style.

Kyushu.

R. wadanum Makino

Leaves rhombic; underside sparsely hairy; petiole densely pilose; corolla rich rose pink or white; ovary densely pilose; style with stalked glands for half its length.

S Honshu.

R. wadanum **in Japan (Y. Kurashige)**

R. sanctum Nakai

(*R. weyrichii* var. *sanctum* Hatus.)

Shrub 2.4–4.5 m (7–15 ft) high; **young shoots** rather densely rufous-pubescent, becoming glabrous later.

Leaves deciduous or semi-evergreen, in whorls of 2 or 3 at the end of the branchlets, broadly ovate or rhombic; **lamina** thinly leathery, 3.8–7.5 cm long, 3–7 cm broad; **upper surface** dark green, shining, hairy or glabrous; **underside** paler, glabrous.

Petiole densely rufose pilose.

Inflorescence 2–3 flowers, flowers opening after the leaves.

Pedicel densely rufous-pilose.

Calyx 0.5 mm long.

Corolla rotate-funnel-shaped, 3–3.6 cm long, 3–5.6 cm across, rose or strong purplish-pink, rarely white.

Hondo (Japan). 275–400 m (900–1,300 ft).

Epithet: Holy, found growing in the sacred area of the Great Shrine of Ise. Hardiness 3.   May–June.

R. tashiroi Maxim.

Shrub or small **tree**; **young shoots** with adpressed flattened brown hairs, soon becoming glabrous.

Leaves monomorphic, persistent, coriaceous in pairs or in threes at the end of the branchlets, lanceolate, oblanceolate or elliptic; **lamina** 3.5–7 cm long, 1.3–2.5 cm broad, apex subacute or acuminate, **both surfaces** at first densely adpressed grey-brown hairy, becoming glabrous above, shining green; **underside** finally glabrous except along the midrib.

Petiole densely adpressed brown-hairy.

Inflorescence 2–5 flowers.

Pedicel densely brown-strigose.

Calyx minute, 1 mm long, densely brown-strigose.

Corolla funnel-campanulate, 2.5–3.5 cm long, pale rose-purple, spotted maroon-purple.

Stamens 10, rarely 12.

S Japan, Liukiu and Kawanabe Islands, Yakushima. 200–500 m (650–1,650 ft).

Epithet: After a Mr. Tashiro, Japanese botanist.   Hardiness 3.   May.

Though, unfortunately, extremely rare in cultivation, this species is included for interest as it is intermediate between Sections Tsutsusi and Brachycalyx. It is evergreen but shares most of the characteristics of species in the latter Section: leaves in whorls of 2–3, pedicel, corolla shape, calyx and stamen number. Some authors have placed it in a separate Subsection.

Related to
R. weyrichii, but:-
the leaves are
lustrous; the corolla
is pink (rarely
white); the
flowering is a
month later.

Leaves deciduous or
semi-evergreen, broadly
ovate or rhombic, in
whorls of 2 or 3 at the
end of the branchlets.

Flowers opening after
the leaves.

Corolla rotate-funnel-
shaped.

R. sanctum at Deer Dell

Leaves
monomorphic,
evergreen,
coriaceous in pairs
of whorls of three
at the end of the
branchlets.

Pedicel densely
brown-strigose.

Corolla funnel-
campanulate, pale
rose-purple.

Calyx minute.

Stamens
normally 10.

R. tashiroi in Japan (Y. Kurashige)

Tsutsusi
(Brachycalyx)

R. weyrichii Maxim.

Shrub or **tree**, 1–5 m (3–16 ft) high; **young shoots** rather densely rufous-pubescent, later glabrous.

Leaves deciduous, in whorls of 2 or 3 at the end of the branchlets, ovate, nearly orbicular or rhombic-ovate; **lamina** chartaceous, 3.2–8 cm long, 2–6 cm broad; **upper surface** pale or dark green, in young leaves rufous-pubescent, soon glabrous; **underside** pale green, glabrous except midrib pubescent.

Petiole densely or moderately rufous-pilose.

Inflorescence 2–4 flowers, flowers opening before or with the leaves.

Pedicel densely or moderately rufous-pilose.

Corolla rotate-funnel-shaped, 3–4 cm long, 3.5–6 cm across, bright red, almost brick-red, spotted within the short tube.

Stamens 6–10.

Japan, Korean. Sea-level up to 1,000 m (3,300 ft).

Epithet: After Dr. Weyrich (1828–1863), a Russian naval surgeon.
Hardiness 2–3.   April–May.

Subgenus Tsutsusi (Sweet) Pojark.
Section Tsutsusi
Obtusum Subseries

Prostrate to occasionally large shrubs, 0.2–3 m (0.7–10 ft).

Leaves with hairs present at maturity, linear to broadly ovate: they are usually dimorphic, the spring leaves larger and deciduous, the summer leaves smaller and evergreen, though in some species the leaves appear to be of one kind and are persistent.

Flowers one or few from a terminal bud, from the lower axils of which the leaves also emerge.

Multiple buds may give rise to an inflorescence of up to 15 flowers. Corolla rotate to tubular campanulate, of all shades except yellow.

Ovary densely strigose, bristly or glandular.

Stamens usually 5–10.

This section represents the evergreen or 'Japanese' azaleas. 'The Rhododendron Handbook' records 40 species as being in cultivation in 1997, and many more have been described by the Chinese. The majority of these are not, however, hardy in northern Europe or across much of the USA. Thousands of hybrids have been developed as house plants (many from the tender R. simsii), and for the garden, to produce floriferous small shrubs with large flowers. In addition, extreme or especially good forms of the species have been selected for centuries, especially in Japan.

Typical forms of the 'pure' species are therefore often almost impossible to identify, and are seldom found, the hybrids predominating even in large collections.

Identification of the species growing in the midst of hybrids is a task that only a dedicated taxonomist would risk undertaking. With this in mind, only species that can be separated easily by the amateur or that are especially garden-worthy are listed here, along with two commonly grown and ancient hybrids.

Rounded near-rhombic leaves, matt, often with some persistent brown hairs.

Corolla bright red, almost brick-red.

Petiole rufous-pilose. Flowers opening before or with the leaves.

New foliage of *R. weyrichii* at Deer Dell

R. weyrichii 'Murasaki-on' at Deer Dell (foliage is *R. glaucophyllum*)

Section *Tsutsusi* (spring and summer leaves) (M.L.A. Robinson)

Tsutsusi
(Brachycalyx/Tsutsusi)

Section Tsutsusi
Species recognisable from the foliage alone

R. macrosepalum 'Linearifolium'
Easily identified from its long twisted linear leaves and mauvish petals. The type species is not as hardy and is rare in cultivation.

R. nakaharae Hayata
The only common truly prostrate creeping species in this section; leaves small, monomorphic, with prominent hairs; flowers late, brick red or rose red, with 6–10 stamens.

R. serpyllifolium Miq.
Very thin branchlets with adpressed hairs; tiny monomorphic leaves which are largely deciduous and glabrous beneath except on the midrib; inflorescence single-flowered; pink or white.

R. taiwanalpinum Ohwi
Leaf coriaceous; rugulose very dark leaves, upper surface hairy; matt, ovate-lanceolate, margin recurved, underside densely reddish brown and hairy; the covering of hairs giving a sheen; leaf margin hairy. Pink flowers. Rare.

Taxa requiring flowers for identification

Stamens 10

R. × 'Mucronatum'
The branchlets densely bristly with spreading bristles. Corolla large, usually white (sometimes rose pink); distinctive long, thinly lanceolate, calyx lobes.

R. yedoense Maxim. var. poukhanense (H.Lev.) Nakai
Almost deciduous. Summer leaf upper side almost glabrous at maturity. Flowers single, pale lilac purple, slightly fragrant.

Stamens 5
(Branches strigose; bud scales not viscid; corolla funnel form; leaves pilose beneath)

R. indicum Sweet Leaves lanceolate; flowers after new leaves, red.

R. kaempferi Planch Tall open habit; flowers before or with new leaves; flowers reddish orange, pink to purple or white.

R. kiusianum Low spreading habit; flowers light to reddish purple, occasionally white.

R. × 'Obtusum' Low twiggy shrub; leaves dimorphic; spring leaves chartaceous; flowers magenta to rosy mauve, crimson, purple red or scarlet.

R. macrosepalum 'Linearifolium' at the Elizabeth Miller garden, Oregon (M.L.A. Robinson)

R. taiwanalpinum at Deer Dell

R. indicum (L.) Sweet

(*R. balsaminiflorum* Carriere)

Shrub, densely branched, 0.5–1.8 m (2–6 ft) high, but usually low and sometimes prostrate; **branchlets** slender, rigid, densely strigose.

Leaves evergreen or semi-evergreen, dimorphic; **spring leaves** narrow-lanceolate, lanceolate or oblanceolate; **lamina** subcoriaceous, 2–4 cm long, 0.3–1.3 cm broad; **upper surface** dark green, glossy, red-brown strigose; margin crenulate, strigose; **underside** paler glaucescent, red-brown strigose; **summer leaves** 1–1.8 cm long, 3–5 mm broad, otherwise same as spring leaves.

Petiole densely brown-strigose.

Inflorescence 1–2 flowers.

Pedicel densely brown-strigose.

Corolla widely funnel-shaped, 3.5–5 cm long, 4.5–6.3 cm across, bright red or scarlet, sometimes rosy-red.

Stamens 5.

Honshu, Kyushu (Japan).

AM 1975 (Wisley), flowers scarlet.

Epithet: Indian.　　Hardiness 2–3.　　June–July.

Much used to produce late-flowering red hybrids.
　　　Cultivated forms, some of which may be hybrids, include:
　　　'Balsaminiflorum'. Flowers double, salmon-red.
　　　'Crispiflorum'. Flowers deep rose or red with wavy margins (frilled petals).
　　　'Laciniatum'. Corolla rich red; 5 strap-like widely separated petals, nearly
　　　　2.5 cm long.
　　　'Variegatum'. Flowers striped red and white.

R. kaempferi Planch.

Shrub 30 cm–3 m (1–10 ft) high; **young twigs** densely brown-strigose, eglandular.

Leaves evergreen or semi-evergreen, chartaceous, dimorphic; **spring leaves** lanceolate, elliptic or ovate, 2–5 cm long, 1–2.5 cm broad, mostly deciduous; **summer leaves** smaller, oblong-obovate, obovate or elliptic, mostly perisistent through the winter; in both types, **upper surface** is glossy green, **underside** is paler, brown-strigose on both surfaces and on the margin.

Inflorescence 2–4 flowers.

Pedicel densely brown-strigose.

Calyx brown-strigose.

Corolla funnel-shaped, 2.6–3.8 cm long, 2.5–4.3 cm across, salmon-red, brick-red or some shade of red.

Stamens 5.

Japan. Sea-level to about 1,600 m (5,250 ft).

AM 1953, FCC 1955 (Windsor) 'Eastern Fire', the flowers camellia rose, darker towards the tips.

Epithet: After E. Kaempfer (1651–1716), who wrote about Japanese plants. Hardiness 3.　　May–June.

There are many selected forms and a multitude of hybrids. Forms with pink, purple or white flowers are cultivated.

Leaves dimorphic, comparatively narrow and dark, margin crenulate. Corolla usually bright red or scarlet. Late flowering, after the new leaves.

Inflorescence 1–2 flowers.

R. indicum in Japan (Y. Kurashige)

Old foliage of a likely pure *R. indicum* at Quarry Hill, California (M.L.A. Robinson)

Tiered upright growth, leaves dimorphic, comparatively large, upper surface glossy green. Flowers almost always a shade of red.

R. kaempferi in the wild in Japan (Y. Kurashige)

Tsutsusi
(Tsutsusi)

R. kiusianum Makino

Shrub, low dense much-branched, often prostrate, sometimes broadly upright, 0.3–1 m (1–3 ft) high; **young shoots** densely brown-strigose.
Leaves evergreen or semi-evergreen, chartaceous, dimorphic; **spring leaves** broadly elliptic, ovate or obovate, small, 0.8–2 cm long, 0.5–1 cm broad; **summer leaves** smaller, obovate, oblanceolate, elliptic or ovate, the uppermost ones are persistent through the winter; in both types **upper surface** is glossy, dark green, **underside** paler, brown-strigose on both surfaces and on the margin.
Inflorescence 2–3 flowers.
Corolla funnel-shaped, small, 1.5–2 cm long, 1.3–2.5 cm across, rose-purple, purple, red or pink, sometimes white.
Stamens 5.
Japan. 600–1,600 m (1,950–5,250 ft).
AM 1977 (Benenden) to a clone 'Chidori', flowers white.
Epithet: From Kyusu, Japan.   Hardiness 3.   May–June.

Var. sataense *Nakai may be a hybrid with* R. kaempferi.

R. macrosepalum 'Linearifolium'

This is a garden form, a sport from *R. macrosepalum* with long, twisted, linear leaves and narrow, mauvish corolla lobes.
Epithet: Linear leaves.   Hardiness 3. April–May.

Photo: page 431.

R. × 'Mucronatum'

Shrub, 1–1.8 m (3–6 ft), or sometimes 3 m (10 ft) high; **branchlets** densely bristly with spreading bristles, and adpressed-bristly, often glandular.
Leaves partly evergreen, dimorphic: **spring leaves** deciduous, rather thin, light green, lanceolate, oblong-lanceolate or oblong, 3–6 cm long, 1–2.5 cm broad, both surfaces rather densely adpressed-hairy; **summer leaves** evergreen, smaller, thicker, dark green, oblanceolate or lanceolate, 1–4 cm long, 0.5–1.2 cm broad, **both surfaces** rather densely adpressed-hairy.
Inflorescence 1–3 flowers.
Pedicel densely bristly, bristles spreading and adpressed-bristly.
Calyx 1–1.6 cm long, glandular.
Corolla widely funnel-shaped, large, 3.8–5 cm long, pure white, fragrant.
Epithet: Pointed, referring to the calyx lobes.   Hardiness 3.   May.

The fragrance is slight to most people. Still found in gardens under the old name of 'Azalea ledifolia'.

A low dense much-branched, often prostrate shrub. Comparatively small leaves, almost deciduous.

Leaves dimorphic usually semi-evergreen, chartaceous, upper surface glossy. Inflorescence 2–3 flowers.

Corolla purple to red, pink, or sometimes white.

R. kiusianum (Davidian)

R. kiusianum lilac form at Deer Dell

Leaves dimorphic: spring leaves deciduous, light green; summer leaves evergreen, smaller, thicker, darker.

The long glandular calyx lobes are distinctive.

Corolla pure white, slightly fragrant.

R. × '*Mucronatum*' at Nymans (E. Daniel)

Tsutsusi
(Tsutsusi)

R. nakaharae Hayata

Low or prostrate twiggy **shrub**, 10–30 cm (0.3–1 ft) high, rarely more, shoots
  densely strigose, eglandular.
Leaves monomorphic, persistent, chartaceous, lanceolate, elliptic or elliptic-obovate;
  lamina somewhat thick, small, 0.3–1.2 cm long, 2–5 mm broad; **upper surface**
  dark green, strigose; margin bristly; **underside** pale green, strigose.
Petiole densely brown-strigose.
Inflorescence 1–3 flowers.
Pedicel densely adpressed-bristly.
Calyx outside densely strigose.
Corolla funnel-shaped, 2 cm long, dark red or scarlet.
Taiwan. 2,000–2,300 m (6,557–7,541 ft).
AM in 1970 (Hydon) 'Mariko', flowers red.
Epithet: After G. Nakahara, Japanese collector.   Hardiness 3.   June–August.

'Mount Seven Star' is a slightly stronger growing form with bright red flowers.

R. × 'Obtusum'

Low **shrub** up to 1 m (3 ft) high, rarely more, sometimes nearly prostrate, twiggy;
  branchlets densely brown-strigose.
Leaves partly evergreen, dimorphic; **spring leaves** (which are formed immediately
  after the flowers open) are deciduous, chartaceous, lanceolate, elliptic-lanceolate
  to ovate, bright green above, pale beneath, **both surfaces** strigose; **summer
  leaves** evergreen, fairly coriaceous, smaller, obovate or oblong-obovate, **both
  surfaces** strigose.
Inflorescence 1–3 flowers.
Pedicel densely strigose.
Calyx 1–3 mm long.
Corolla funnel-shaped, 2.4–2.8 cm long, bright red, scarlet, crimson, salmon or
  salmon-red.
Japan.
AM 1898 (Basing Park, Alton), flowers orange-scarlet.
Epithet: Blunt.   Hardiness 3.   May–June.

*Small flowers compared to most modern hybrids, about the size of those in the
'Wilson 50', and with small foliage to match. Very floriferous.*

Clones of *R.* × *'Obtusum'*
'Ramentaceum' (*Obtusum album*). Flowers white, spring leaves broadly
elliptic.
'Amoenum' (*Azalea amoena*). Taller; 'hose-in-hose' flowers, rosy-purple or
rich crimson-purple. *Still one of the most widely grown azaleas in the UK.*
'Amoenum coccineum'. A branch-sport of 'Amoenum' with light
carmine-red, 'hose-in-hose' flowers.
'Splendens'. Rose flowers. AM 1965 (Knaphill).

Markedly prostrate twiggy growth – more so than any of its hybrids so far.

Corolla dark red or scarlet.

Late flowering.

Leaves monomorphic, small; underside pale green, strigose.

Petiole densely brown-strigose.

Inflorescence 1–3 flowers.

Corolla dark red or scarlet.

R. nakaharae (Davidian)

Leaves dimorphic: small spring leaves deciduous, chartaceous and formed immediately after the flowers open; summer leaves evergreen, coriaceous.

Corolla fairly small, bright red, scarlet, crimson, salmon or salmon-red.

R. × *'Obtusum'* 'Natsugiri' at RBG Edinburgh (Davidian)

R. × *'Obtusum'* 'Amoenum' (B. Starling)

R. serpyllifolium (A.Gray) Miq.

Shrub, low and much-branched, 60–90 cm (2–3 ft) high; **shoots** slender, densely brown-strigose.

Leaves deciduous, thin, mostly crowded at the end of short branchlets, monomorphic, obovate, oblong-obovate, oblong or elliptic, very small, 0.3–1.6 cm, usually 0.3–1.1 cm long, 2–8 mm broad; **upper surface** bright green, brown-strigose; margin strigose; **underside** pale green, glabrous or brown-strigose.

Inflorescence 1 or rarely 2 flowers.

Corolla funnel-shaped, small, 1–1.2 cm long, 1.1–1.8 cm broad, pale rose, rosy-pink or bright rose-red.

Stamens 5.

Central and S Japan. 150–800 m (450–2,600 ft).

Epithet: With leaves like *Thymus serpyllum*.   Hardiness 2–3.   April–May.

A delightful form with white flowers is in cultivation.

R. taiwanalpinum Ohwi

Low **shrub**.

Leaves monomorphic, semi-evergreen, coriaceous, ovate-lanceolate, 2–3 cm long; **upper surface** densely hairy, margin revolute; **underside** densely reddish-brown hairy.

Inflorescence 1–3 flowers.

Pedicel short.

Corolla widely funnel-shaped, 2 cm long, red.

Stamens 9–10, longer than corolla.

N Taiwan. 2,800–3,000 m (9,000–9,800 ft).

Epithet: Alpine from Taiwan.   Hardiness 3.   April–May.

Rare in cultivation.

Note the epithet.
Leaves very small,
monomorphic,
largely deciduous, on
thin twigs.
Small flowers.
Inflorescence 1 or rarely
2 flowers.
Corolla pale rose, rosy-
pink or bright rose-red.

R. serpyllifolium at Deer Dell

Leaves
monomorphic,
semi-evergreen,
coriaceous, hairs on
the underside with a
silky-sheen-like
appearance.
Pedicel short.
**Corolla widely
funnel-shaped, red.**

R. taiwanalpinum at Deer Dell

Tsutsusi
(Tsutsusi)

R. yedoense Maxim.

The name **R. yedoense** was given by Maximowicz in 1886 to a double-flowered garden (cultivated) form of an azalea that grows wild in Korea. It is unfortunate that the latter, the single-flowered **var. poukhanense**, is treated as a variety of a double-flowered cultivated form.

The double-flowered azalea is also known as 'Yodogawa'.

AM 1961 (Benenden), flowers mauve with darker spots.

R. yedoense var. poukhanense (H.Lev.) Nakai

Shrub 0.6–1.8 m (2–6 ft) high, sometimes low or prostrate; **shoots** densely brown or grey-strigose, eglandular.

Leaves dimorphic, deciduous or semi-deciduous; **spring leaves** chartaceous, oblanceolate, lanceolate, oblong-lanceolate or ovate-lanceolate; **lamina** 2.8–7.8 cm long, 0.5–2 cm broad; **upper surface** dark green, grey or brown strigose, margin strigose, **underside** paler, grey or brown-strigose; **summer leaves** thicker, oblanceolate or lanceolate, smaller, 0.4–1 cm long; **upper surface** soon glabrescent, otherwise same as spring leaves.

Petiole densely brown-strigose.

Inflorescence 2–4 flowers, flowers expanding with or shortly before the leaves.

Pedicel brown-strigose.

Calyx outside brown-strigose.

Corolla widely funnel-shaped, large, 3.2–4.2 cm long, 3.2–5 cm across, rose, rosy-purple or pale lilac-purple, with purple-brown spots, fragrant.

Korea, and Tsushima (Japan). Sea-level up to 1,800 m (5,900 ft).

Epithet: From Mt Poukhan-san, Korea.   Hardiness 3.   May.

Double flowers.

R. yedoense at Trewidden (Davidian)

Leaves dimorphic, almost deciduous. Summer leaf upper side almost glabrous at maturity.

Calyx and ovary eglandular.

Corolla rosy-purple or pale lilac-purple, with purple-brown spots, somewhat fragrant.

Flowers expanding with or shortly before the leaves.

R. yedoense var. *poukhanense* in the Valley Gardens

Subgenus Candidastrum
Albiflorum Series
A monotypic subgenus.

Recent molecular studies (Goetsch et al. 2005) have led to the suggestion that this species should be included in Section Sciadorhodion.

R. albiflorum Hook.f.

Shrub broadly upright, 0.75–1.8 m (2.5–6 ft) high; **branchlets** moderately or rather densely hairy with long, adpressed brown hairs.

Leaves deciduous, oblong, oblong-elliptic, oblanceolate, lanceolate or obovate; **lamina** thin, chartaceous, 3.2–7.5 cm long, 1.3–2.5 cm broad; **upper surface** bright green, adpressed-hairy.

Petiole with long, adpressed brown hairs.

Inflorescence 1–2 flowers from axillary buds, flowers appearing after the leaves develop, pendulous.

Corolla rotate-campanulate or campanulate, 1.3–2 cm long, white, creamy-white or greenish-white.

Rocky Mountains from British Columbia and Alberta to Washington, Oregon and Colorado. 1,350–2,200 m (4,400–7,200 ft).

Epithet: With white flowers.   Hardiness 3.   June–July.

Extremely rare in cultivation: difficult to grow. Hard to recognise as a rhododendron.

Flowers not terminal, appearing after the leaves develop, pendulous all down the stems.

Branchlets hairy with long, adpressed brown hairs.

Leaf upper surface adpressed-hairy.

Inflorescence from axillary buds.

R. albiflorum (B. Starling)

Subgenus Mumeazalea
Semibarbatum Series

Recent molecular studies (Goetsch et al. 2005) have led to the suggestion that this species should be included in Section Tsutsusi.

R. semibarbatum Maxim.

Shrub 0.6–3 m (2–10 ft) high; **branches** slender; **branchlets** pubescent, glandular with long-stalked glands.

Leaves deciduous, elliptic or ovate; **lamina** thin, chartaceous, 2–5 cm long, 1–2.5 cm broad; **upper surface** dark green, glabrous except midrib pubescent; margin crenulate; **underside** pale green, setulose.

Petiole densely pubescent, setulose-glandular.

Inflorescence 1 flower, axillary from lateral buds crowded at the ends of the branchlets, flowers appear after the unfolding of the leaves.

Pedicel setulose-glandular.

Calyx margin setulose-glandular.

Corolla rotate, with a short tube and spreading lobes, 1.5–2 cm across, white, white flushed pink, or yellowish-white, with red spots.

Stamens 5, very unequal.

Honshu, Shikoku and Kyushu (Japan).

Epithet: Partially bearded.   Hardiness 3.   June.

Extremely rare in cultivation.

More like a *Menziesia* species than a *Rhododendron*.

Flowers insignificant, whitish, appearing after the leaves.

Stamens 5, very unequal.

Branchlets pubescent, glandular.

Leaves deciduous, chartaceous, margin crenulate.

Petiole pubescent, glandular.

Inflorescence with 1 flower, from axillary buds.

Pedicel setulose glandular.

Calyx margin setulose glandular.

R. semibarbatum (S. Hootman)

Leaves of *R. semibarbatum* at Deer Dell showing the crenulate margins

Mumeazalea

Subgenus Therorhodion (Maxim.) A.Gray
Camtschaticum Series

**A small Subsection of low or prostrate elepidote shrubs.
Leaves emerge from separate buds below the flowers.**

*The only species of this subgenus in general cultivation is
R. camtschaticum itself, which is rare, being extremely hard to
please in climates warmer than the south of Scotland.*
Pedicel a leaf-like bract.

R. camtschaticum Pall.
(Rhodothamnus camschaticus [Pall.] Lindl.)

Shrub, low-growing or prostrate or a cushion, 10–30 cm (0.3–1 ft) high, and
spreads by means of underground suckers; young **branchlets** hairy with very
long hairs, glandular or eglandular.
Leaves deciduous, short-petioled or sessile, obovate or spathulate-obovate;
lamina thin, chartaceous, 1.6–5 cm long, 0.8–1.8 cm broad; margin crenulate,
bristly; **underside** bristly.
Inflorescence 1–2 flowers produced at the ends of the young leaf-like bracts.
Corolla rotate, 1.8–2.5 cm long, tube split to the base, carmine-purple, reddish-
purple or pink, spotted, outside pubescent.
Stamens 10.
From Kamtschatka, the Aleutian Islands, Alaska, southward to Saghalien, the
Kurile Islands and N Japan.
AM 1908 (Reuthe), as *Rhodothamnus kamtschaticus*, flowers purplish-red.
Epithet: From Kamtschatka.   Hardiness 3.   May–June.

R. camtschaticum var. *albiflorum* Koidzumi
The variety differs from the species type in its white flowers.
Japan; Kodiak Island, Alaska.
Epithet: With white flowers.   Hardiness 3.   May–June.

*Work subsequent to Davidian's description has shown that, although the
inflorescence appears to be on the end of a leafy shoot, this is in fact the pedicel,
which is a **leaf-like bract** and not a true leaf. The flowers are therefore borne on
a terminal bud as with most rhododendrons. Where grown in the UK, the plants
appear to be deciduous.*

Leaves deciduous, short-petioled or sessile.

Inflorescence, terminal, 1–2 flowers, produced at the ends of the young leaf-like bract.

Low-growing or prostrate or cushion shrub.

Corolla rotate, tube split to the base, carmine-purple, reddish-purple or pink.

R. camtschaticum at Deer Dell

R. camtschaticum (Davidian)

Subgenus *Rhododendron*
Section *Pogonanthum* G.Don
Anthopogon Series

Shrubs, usually dwarf.
Leaf usually small, aromatic; underside densely scaly with overlapping lacerate scales.
Inflorescence dense and has many flowers.
Flower-bud scales always fringed with large dendroid hairs.
Corolla narrowly tubular with spreading lobes, similar to a daphne species.
Stamens 5–10, included in the corolla tube.

R. anthopogon ssp. anthopogon var. anthopogon

R. anthopogon D.Don

Shrub prostrate, rounded compact, spreading or upright, 0.15–1.5 m (0.5–5 ft) high; **branchlets** densely scaly; **leaf-bud scales** often deciduous.
Leaves orbicular, oval, elliptic or oblong; aromatic; **lamina** 1.3–3.8 cm long, 0.8–2.5 cm broad; **upper surface** dark green or pale green; **underside** densely covered with overlapping lacerate scales.
Inflorescence capitate, 5–9 flowers.
Corolla narrowly tubular with spreading lobes, 1.2–1.9 cm long, pink deep rose, reddish or rarely crimson.

Himalayan range extending from Kashmir, Nepal, Sikkim, Bhutan to Assam and Tibet. 2,750–5,050 m (9,000–16,500 ft).
AM 1955 (Nymans), flowers pink.
AM 1969 (Glendoick) 'Betty Graham', flowers deep pink.
Epithet: With bearded flowers, referring to hairs in floret-tube.   Hardiness 3. April–May.

Habit very variable. P.A. and K.N.E Cox report a population that has uniformly yellow flowers and deciduous leaf-bud scales in Nepal.

R. anthopogon ssp. anthopogon var. album Davidian

Compact or broadly upright **shrub**, 30–90 cm (1–3 ft) high.
Nepal, Bhutan, Tibet. 3,150–4,600 m (10,500–15,000 ft).
Epithet: With white flowers.   Hardiness 3.   April–May.

Foliage of Section Pogonanthum at Deer Dell. Top row
(left to right, 2 leaves each): *R. anthopogon* 'Betty Graham'
AM and *R. anthopogon* ssp. *hypenanthum*. Bottom row (left
to right, 2 leaves each): *R. primuliflorum* 'Doker-La',
R. sargentianum and *R. trichostomum* var. *radinum*. Bar = 2 cm

**Leaf-bud scales
often deciduous.
Leaf underside
cinnamon or rust-
coloured.**

Corolla pink, deep rose
or reddish.

R. anthopogon ssp. *anthopogon* var. *anthopogon* L&S1091 at
RBG Edinburgh (Davidian)

R. anthopogon ssp. hypenanthum Cullen

R. hypenanthum Balf.f.

Shrub 15–90 cm (0.5–3 ft) high; **branchlets** densely scaly, **leaf-bud scales** persistent.
Leaves oval, oblong-oval, elliptic, oblong or orbicular; aromatic; **lamina** 1–4.2 cm long, 0.8–2 cm broad; **upper surface** green, not scaly or sometimes scaly; **underside** densely covered with reddish-brown, cinnamon, rust-coloured or dark brown overlapping lacerate scales.
Inflorescence capitate, 5–10 flowers.
Corolla narrowly tubular with spreading lobes, 1.1–1.9 cm long, yellow, pale creamy-yellow or lemon-green.
Stamens 5–8, included in the corolla-tube.
Style very short, as long as the ovary or slightly longer.
Kashmir, Punjab, Nepal, Sikkim, Bhutan. 3,150–4,250 m (10,500–14,000 ft).
AM 1974 (Glendoick) 'Annapurna', flowers yellow (SS&W9090).
Epithet: Bearded flowers.   Hardiness 3.   April–May.

R. anthopogonoides Maxim.

Inflorescence capitate, 10–20 flowers: very dense, with short-lobed corollas.
Corolla yellow, greenish-yellow, whitish-pink or white.
Pedicel and **calyx** not scaly.

Extremely rare, and almost certainly not introduced into cultivation before 2000, older plants being incorrectly labelled. Said to be fragrant.

R. cephalanthum ssp. cephalanthum

R. cephalanthum Franch.

Shrub 0.05–1.5 m (0.16–5 ft) high; **branchlets** densely scaly, **leaf-bud scales** persistent.
Leaves oblong-elliptic, elliptic, oblong, oblong-oval or oval; aromatic; **lamina** 1–3.5 cm long, 0.5–1.6 cm broad; **upper surface** dark green, shiny, not scaly; **underside** densely covered with fawn reddish or brown overlapping lacerate scales.
Inflorescence capitate, 5–10 flowers.
Corolla narrowly tubular with spreading lobes, white, rose, deep rose, pink or rose-crimson; outside not scaly.
Stamens 5–8, included in the corolla-tube.
Mid-W and NW Yunnan, SW Sichuan, E and SE Tibet, Upper Burma, Assam. 2,600–4,600 m (8,500–15,000 ft).
AM 1934 (Exbury), flowers white, tinged yellow.
AM 1979 (Mrs. K. Dryden, Sawbridgeworth), Winifred Murray.
Epithet: With flowers in a head.   Hardiness 3.   April–May.

R. cephalanthum ssp. *platyphyllum* has wider leaves and white flowers.

Similar to
R. anthopogon,
but with persistent
leaf-bud scales.

Leaf underside
densely covered
with reddish-brown
scales.

Corolla yellow, pale
creamy-yellow or
lemon-green.

R. anthopogon ssp. *hypenanthum* at RBG Edinburgh (Davidian)

R. cephalanthum ssp. *cephalanthum* at Deer Dell

Leaf-bud scales
persistent,
conspicuous.

Under tier of scales
golden yellow,
contrasting with the
darker top layer.

Stamens usually 5.

Corolla narrowly tubular
with spreading lobes,
white, rose, deep rose,
pink or rose-crimson.

R. cephalanthum ssp. *platyphyllum* at Glendoick (M.L.A. Robinson)

Rhododendron
(Pogonanthum)

R. cephalanthum ssp. cephalanthum
Crebreflorum Group
R. crebreflorum Hutch. & Ward

Shrub prostrate and spreading or very compact, 5–25 cm (2–10 inches) high; **branchlets** densely or moderately scaly; **leaf-bud scales** persistent.

Leaves broadly elliptic, oval or oblong-elliptic; aromatic; **lamina** 1.3–2.5 cm long, 0.6–1.7 cm broad; **upper surface** dark green, shiny; **underside** densely covered with dark brown or reddish-brown, flaky, overlapping lacerate scales.

Inflorescence capitate, 5–12 flowers.

Corolla narrowly tubular with spreading lobes, 1.6–2 cm long, pale pink, white tinged pink, deep rose or reddish.

Stamens 6, inducted in the corolla-tube.

Style very short, straight, shorter than or as long as the ovary.

Upper Burma, Assam. 3,950 m (13,000 ft).

AM1934 (Nymans).

Epithet: Densely flowers.   Hardiness 3.   April–May.

R. cephalanthum ssp. cephalanthum Nmaiense Group
R. nmaiense Balf.f. & Kingdon Ward

Submerged into subspecies cephalanthum, *from which it differs mainly in the yellow flower colour. Not so tall growing.*

R. collettianum Aitch. & Hemsl.

Shrub upright or lax and somewhat rounded, up to 1 m (3 ft) high; **branchlets** densely scaly, **leaf-bud scales** deciduous.

Leaves lanceolate or oblong-lanceolate, aromatic; **lamina** 3–5.6 cm long, 1.1–2.3 cm broad; **upper surface** bright green; **underside** densely covered with creamy-yellow, overlapping, lacerate scales.

Inflorescence capitate, 5–12 flowers.

Pedicel 3–4 mm long.

Corolla narrowly tubular-funnel shaped, 1.5–2.4 cm long, white or white-tinged rose, outside not scaly.

Stamens 10, included in the corolla tube.

Afghanistan, Pakistan. 3,050–4,000 m (10,000–13,000 ft).

Epithet: After General Sir Henry Collett (1836–1901).   Hardiness 3.
April–May.

Very similar to
R. cephalanthum, but
prostrate, spreading or
very compact and may
be stoloniferous.
Corolla white tinged
pink, pink, deep rose or
reddish.
Stamens 6.

R. cephalanthum ssp. *cephalanthum* Crebreflorum Group at Eckford

**Leaf-bud scales
deciduous.
Leaf larger and
more lanceolate
than those of
R. anthopogon.
Scales on leaf
underside creamy
yellow.
Stamens 10.**

Corolla white or white
tinged rose.

R. collettianum Wendlebo 8975 in R. White's garden, Sarum, Surrey

R. kongboense Hutch.

Shrub 0.15–2.4 m (0.5–8 ft) high; **branchlets** densely scaly, **leaf-bud scales** deciduous.

Leaves oblong, oblong-lanceolate or oblong-oval; aromatic; **lamina** 0.9–2.8 cm long, 0.4–1.1 cm broad; **upper surface** dark green, matt, densely scaly; **underside** densely covered with cinnamon, rust-coloured or brown, overlapping lacerate scales.

Inflorescence capitate, 6–12 flowers.

Calyx 3–5 mm long, outside densely scaly down the middle of the lobes.

Corolla narrowly tubular with spreading lobes, 0.8–1.1 cm long, rose, pink, strawberry-red or deep red; tube densely hairy outside.

Stamens 5, included in the corolla-tube.

Style very short, shorter or longer than the ovary.

Tibet–Bhutan Border, SE Tibet. 3,150–4,600 m (10,500–15,000 ft).

Epithet: From Kongbo, SE Tibet.   Hardiness 3.   April–May.

R. laudandum var. laudandum

R. laudandum Cowan

Leaf bud scales deciduous.

Leaves oblong-oval, oval, elliptic or oblong; aromatic; **lamina** 0.9–1.8 cm long, 0.4–1 cm broad; **upper surface** pale green or dark green (in young leaves, densely scaly); **underside** densely covered with dark reddish-brown, chocolate or cinnamon-coloured, overlapping lacerate scales.

Corolla pale pink, pink, white or rarely creamy-yellow, outside not scaly, tube densely or moderately hairy.

S Tibet, Bhutan. 4,250–4,575 m (14,000–15,000 ft).

Epithet: Praiseworthy.

Field studies of this taxon permit the suspicion that it is a form of the variable R. primuliflorum. *Probably not in cultivation.*

R. laudandum var. temoense Kingdon Ward ex Cowan & Davidian

Leaf-bud scales usually persistent.

Corolla usually white.

Ovary glabrous, densely scaly.

Tibet–Bhutan border. SE Tibet. 2,900–4,750 m (9,500–15,000 ft).

Epithet: From Temo La, S Tibet.   Hardiness 3.   April–May.

Similar to
R. primuliflorum,
and hard to identify
out of flower, but
usually upright.
Leaves very
aromatic.
Calyx large.
Corolla short, rose,
pink, strawberry-red
or deep red, tube
densely hairy
outside.
Stamens 5.

Leaf of *R. kongboense* (J.C. Birck). Bar = 1 cm

R. kongboense at Deer Dell

Leaf bud scales
deciduous.
Leaf round;
underside dark
chocolate brown
or cinnamon.
Corolla usually
pink, lavender
forms being
hybrids.
Ovary densely
hairy over at least
the lower half.

R. laudandum var. *temoense* (J.C. Birck)

Rhododendron
(Pogonanthum)

R. primuliflorum Bur. & Franch.
(*R. cephalanthoides* Balf.f. & W.W.Sm)

Shrub 0.05–1.8 m (0.16–6 ft) or rarely 2.5–2.75 m (8–9 ft) high; **branchlets** densely scaly; **leaf-bud scales** deciduous.

Leaves oblong, elliptic, oblong-elliptic or oblong-lanceolate; aromatic; **lamina** 0.8–3.4 cm long, 0.5–1.5 cm broad; **upper surface** dark green, glabrous; **underside** densely covered with brown or rust-coloured, overlapping, lacerate scales.

Inflorescence capitate, 5–10 flowers.

Corolla narrowly tubular with spreading lobes, 1–1.6 cm long, white, pink, creamy-yellow or yellow.

Stamens 5, included in the corolla-tube.

Style very short, as long as the ovary or shorter.

E Siberia, SE and E Tibet, Gansu, SW Sichuan, NW and mid-W Yunnan. 3,050–5,200 m (10,000–17,000 ft).

AM 1980 (Glendoick) 'Doker-La'.

Epithet: Primrose flowers.   Hardiness 3.   April–May.

Flowers a month earlier than R. trichostomum. R. fragrans *is similar, having yellowish scales on the leaf underside and pink flowers, but is almost never found in cultivation.*

R. primuliflorum Bur. & Franch.
R. primuliflorum var. cephalanthoides Cowan & Davidian

Corolla tube outside moderately or densely puberulous.

NW Yunnan, SE Tibet. 3,660 m (12,000 ft).

Epithet: Like *R. cephalanthum.*   Hardiness 3.   April–May.

Submerged into R. primuliflorum.

R. sargentianum Rehd. & E.H.Wilson

Compact **shrub** 30–60 cm (1–2 ft) high; **branchlets** short, twiggy, densely scaly; **leaf-bud scales** persistent.

Leaves broadly elliptic, oblong-elliptic or oval; aromatic; **lamina** 0.8–2 cm long, 0.5–1.1 cm broad; **upper surface** dark green, shiny, not scaly; **underside** densely scaly, scales lacerate, overlapping, rust-coloured or brown.

Inflorescence capitate, 5–7 or rarely 12 flowers.

Corolla narrowly tubular with spreading lobes, 1.2–1.6 cm long, lemon-yellow, pale yellow or white, outside scaly.

Stamens 5.

W Sichuan. 3,000–4,300 m (9,800–14,100 ft).

AM 1923 (Bodnant), flowers pale yellow.

AM 1966 (Glendoick) 'Whitebait', flowers nearly white.

Epithet: After G.S. Sargent, Director of the Arnold Arboretum, Massachusetts. Hardiness 3.   April–May.

Relatively common in cultivation.

Very similar to
R. cephalanthum,
except for the
stamens (5) and
leaf-bud scales
(deciduous).

Scales on the leaf
underside are brown
or rust-coloured.
Corolla white, pink,
creamy-yellow
or yellow.

Bush of *R. primuliflorum* at Glendoick

R. primuliflorum 'Doker-La' AM at Deer Dell

Dwarf dense habit.
Leaf-bud scales
persistent.
Corolla shades of
yellow, and densely
scaly outside.
Stamens 5.

Scales on the leaf
underside brown or
rust-coloured.

R. sargentianum (J.C. Birck)

Rhododendron
(Pogonanthum)

R. trichostomum Franch.

(*R. ledoides* Balf.f. & W.W.Sm., *R. radinum* Balf.f. & W.W.Sm., *R. sphaeranthum* Balf.f. & W.W.Sm.)

Shrub 0.2–1.2 m (0.6–4 ft) or rarely 2 m (6 ft) high; **branchlets** densely scaly, **leaf-bud scales** deciduous.

Leaves narrowly oblanceolate, linear or linear-lanceolate; aromatic; **lamina** 0.8–3 cm long, 2–8 mm broad; **upper surface** pale green or dark green; matt, scaly or not scaly; margin recurved; **underside** densely covered with overlapping lacerate scales.

Inflorescence capitate, globose, 8–20 flowers.

Corolla narrowly tubular with spreading lobes, 0.8–1.6 cm long, white, pink, rose or deep rose; inside densely hairy within the tube.

Stamens 5, included in the corolla-tube.

Style very short.

W Yunnan, W and SW Sichuan. 2,500–4,300 m (8,200–14,100 ft).

AM 1925 (Ness) as *R. ledoides*.

AM 1971 (Quarry Wood, Newbury) as *R. trichostomum* var. *ledoides* 'Quarry Wood', flowers white flushed with deep rose.

AM 1972 (Windsor) as *R. trichostomum* var. *ledoides* 'Lakeside', flowers white flushed with deep rose.

FCC 1976 (Rosemoor) as *R. trichostomum* var. *ledoides* 'Collingwood Ingram', flowers white flushed with deep rose.

Epithet: Hairy-mouthed.   Hardiness 3.   May–June.

Submerged into R. trichostomum *are:-*

R. trichostomum var. hedyosmum Cowan & Davidian

Leaves oblong.

Corolla large, 1.9–2.1 cm long, with a long tube.

W Sichuan.

Epithet: Sweet scented.   Hardiness 3.   May–June.

Known only in cultivation.

R. trichostomum var. radinum Cowan & Davidian

Leaf upper side often densely scaly.

Corolla scaly.

W Yunnan. 2,600–4,450 m (8,500–14,500 ft).

AM 1960 (Windsor) 'Sweet Bay', flowers pink.

AM 1972 (Quarry Wood, Newbury), flowers deep rose.

Epithet: Slender.   Hardiness 3.   May–June.

Leaf-bud scales deciduous.

Leaf narrow (narrowly oblanceolate, linear or linear-lanceolate) with recurved margins.

Flowers a month later than _R. primuliflorum._

Scales on the leaf underside brown or rust-coloured.

Inflorescence globose, 8–20 flowers.

Corolla white, pink, rose or deep rose.

R. trichostomum 'Collingwood Ingram' FCC at Deer Dell

Leaf upper side often densely scaly.

R. trichostomum var. *radinum* at Deer Dell

Subgenus *Rhododendron*
Section *Rhododendron*
Subsection *Afghanica* Cullen
Triflorum Series, Hanceanum Subseries

R. afghanicum Aitch. & Hemsl.

Shrub low-growing or straggly or creeping, poisonous, up to 50 cm (1.5 ft) high; **branchlets** scaly.

Leaves lanceolate, oblong-lanceolate or oblong; **lamina** 3–8 cm long, 1–2.3 cm broad; **upper surface** olive-green or bluish-green, sparsely scaly or not scaly; **underside** scaly, scales nearly contiguous or their own diameter apart.

Inflorescence distinctly racemose, 8–15 flowers; **rhachis** 1.3–5 cm long.

Corolla campanulate, small, 0.8–1.3 cm long, whitish-green, creamy-yellow or white.

Style sharply bent or deflexed or straight.

Afghanistan. 2,150–3,000 m (7,000–9,850 ft).

Epithet: From Afghanistan.   Hardiness 3.   April–May.

Extremely rare in cultivation. Davidian placed this plant with R. hanceanum, *which was then in the* Triflorum *Series.*

Subgenus *Rhododendron*
Section *Rhododendron*
Subsection *Baileya* Sleumer
Lepidotum Series, Baileyi Subseries

The crenulate (scalloped) scales have led *R. baileyi* being placed in its own Subsection.
Not easily recognised out of flower without a microscope, but the colour of the corolla is a striking reddish-purple, deep purple or crimson-purple.

Leaf underside densely scaly, usually cinnamon rust-coloured, overlapping. Style short, stout and sharply bent.

The comparatively large and smooth recurved lepidote leaves are unusual.

Distinctive only in flower from its long rhachis and racemose inflorescence.

Corolla campanulate, small, whitish-green, creamy-yellow or white.

Leaves of *R. afghanicum* (A. Dome). Bar = 1 cm

***R. baileyi* LS&H17359 at Deer Dell**

Rhododendron
(Afghanica/Baileya)

R. baileyi <small>Balf.f.</small>

Shrub 0.6–1.8 m (2–6 ft) high; **branchlets** rather densely scaly.

Leaves obovate, elliptic, oblong-oval or oval; **upper surface** densely or
sometimes moderately scaly; **underside** densely scaly, scales crenulate,
cinnamon, rust-coloured or brown, overlapping.

Inflorescence 4–9 (rarely 18) flowers.

Corolla rotate, 0.8–1.6 cm long, reddish-purple, deep purple or crimson-purple;
outside rather densely or moderately scaly.

Style short, stout and sharply bent.

S and SE Tibet, Sikkim, Bhutan. 2,440–4,270 m (8,000–14,000 ft).

AM 1960 (Glenarn), flowers deep purple.

Epithet: After Lt. Col. F.M. Bailey (1882–1967), traveller in Tibet.   Hardiness 3.
April–May.

Davidian placed this plant in the Lepidotum *Series, as it resembles* R. lepidotum *in
leaf, but* R. baileyi *is larger in all its parts.*

Subgenus *Rhododendron*
Section *Rhododendron*
Subsection *Boothia* <small>Sleumer</small>
Boothii Series, Boothii and Megeratum Subseries

Small or straggly shrubs.
Branchlets often bristly.
Leaf underside almost always glaucous, densely scaly.
Corolla campanulate to almost rotate, yellow or white.
Stamens 10.
**Ovary densely lepidote tapering into the style, which is short, stout
and sharply bent.**

*For ease of identification, this Subsection can be split into two groups
depending on the detail of the scales (vesicular or not), which is visible
using a magnifying glass:*

Group A. Scales pale, vesicular, lacking a rim, bladder-like
R. leucaspis *and* R. megeratum.

Group B. Scales darker, not vesicular
R. boothii, R. boothii *Mishmiense Group,* R. chrysodoron,
R. dekatanum *and* R. sulfureum.

Leaf underside densely
scaly.

Scales crenulate,
cinnamon, rust-coloured
or brown, overlapping.

Corolla reddish-purple,
deep purple or crimson-
purple.

Style stout and sharply
bent.

R. baileyi (S. Hootman)

R. sulfureum (Muncaster Form) at Deer Dell

Rhododendron
(Baileya/Boothia)

R. leucaspis Tagg   Boothia Gp A

Shrub sometimes epiphytic, 0.3–1.5 m (1–5 ft) high; **branchlets** scaly, bristly.

Leaves nearly orbicular, oval, obovate or elliptic; **lamina** 2.8–5 cm long, 1.5–3.4 cm broad; **upper surface** dark green, shiny, bristly, margin recurved, bristly; **underside** glaucous, densely scaly, scales bladder-like (vesicular), one-half to their own diameter apart.

Petiole scaly, bristly.

Inflorescence 1–2 (or rarely 3) flowers.

Corolla rotate or almost saucer-shaped, 2.3–3.2 cm long, white, outside scaly.

SE Tibet. 2,000–3,050 m (8,000–10,000 ft).

AM 1929 (Exbury), flowers white (KW6273).

FCC 1944 (Exbury) (KW7171).

Epithet: White shield.   Hardiness 2–3.   February–April.

R. megeratum Balf.f. & Forrest   Boothia Gp A

Shrub often epiphytic, 0.08–1 m (0.25–3 ft) or rarely 1.5–1.8 m (5–6 ft) high; **branchlets** scaly, moderately or densely bristly.

Leaves elliptic, obovate-elliptic or oval; **lamina** 1.3–4 cm long, 0.8–2 cm broad; **upper surface** dark green, shiny, not scaly, not bristly; margin recurved often bristly; **underside** very glaucous, densely scaly, scales bladder-like (vesicular), submerged in pits, 1–1.5 times their own diameter apart.

Petiole bristly.

Inflorescence 1–2 or rarely 3 flowers.

Pedicel bristly.

Calyx 0.6–1 cm long.

Corolla campanulate or rotate-campanulate, 1.6–2.5 cm long, yellow, deep yellow, pale lemon-yellow or rarely white, outside scaly.

Style short, stout and sharply bent.

NW Yunnan, NE Upper Burma, S, SE and E Tibet, Assam. 2,440–4,000 m (8,000–13,100 ft).

AM 1935 (Townhill Park), flowers deep yellow.

AM 1970 (Bodnant), flowers yellowish-green.

Epithet: Passing lovely.   Hardiness 2–3.   March–April.

Easily recognised in flower, and out of flower by the bristly upper leaf surface and the growth habit — a compact rounded shrub.

Corolla rotate or almost saucer-shaped, white.

Stamens chocolate-coloured.

Branchlets bristly.

Leaf underside glaucous.

Petiole scaly, bristly.

Inflorescence usually 1–2 flowers.

R. leucaspis KW7171 FCC at Deer Dell

New foliage of *R. leucaspis* at Deer Dell

Usually low-growing, but may be compact or spreading.

In cultivation, has smaller less hairy leaves than *R. leucaspis*, with which it shares the bladder-like scales.

Usually has yellow flowers.

Branchlets moderately or densely bristly.

Leaf underside very glaucous.

Petiole and pedicel bristly.

Inflorescence usually 1–2 flowers.

R. megeratum at Glenarn

Rhododendron
(Boothia Group A)

R. boothii Nutt.

Boothia Gp B

Shrub usually epiphytic, 1.5–2.4 m (5–8 ft) or sometimes up to 3 m (10 ft) high; **branchlets** scaly, densely hairy.

Leaves ovate, ovate-elliptic or elliptic; **lamina** coriaceous, very thick, 8.2–12.6 cm long, 3.5–6.2 cm broad, apex acutely acuminate; **upper surface** bright green, not scaly; **underside** pale glaucous green, densely scaly, scales their own diameter apart.

Petiole densely woolly.

Inflorescence 7–10 flowers.

Pedicel densely scaly, bristly.

Calyx large, 0.8–1.3 cm long.

Corolla broadly campanulate, 2–3.1 cm long, bright lemon-yellow or sulphur-yellow, without spots.

Style short, stout and sharply bent.

Assam, SE Tibet. 1,525–3,050 m (5,000–10,000 ft).

Epithet: After T.J. Booth, who first collected it in 1849.   Hardiness 1–2. April–May.

Probably lost to cultivation in the UK, but re-introduced very recently by P.A. and K.N.E. Cox.

R. boothii Mishmiense Group

Boothia Gp B

R. mishmiense Hutch & Kingdon Ward

Branchlets densely bristly.

Leaf apex acuminate.

Corolla yellow with reddish spots.

Calyx large, 0.4–1.3 cm long.

Leaf coriaceous, very thick, underside pale green or glaucous green.

Petiole bristly.

Pedicel woolly.

Inflorescence 3–4 flowers.

Overlaps with R. boothii geographically. It may deserve varietal status.

Branchlets densely hairy.

Leaf margins and lower midrib hairy, apex acutely acuminate.

Corolla yellow, without spotting.

Calyx large, 0.8–1.3 cm long.

Flowers small compared to the hanging dark leaves.

Leaf coriaceous, very thick, underside pale glaucous green.

Petiole woolly.

Pedicel bristly.

Inflorescence 7–10 flowers.

R. boothii (A. Dome)

Young foliage of *R. boothii* (S. Hootman)

Rhododendron
(Boothia Group B)

R. chrysodoron Tagg ex Hutch.   Boothia Gp B

Shrub sometimes epiphytic, 0.2–1.8 m (0.6–6 ft) high; **branchlets** often bristly with slender bristles.

Leaves broadly elliptic, obovate or oblong-elliptic; **lamina** 4.8–8 cm long, 2–4.1 cm broad; **upper surface** bright green or dark green, shiny; **underside** pale glaucous green, densely scaly, scales 1–2 times their own diameter apart.

Inflorescence 3–6 flowers.

Corolla broadly campanulate, 3–4 cm long, bright canary yellow or yellow.

Style short, stout and sharply bent.

W Yunnan, NE Upper Burma. 6,500–8,500 ft.

AM 1934 (Bodnant), flowers clear yellow.

Epithet: Golden gift, referring to the yellow-flowered plant given by Lord Stair to the Royal Botanic Garden, Edinburgh.   Hardiness 1–2.   February–April.

R. dekatanum Cowan   Boothia Gp B

Shrub 0.6–1.2 m (2–4 ft) high; **branchlets** scaly, not bristly, leaf-bud scales deciduous.

Leaves oval, obovate, oblong-oval, oblong-obovate or broadly oblong; 3–6 cm long, 2–3.5 cm broad; apex rounded, mucronate; base rounded or obtuse; **upper surface** olive-green, matt, not scaly; margin slightly recurved, glabrous; **underside** pale glaucous, densely scaly, the scales markedly unequal in size, large and medium-sized, dark brown and brown, contiguous.

Petiole 5–6 mm long, densely scaly, not bristly.

Inflorescence terminal, 2–3 flowers, **flower-bud scales** deciduous.

Pedicel 6–8 mm long, thick, densely scaly, glabrous.

Calyx 6–8 mm long.

Corolla broadly campanulate, 2–2.6 cm, bright lemon-yellow.

Stamens 10.

Style short, stout and sharply bent, scaly at the base, glabrous.

S Tibet. 3,500 m (11,500 ft).

Epithet: After Mrs Dekat.   Hardiness 2–3.   March–April.

Recently recognised as being in cultivation by P.A. Cox.

Bushy or upright growing shrub.
Reddish bark.
Long slender bristles at first on the branchlets.
Corolla large and deep yellow.
Calyx minute.
Leaf underside pale glaucous green.

R. chysodoron at Arduaine

Similar to *R. sulfureum* but: leaves are usually oval;
has markedly unequally sized scales on the leaf underside;
and the calyx (6–8 mm) and corolla (2–2.6 cm) are larger.

Leaf underside pale, glaucous.
Inflorescence 2–3 flowers.
Corolla lemon-yellow.

R. dekatanum (S. Hootman)

Foliage of *R. dekatanum* showing the unequal scales (S. Hootman)

R. sulfureum Franch.   Boothia Gp B

Shrub often epiphytic, 0.3–1.5 m (1–5 ft) high; **branchlets** scaly.
Leaves obovate, oblong-obovate, elliptic or oval; **lamina** 2.6–8.6 cm long,
    1.3–4.2 cm broad; **upper surface** bright or dark green; **underside** glaucous,
    densely scaly, scales one-half to their own diameter apart.
Inflorescence 4–8 flowers.
Calyx 3–6 mm long.
Corolla campanulate, 1.3–2 cm long, bright or deep yellow, bright or deep
    sulphur-yellow, or greenish-yellow, outside scaly.
Style short, stout and sharply bent.
W and NW Yunnan, NE Upper Burma. 2,150–3,850 m (7,000–12,620 ft).
AM 1937 (Lochinch), as *R. commodum*, flowers sulphur-yellow.
Epithet: Sulphur-coloured.   Hardiness 2–3.   March–April.

Large calyx but small flowers.

Subgenus *Rhododendron*
Section *Rhododendron*
Subsection *Camelliiflora*
Camelliiflorum Series

R. camelliiflorum Hook.f.

Shrub 0.6–1.8 m (2–6 ft) high; **branchlets** rather densely scaly.
Leaves oblong, oblong-lanceolate or oblong-elliptic; **lamina** 4.8–12.3 cm long,
    1.5–3.8 cm broad; **upper surface** dark green, not scaly; **underside** densely scaly,
    scales contiguous or slightly overlapping or one-half their own diameter apart.
Petiole densely scaly.
Inflorescence terminal, 1–2 flowers.
Pedicel densely scaly.
Calyx 0.5–1.1 cm long.
Corolla campanulate with 5 spreading lobes, 1.5–2.5 cm long, white, white
    tinged pink, red or deep wine-red; outside scaly.
Stamens 12–16.
Style short, stout and sharply bent.
Sikkim, Nepal, Bhutan. 2,750–3,650 m (9,000–12,000 ft).
Epithet: With *Camellia*-like flowers.   Hardiness 1–2.   May–July.

Rare in cultivation. Flowers often hidden beneath the leaves.

Very variable in growth habit, but often straggly.

Leaves obovate, oblong-obovate, elliptic or oval.

Leaf underside with small, brown or dark brown, scales with up-turned rims.

Calyx (3–6 mm) and corolla (1.3–2 cm), smaller than those of *R. dekatanum*.

Leaf underside glaucous.

Inflorescence 4–8 flowers.

Corolla shades of yellow.

R. sulfureum F24235 at Brodick Castle

The small flowers are reminiscent of Subsection *Boothia*.

Leaves and growth habit similar to those of Subsection *Maddenia*.

Distinguished from most *Maddenia* species by the intensely lepidote leaves.

Corolla with 5 spreading oval lobes.

Style short, stout and sharply bent.

Leaf underside with scales contiguous or slightly overlapping.

Pedicel densely scaly.

Inflorescence 1–2 flowers.

Corolla white, or pink to deep wine red.

r. camelliiflorum S&L5696 at Benmore

Rhododendron
(Boothia Group B/Camelliflora)

Subgenus *Rhododendron*
Section *Rhododendron*
Subsection *Campylogyna* Sleumer
Campylogynum Series

Prostrate, mounding or broadly spreading shrubs.
Leaves small.
Leaf margins crenulate (notched with rounded teeth).
Scales vesicular (lacking a rim, bladder-like), often only present on the margin, with most of the lamina glabrous because the scales are shed as the leaf matures.
Inflorescence 1–2 flowers.
Corolla small, campanulate, with a very long pedicel.

The varieties described by Davidian are now considered to be cultivated variants of one species and have been reduced to Group status.

R. campylogynum Franch.

Shrub dwarf, prostrate cushion, compact or small erect, usually 2.5–4.5 cm (1–2 inches) high; **branchlets** short scaly.

Leaves obovate or oblanceolate; **lamina** 0.6–2.5 cm long, 0.3–1.8 cm broad; **upper surface** dark green, shiny, not scaly; margin crenulate; **underside** glaucous or pale green, scaly, scales vesicular, 1–6 times their own diameter apart.

Inflorescence 1–3 (or rarely up to 5) flowers.

Pedicel very long, 1.8–5 cm in length, scaly.

Corolla campanulate, 1.4–1.8 cm long, nodding; pale rose-purple, salmon-pink, carmine to deep plum-purple or almost black-purple; not scaly outside.

Style thick, bent.

W and NW Yunnan, E and NE Upper Burma, SE Tibet. 3,650–4,000 m (12,000–13,000 ft).

AM 1966 (Collingwood Ingram) 'Thimble', flowers salmon-pink.

AM 1973 (Collingwood Ingram) 'Baby Mouse', flowers deep plum-purple.

Epithet: With bent ovary.   Hardiness 3.   May–June.

R. campylogynum (pink form) (J.C. Birck)

Leaf dark and shiny,
margin crenulate.
Pedicel very long,
1.8–5 cm.
Corolla with a
bloom, campanulate,
nodding.
Corolla pale rose
purple, salmon-pink,
carmine, or deep
purple shades.

R. campylogynum 'Claret' at Deer Dell

R. campylogynum Celsum Group

R. campylogynum Franch. var. *celsum* Davidian

Growth habit erect, height 0.45–1.8 m (1.5–6 ft).
Corolla plum purple.

R. campylogynum Charopoeum Group

R. campylogynum Franch. var. *charopoeum* Davidian

Shrub, very compact, spreading, up to 45 cm (1.5 ft) high.
Corolla plum-purple or pale rose.

R. campylogynum Myrtilloides Group

R. campylogynum Franch. var. *myrtilloides* Davidian

Shrub up to 30 cm (1 ft): dwarf compact or somewhat spreading.
Flowers smaller, 0.8–1.4 cm long.
NE Upper Burma, mid-W and NW Yunnan, SE Tibet, Assam. 2,450–4,750 m
  (8,000–15,500 ft).
AM 1925 (Exbury).
FCC 1943 (Exbury), flowers rose-purple.
Epithet: Myrtle-like.   Hardiness 3.   May–June.

R. campylogynum Cremastum Group

R. cremastum Balf.f. & Forrest

Shrub 0.6–1.8 m (2–6 ft) high; **branchlets** scaly.
Leaves obovate or oblong-obovate; **lamina** 2–3.7 cm long, 1.2–1.6 cm
  broad; **upper surface** pale green, margin crenulate; **underside** pale
  green, not glaucous, scaly, scales 3–5 times their own diameter apart.
Inflorescence 1–4 flowers.
Pedicel very long, 2.5–4.5 cm long.
Corolla campanulate, 1.4–2 cm long, nodding; light plum-rose or bright red
  or deep wine-red.
Stamens 10.
Style thick, sharply bent or bent.
NW Yunnan, SE Tibet. 3,350 m (11,000 ft).
AM 1971 (Windsor) 'Bodnant Red'.
Epithet: Suspended.   Hardiness 3.   May–June.

'Leucanthum' *AM (Benenden) 1970, with openly campanulate white flowers,
has a similar overall appearance, and is a selected form of this species.*

R. campylogynum 'Leucanthum' AM at Deer Dell

Very small leaves and flowers.
Corolla usually plum-purple with a bloom.

R. campylogynum F25357 Myrtilloides Group at Deer Dell

More upright growing.
Leaves pale green and matt above, pale green not glaucous below, and are usually larger than those of the type of *R. campylogynum*.
In cultivation, the flowers are red or reddish.

R. campylogynum Cremastum Group at Deer Dell

Rhododendron
(Campylogyna)

Subgenus *Rhododendron*
Section *Rhododendron*
Subsection *Caroliniana*
Carolinianum Series

Shrubs.
Scales on the leaf underside dense, not more than their own diameter apart.
Inflorescence terminal, or terminal and axillary in the upper one or two leaves.
Corolla usually white to pale pink or pale rose.

All species here have now been grouped under R. minus.

R. minus var. minus

R. minus Michaux

Shrub 1.2–6 m (4–20 ft) or rarely 9 m (30 ft) high; **branchlets** scaly.
Leaves elliptic, ovate-elliptic or oblong-lanceolate; **lamina** 5.8–11.8 cm long, 2.3–6.5 cm broad; **upper surface** dark green or paler green, scaly or not scaly; **underside** densely scaly, scales one-half to all of their own diameter apart.
Inflorescence 6–12 flowers.
Corolla narrowly tubular-funnel-shaped, 2.6–3.4 cm long, pink, rose, pinkish-purple or sometimes white, with or without brown or greenish spots.

E and SE USA, from N and S Carolina to Georgia and Alabama.
Epithet: Smaller.   Hardiness 3.   May–June.

R. minus var. *minus* (pink form) in the Valley Gardens

Leaves elliptic, ovate-elliptic or oblong-lanceolate; usually larger and smoother than leaves of *R. minus* var. *chapmanii*.

Leaf apex acuminate or acute.

Corolla narrowly tubular-funnel-shaped, pink to purplish pink, rose or white.

Scales on leaf underside not contiguous.

R. minus var. *minus* in the Valley Gardens

Rhododendron
(Caroliniana)

R. minus var. minus Carolinianum Group

R. carolinianum Rehder

Shrub, compact or somewhat compact, 1–2.4 m (3–8 ft) high.
Leaves ovate, elliptic, ovate-lanceolate or lanceolate; **lamina** 4.6–10.8 cm long, 1.8–4.6 cm broad; **upper surface** dark green or pale green, scaly or not scaly; **underside** densely scaly, scales contiguous to their own diameter apart.
Inflorescence 4–12 flowers.
Corolla widely funnel-shaped, 2–3 cm long, pink or pale rosy-purple, with or without faint spots, outside scaly; tube shorter than, or as long as, the lobes.

Eastern USA, N and S Carolina, Tennessee.
AM 1968 (Colville, Launceston).
Epithet: From Carolina.   Hardiness 3.   May–June.

> *R. carolinianum* var. *album* Rehder
>
> **Flowers** a white or whitish, with or without greenish spots.
> **Epithet:** With white flowers.   Hardiness 3.   May–June.
>
> *Correctly, no longer has separate status.*

R. minus var. chapmanii (A.Gray) W.H.Duncan & Pullen

R. chapmanii A.Gray

Leaves oval or oblong-oval, markedly curved (revolute), and usually rugose in cultivation.
Corolla tubular-funnel-shaped, pink or rose.
Scales on leaf underside sometimes contiguous, and never more than half their diameter apart.

W Florida, USA.
Epithet: After A.W. Chapman, American botanist (1809–1899).   Hardiness 3.   April–May.

Even coming from Florida, it is now considered hardy, but it requires summer heat to ripen the wood.

Generally smaller growing and with a wider corolla than is typical for *R. minus* var. *minus*.

Scales on leaf underside sometimes contiguous. Corolla pink or pale rosy-purple.

R. minus var. *minus* Carolinianum Group in the Valley Gardens

R. minus var. *chapmanii* at Deer Dell

Rhododendron
(Caroliniana)

Subgenus *Rhododendron*
Section *Rhododendron*
Subsection *Cinnabarina* Sleumer
Cinnabarinum Series

Upright shrubs.
Foliage evergreen (except *R. tamaense*).
Leaves often glaucous above and below; densely scaly below.
Corolla tubular or tubular campanulate, usually pendulous, outside not scaly.

R. cinnabarinum ssp. cinnabarinum
R. cinnabarinum Hook.f.

Shrub 1.2–5.5 m (4–18 ft) high; **branchlets** moderately or sparsely scaly.
Leaves evergreen, obovate-elliptic, oblong-obovate, elliptic, oblong or oblong-lanceolate; **lamina** 3.2–9 cm long, 1.8–5 cm broad; **upper surface** bluish-green, olive-green or greyish-green, moderately or slightly glaucous, often not scaly; **underside** densely scaly, scales one-half their own diameter apart or nearly contiguous or sometimes their own diameter apart.
Calyx 1–2 mm long.
Corolla pendulous, tubular, slightly widened towards the top, the lobes spreading, or tubular-campanulate, 2.6–3.8 cm long, cinnabar-red, salmon-pink suffused with yellow, salmon-pink, yellow or rarely orange; outside not scaly.

Nepal, Sikkim, Bhutan, S and SE Tibet. 2,150–4,150 m (7,000–13,500 ft).

AM 1954 (Collingwood Ingram) 'Cuprea', with coral flowers, bright red outside (LST6560).
Epithet: Cinnabar-red.   Hardiness 3.   April–July.

R. cinnabarinum ssp. *cinnabarinum* KW8239 at Deer Dell

Leaves narrower than those of *R. cinnabarinum* ssp. *xanthocodon.*
Upper leaf surface usually not scaly, moderately or slightly glaucous.

Leaf upper surface bluish-green, olive-green or greyish-green.

Corolla pendulous, tubular or tubular-campanulate, reddish, salmon-pink or yellow or rarely orange.

Foliage of *R. cinnabarinum* ssp. *cinnabarinum* BLM234 at Nymans (E. Daniel)

R. cinnabarinum ssp. *cinnabarinum* yellow form in the Valley Gardens

Rhododendron
(Cinnabarina)

The following variations of R. cinnabarinum *ssp.* cinnabarinum *are noteworthy and distinct in cultivation:*

R. cinnabarinum ssp. cinnabarinum

Blandfordiiflorum Group

R. cinnabarinum var. *blandfordiiflorum* W.J.Hooker

This variety differs from the *R. cinnabarinum* ssp. *cinnibarinum* type in that the tubular **corolla** is usually about twice the length, 5–6.3 cm (2–2.5 inches), it is red outside, yellow or greenish-yellow within.

Sikkim. 3,050–3,650 m (10,000–12,000 ft).

AM 1945 (Bodnant).

Epithet: Blandfordia flowers.   Hardiness 3.   May–July.

R. cinnabarinum ssp. cinnabarinum Roylei Group

R. cinnabarinum var. *roylei* (Hook.f.) Hutch.

Nepal, Sikkim, Bhutan. 2,900–3,800 m (9,500–12,500 ft).
Epithet: After Dr. Royle.   Hardiness 3.   May–July.

R. cinnabarinum var. roylei f. magnificum W.Watson

Corolla deep plum-crimson, exceptionally large, 4–5.3 cm long, with a broad tube.

AM 1918 (Reuthe), flowers dark crimson on the outside, orange-red inside.

AM 1953 (Windsor) 'Vin Rose', flowers currant-red outside, blood-red inside.

Epithet: Magnificent.   Hardiness 3.   June–July.

Corolla bicoloured, usually about twice as long as that of the *R. cinnabarinum* ssp. *cinnibarinum* type; red outside, yellow or greenish-yellow within.

R. cinnabarinum ssp. *cinnabarinum*
**Blandfordiiflorum Group
at Benmore**

R. cinnabarinum
ssp. *cinnabarinum*
**Blandfordiiflorum Group –
fallen corollas at Benmore**

Corolla deep plum-crimson, with a glaucous sheen.

R. cinnabarinum ssp. *cinnabarinum* **Roylei Group at Benmore**

Rhododendron
(Cinnabarina)

R. cinnabarinum ssp. cinnabarinum
Breviforme Group
R. cinnabarinum ssp. *cinnabarinum* var. *breviforme* Davidian

Short tubular **corolla** 3–3.5 cm long, red outside, yellowish at the margins and within.

NE Bhutan. 3,150 m (10,500 ft).

AM 1977 (Hydon) 'Nepal' (LS&H21283 – the Type No.)

Epithet: A form with shorter corolla.   Hardiness 3.   June–July.

In cultivation, the plant has been known as R. cinnabarinum *'var. Nepal'.*

The remaining variation, although submerged by modern botanists, may have some botanical significance.

R. cinnabarinum Hook.f. **var. aestivale** Hutch.

Epithet: Summer-flowering.   Hardiness 3.   July.

R. cinnabarinum ssp. tamaense (Davidian) Cullen
R. tamaense Davidian

Shrub often epiphytic, 1–1.8 m (3–6 ft) high, or a small **tree**; **branchlets** scaly.

Leaves deciduous or semi-deciduous, oblong-oval, elliptic, oblong or oblong-lanceolate; **lamina** 2.5–5.8 cm long, 1.5–2.6 cm broad; **upper surface** green, not scaly; **underside** pale glaucous green or brown, scaly, scales 2–5 times their own diameter apart.

Inflorescence 2–5 flowers.

Calyx minute, 1 mm long.

Corolla tubular or tubular-campanulate, 3–4.6 cm long, deep royal purple, purple or pale lavender; outside scaly.

N Burma. 2,750–3,200 m (9,000–10,500 ft).

AM 1978 (Sandling Park) 'Triangle', flowers purple with red-purple spots.

Epithet: From Tama, North Triangle, N Burma.   Hardiness 3.   April–May

Leaves narrow, oblong-
lanceolate.

Later flowering in July.

R. cinnibarinum var. *aestivale* (J.C. Birck)

**Leaf underside less
scaly, scales 2–5
times their own
diameter apart.**

**No persistent leaves
from previous years.**

Corolla deep royal
purple, purple or pale
lavender.

R. cinnabarinum ssp. *tamaense* KW21003 in the Valley Gardens

R. cinnabarinum ssp. *tamaense* KW21003 showing autumn leaf
colour at Deer Dell

R. cinnabarinum ssp. xanthocodon

(Hutch.) Cullen

R. xanthocodon Hutch.

Shrub or **tree** 1.8–7.6 m (6–25 ft) high; **branchlets** scaly.

Leaves elliptic, oblong-elliptic, obovate or oval; **lamina** 3–7.5 cm long, 1.6–4.5 cm broad; **upper surface** olive-green, shiny, scaly; **underside** pale glaucous green or pale green, densely scaly, scales one-half or their own diameter apart.

Inflorescence 2–6 or rarely 8 flowers.

Calyx 1–2 mm long.

Corolla tubular-campanulate, 2.5–3.2 cm long, creamy-yellow, without spots.

Tibet, Bhutan. 3,350–3,650 m (11,000–12,000 ft).

AM 1935 (Exbury), flowers yellow (KW6026).

Epithet: Yellow bell.   Hardiness 3.   May.

R. cinnabarinum ssp. xanthocodon

Concatenans Group

R. concatenans Hutch.

Shrub 1.5–2 m (5–7 ft) high; **branchlets** scaly.

Leaves elliptic or oblong-elliptic; **lamina** 4–8.7 cm long, 2–5 cm broad; **upper surface** bluish-green, glaucous (in young leaves, markedly bluish-green, glaucous, matt), not scaly; **underside** tinged purple or purplish-brown (in young leaves, glaucous), densely scaly, scales nearly contiguous.

Inflorescence 3–8 flowers.

Corolla tubular-campanulate, 3–4 cm long, apricot, faintly tinged pale purple outside or not tinged, not scaly outside.

SE Tibet. 3,650–3,950 m (12,000–13,000 ft).

FCC 1935 (Nymans) 'Orange Bill', flowers apricot, flushed rose outside (KW5874).

AM 1950 (Collingwood Ingram) 'Copper' (LS&T6560).

Epithet: Linking together.   Hardiness 3.   April–May.

Many plants so labelled are hybrids or wrongly identified.

Upper leaf surface not usually glaucous, but very close to that of *R. cinnibarinum* ssp. *cinnibarinum*.

Corolla always tubular-campanulate, creamy yellow.

R. cinnabarinum ssp. *xanthocodon* in R. White's garden (E. Daniel)

Relatively compact growth habit, with quite large, broad, very glaucous leaves.

Underside of leaves tinged purple or purplish-brown (in young leaves, glaucous). Corolla apricot.

R. cinnabarinum ssp. *xanthocodon* Concatenans Group at Borde Hill, showing glaucous young foliage

Rhododendron (Cinnabarina)

R. cinnabarinum ssp. xanthocodon
Purpurellum Group

R. cinnabarinum var. *purpurellum* Cowan

S Tibet. 3,050 m (10,000 ft).
AM 1951 (Collingwood Ingram) (LS&T6349A).
Epithet: Purple.   Hardiness 3.   May–July.

This plant is intermediate between Subsection Cinnibarina and R. oreotrephes (Subsection Triflora). The corolla is short campanulate, and, because of this, the plant is better classified as a variety of R. xanthocodon. It does not seem to suffer from powdery mildew as do many others in this Subsection. (This may be evidence of its possible hybridity with R. oerotrephes, a species that is not much affected by powdery mildew.)

R. keysii Nutt.
(R. igneum Cowan)

Shrub or **tree**, lax and upright, sometimes epiphytic, 0.6–6 m (2–20 ft) high; **branchlets** scaly.

Leaves lanceolate, oblong or oblong-lanceolate; **lamina** 5–15.5 cm long, 1.5–3.6 cm broad; **upper surface** dark green, shiny, scaly; **underside** densely scaly, scales half or their own diameter apart.

Inflorescence terminal and axillary in the uppermost 1–3 leaves, 2–6 flowers.

Calyx minute, 1 mm long.

Corolla tubular, slightly ventricose, 1.5–2.5 cm long, lobes erect, small; orange, coral, salmon-pink or deep scarlet, with yellow or yellowish lobes.

Bhutan, Assam, S and SE Tibet. 2,450–3,650 m (8,000–12,000 ft).

Epithet: After a Mr. Keys.   Hardiness 3.   June–July.

R. keysii Nutt. var. unicolor Hutch. ex Stearn.

Corolla a uniform deep red.
The short erect lobes of the corolla are rarely tipped with a yellowish tinge.

AM 1933, flowers red, tips of corolla-lobes slightly yellowish (KW6257).
Epithet: Uniform flower colour.   Hardiness 3.   June–July.

This variety is of no botanical significance.

Leaves usually smaller. Corolla short, campanulate; usually rich plum-purple, sometimes bright pinkish-mauve.

R. cinnabarinum ssp. *xanthocodon* Purpurellum Group at Deer Dell

Leaves often larger, narrower, and thinner in texture than those of *R. cinnibarinum*. Flowers thin tubes in clusters; usually orange, salmon pink or scarlet.

Inflorescences terminal and axillary.

R. keysii in Philip Urlwin-Smith's garden

Rhododendron (Cinnabarina)

Subgenus *Rhododendron*
Section *Rhododendron*
Subsection *Edgeworthia* Sleumer
Edgeworthii Series

Branchlets densely woolly.
Leaf underside densely woolly and scaly.
Petiole and pedicel densely woolly.
Stamens 10.
Capsule woolly.

R. edgeworthii Hook.f.
(*R. bullatum* Franch.)

Shrub often epiphytic, 0.3–3.5 m (1–12 ft) high; **branchlets** densely woolly with brown, rust-coloured or whitish wool.

Leaves ovate-lanceolate, ovate, ovate-elliptic or oblong-lanceolate; **lamina** 4–14 cm long, 2–5.6 cm broad; **upper surface** strongly bullate, glabrous; **underside** densely woolly with brown or rust-coloured wool, densely scaly.

Petiole densely woolly.

Inflorescence 1–3 (or sometimes 4) flowers.

Calyx 0.6–1.9 cm long.

Corolla campanulate or funnel-campanulate, very fragrant, 3.2–7.6 cm long, white or white-tinged pink, or rose, with or without a yellow blotch, outside scaly.

Nepal, Sikkim, Bhutan, Assam, S and E Tibet, NE Upper Burma, NW Yunnan, SW Sichuan. 1,850–4,000 m (6,000–13,000 ft).

AM 1923 (Sunninghill) as *R. bullatum*, flowers white (Farrer842).

AM 1946 (Bodnant) as *R. bullatum*, flowers white flushed rose.

FCC 1933 (Nymans), flowers white.

FCC 1937 (Exbury) as *R. bullatum*, flowers white (F26618).

FCC 1981 (Leonardslee) 'Red Collar' (KW20840).

Epithet: After H.P. Edgeworth, Bengal Civil Service (1812–1881).   Hardiness 2–3.
April–May.

Hardier forms have been introduced recently.

R. pendulum L&S6660 at Deer Dell

Easy to identify because of the bullate leaves with a woolly underside.
Corolla very fragrant, usually white or white suffused pink.

Branchlets, petiole and pedicel densely woolly.

R. edgeworthii (Stronachulin form) at Deer Dell

R. pendulum Hook.f.

Shrub sometimes epiphytic, 0.3–1.2 m (1–4 ft) high; **branchlets** densely woolly with brown or fawn wool.

Leaves oblong, elliptic or oblong-elliptic; **lamina** 2.3–5 cm long, 1.2–2.5 cm broad; **upper surface** convex, not bullate, glabrous, midrib tomentose; **underside** densely woolly with brown or fawn wool, densely scaly.

Petiole densely woolly.

Inflorescence 1–3 flowers.

Pedicel densely woolly.

Calyx 5–9 mm long.

Corolla rotate-campanulate, 1.5–2.2 cm long, white or white suffused with pink or pale yellow, outside scaly.

Nepal, Sikkim, Bhutan, S and SE Tibet. 2,300–3,650 m (7,500–12,000 ft).
Epithet: Hanging.   Hardiness 1–3.   April–May.

Rare in cultivation.

R. seinghkuense Kingdon Ward

Shrub usually epiphytic, 0.3–1 m (1–3 ft) high; **branchlets** densely or moderately woolly, scaly or not scaly.

Leaves ovate, ovate-lanceolate or oblong-elliptic; **lamina** 3–8 cm long, 1.6–4 cm broad, **upper surface** bullate, glabrous; **underside** densely brown woolly, scaly, scales very small, 1–2 times their own diameter apart.

Petiole densely woolly.

Inflorescence 1 (or rarely 2) flowers.

Pedicel densely woolly.

Calyx densely woolly.

Corolla rotate-campanulate, 2–2.5 cm long, sulphur-yellow, outside rather densely scaly.

Style short, sharply bent.

Upper Burma, Burma–Tibet Frontier, NW Yunnan. 1,850–3,050 m (6,000–10,000 ft). AM 1953, flowers sulphur-yellow.
Epithet: From the Seinghku Valley, Upper Burma.   Hardiness 1.   March–April.

Rare in cultivation. Likely to be hardier than stated above.

An open or fairly
compact shrub.

Leaves convex, not
bullate, somewhat
rugulose, underside
densely woolly.

Calyx red.
Flowers openly
campanulate.

Corollas white,
white-tinged pink or
pale yellow.

Branchlets, petiole and
pedicel densely woolly.

R. pendulum in June in J. Sinclair's garden, Seattle (M.L.A. Robinson)

In cultivation, similar
to *R. edgeworthii*,
but smaller in all
its parts.

Corolla sulphur-
yellow.

Branchlets moderately
or densely woolly.

Petiole and pedicel
densely woolly.

R. seinghkuense KW9254 at Deer Dell

Subgenus *Rhododendron*
Section *Rhododendron*
Subsection *Fragariiflora* Cullen

Lapponicum Series

Small shrublets.
Inflorescence terminal.
Calyx conspicuous.

R. fragariiflorum *and* R. setosum *are very similar in the appearance of their foliage. They differ in that* R. fragariiflorum *is somewhat ciliate whereas* R. setosum *is somewhat bristly. For identification purposes, these two species have been placed together, and it is likely that they are very closely related botanically.*

R. fragariiflorum Kingdon Ward

Dwarf **shrub** 10–30 cm (0.33–1 ft) high; **branchlets** very short, scaly, rather densely puberulous.

Leaves obovate, elliptic, oval or oblong; **lamina** 0.5–1.6 cm long, 3–7 mm broad; **upper surface** dark green or bright green, shiny, rather densely scaly; margin recurved, crenulate; **underside** scaly, scales large, dark brown and yellow, 1.5–6 times their own diameter apart.

Inflorescence 2–6 flowers.

Pedicel scaly, minutely puberulous.

Calyx 3–8 mm long, crimson or pink.

Corolla widely funnel-shaped or almost rotate, 1–1.7 cm long, purple or 'crushed strawberry' colour, plum-purple or purplish-crimson.

S and SE Tibet, Assam, Bhutan. 3,650–4,600 m (12,000–15,000 ft).

Epithet: Strawberry flowers.   Hardiness 3.   May–June.

Formerly in Subsection Lapponica, but placed in its own Subsection on account of the leaf margins and the scales. Related to R. setosum, *which has vesicular (and other) scales, and which remains in Subsection Lapponica. There seems to be some difference between the authorities as to the nature of the scales on* R. fragariiflorum.

R. fragariiflorum at Tsari, SE Tibet (P. Evans)

Dwarf shrub,
10–30 cm.

Branchlets
puberulous not
bristly.

Prominent golden
scales on the upper
leaf surface.

Recurved, crenulate
leaf margins bristly
or not bristly.

Leaf underside with
bicolourous scales,
dark brown and
yellow.

Petiole not bristly or
rarely bristly.

Calyx usually crimson.
Corolla purple, 'crushed
strawberry' colour,
plum-purple or purplish-
crimson.

R. fragariiflorum (S. Hootman)

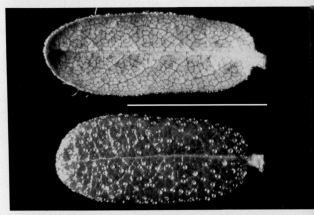

Leaves of *R. fragariiflorum* showing the scales (Rhododendron
Species Foundation). Bar = 1 cm

Rhododendron
(Fragariiflora)

R. setosum D.Don.
(formerly in Subsection *Lapponica*)

Shrub 0.15–1.2 m (0.5–4 ft) high; **branchlets** densely or moderately bristly, often puberulous.
Leaves obovate, oblong-obovate, oblong-oval or oblong-elliptic; **lamina** 0.6–1.7 cm long, 4–7 mm broad; **upper surface** pale greyish-green, matt, densely scaly; margin recurved, bristly; **underside** scaly, scales 1–4 times their own diameter apart.
Petiole bristly.
Inflorescence 3–6 flowers.
Pedicel not bristly or bristly.
Calyx 4–6 mm long, reddish or crimson-purple.
Corolla widely funnel-shaped, 1.3–2 cm long, reddish-purple, reddish-pink, pink or wine-red, outside not scaly.
Nepal, Sikkim, Bhutan, S Tibet. 2,750–5,000 m (9,000–16,500 ft).
Epithet: Bristly.   Hardiness 3–4.   April–May.

Subgenus *Rhododendron*
Section *Rhododendron*
Subsection *Genestieriana* Sleumer
Glaucophyllum Series, Genestierianum Subseries

R. genestierianum Forrest

Shrub or small **tree** 1.2–4.5 m (4–15 ft) high; **branchlets** purplish or reddish-brown (young growths reddish-purple, glaucous).
Leaves lanceolate or oblanceolate; **lamina** 5.5–15.3 cm long, 1.4–4.5 cm broad, apex acutely acuminate; **upper surface** olive-green or dark green, not scaly; **underside** markedly glaucous, laxly scaly, scales small, 4–6 times or rarely 10 or more times their own diameter apart.
Inflorescence racemose, 6–15 flowers.
Pedicel slender, 1.6–3 cm long.
Corolla tubular-campanulate, 1.2–1.8 cm long, fleshy, deep plum-purple, covered with a glaucous bloom.
Stamens 8–10.
NE Upper Burma, mid-W Yunnan, E and SE Tibet. 2,000–4,400 m (6,500–14,500 ft).
Epithet: After Pere A. Genestier (b. 1858) of the French R.C. Tibetan Mission, friend of G. Forrest.   Hardiness 1–2.   April–May.

Only seen in the milder areas.

Branchlets densely or moderately bristly.

Densely golden scaly upper leaf surface.

Leaf margin and petiole bristly.

Leaf underside with bicolourous scales.

Calyx crimson-purple.

Corolla reddish-purple, pink or wine-red.

R. setosum (B. Starling)

Foliage of *R. setosum* showing the setose hairs and very scaly upper leaf (M.L.A. Robinson)

The peeling purplish or reddish bark is noticeable.

Unmistakable with its somewhat hanging pointed leaves, very glaucous beneath, without dark scales, and the lax pendant truss.

Corolla deep plum-purple, covered with a glaucous bloom.

R. genestierianum at Brodick

Rhododendron
(Fragariiflora/Genestieriana)

Subgenus *Rhododendron*
Section *Rhododendron*
Subsection *Glauca* Sleumer
Glaucophyllum Series

Small to medium-sized shrubs, rarely small trees; bark smooth, peeling.

Leaf aromatic; upper surface at least somewhat shiny; undersides markedly glaucous with scales of two kinds: smaller pale yellow and widely scattered large dark brown.

Stamens 10 or rarely 8–9.

Calyx deeply 5-lobed.

Capsule often enclosed by the persistent calyx lobes.

It should be noted that in some species the glaucescence may decrease with the age of the leaf, and may vary with the local environment.

R. brachyanthum ssp. brachyanthum

R. brachyanthum Franch.

Shrub 0.3–1.8 m (1–6 ft) high; **stem** and **branches** with smooth brown flaking bark; **branchlets** scaly.

Leaves oblong, oblong-obovate or lanceolate, very aromatic; **lamina** 2–6.5 cm long, 1–2.6 cm broad; **upper surface** bright green, shiny; **underside** markedly glaucous, laxly scaly, scales 4–10 times their own diameter apart.

Inflorescence 3–10 flowers.

Pedicel slender, 1.4–4 cm long, longer than the corolla.

Calyx large and leafy 3–8 mm long.

Corolla campanulate, 0.8–1.5 cm long pale or deep yellow, or greenish-yellow.

Style short, stout and sharply bent.

Yunnan, S and SE Tibet. 3,000–3,350 m (9,800–11,000 ft)

AM 1966 (Collingwood Ingram) 'Jaune', flowers primrose-yellow.

Epithet: With short flowers.   Hardiness 3.   May–July.

Rare in cultivation.

R. glaucophyllum ssp. *glaucophyllum* at Deer Dell

Leaf underside glaucous, laxly scaly, the scales being partly or even completely deciduous.

Pedicel slender, 1.4–4 cm long, longer than the small yellow corolla.

Leaves very aromatic.

Corolla campanulate, small, long pale or deep yellow, or greenish-yellow.

Style short, stout and sharply bent.

R. brachyanthum ssp. *brachyanthum* (S. Hootman)

R. brachyanthum ssp. hypolepidotum

(Franch.) Cullen
R. brachyanthum var. *hypolepidotum* Franch.

The variety differs from the species in that the underside of the **leaves** is densely scaly, pale yellow scales being contiguous to 3 times their own diameter apart, and brown scales widely or closely separated.

SE Tibet, NW Yunnan, NE Upper Burma. 2,750–4,400 m (9,000–14,500 ft).
AM 1951 (Windsor) 'Blue Light'.
Epithet: Scaly beneath.   Hardiness 3.   May–July.

This variety is much more common in cultivation.

R. charitopes ssp. charitopes

R. charitopes Balf.f. & Farrer

Shrub somewhat compact, rounded, 0.2–1.5 m (0.75–5 ft) high; **branchlets** scaly.
Leaves obovate or oblong-obovate, aromatic; **lamina** 2.6–7 cm long, 1.3–2.9 cm broad; **upper surface** dark green, somewhat shiny; **underside** markedly glaucous, densely scaly, scales of two kinds, pale yellow scales 0.5–2 times their own diameter apart, larger brown scales widely separated.
Inflorescence 2–6 flowers.
Corolla broadly campanulate or almost saucer-shaped, 1.8–2.6 cm long, apple-blossom pink, speckled with crimson.

NE Upper Burma, NW Yunnan. 3,150–4,250 m (10,500–14,000 ft).
AM 1979 (Windsor) 'Parkside'.
Epithet: Graceful of aspect.   Hardiness 3.   April–May.

Leaf underside
densely scaly, with
pale yellow scales
contiguous to 3
times their own
diameter apart.

Flowers
indistinguishable from
ssp. *brachyanthum*.

R. brachyanthum ssp. *hypolepidotum* at Deer Dell

Corolla: pink, broadly
campanulate, almost
saucer-shaped.

Leaf underside
markedly glaucous,
densely scaly.

The scales on the
leaf underside,
though dense, are
not touching as they
are in *R. prunifolium*,
which has a similar
leaf shape (obovate
or oblong-obovate).

Larger calyx than that
of *R. tsangpoense*.

Corolla apple blossom
pink, speckled with
crimson.

R. charitopes ssp. *charitopes* F25570 at Deer Dell

R. charitopes ssp. tsangpoense (Kingdon Ward) Cullen

R. tsangpoense Kingdon Ward

Shrub 0.3–1.2 m (1–4 ft) high; **stem** and **branches** with smooth, brown, flaking bark; **branchlets** scaly.

Leaves obovate, oblong-obovate, elliptic, oblong or oval, aromatic; **lamina** 1.3–5.2 cm long, 0.6–3.2 cm broad; **upper surface** dark green, shiny; **underside** markedly glaucous, scaly, the smaller pale yellow scales 1–6 times or rarely 10 times their own diameter apart, the larger brown scales widely or closely separated.

Inflorescence 2–6 flowers.

Calyx 2–7 mm long.

Corolla campanulate, 1.3–2.6 cm long, reddish-purple, pinkish-purple, pink or rarely cerise.

SE Tibet. 2,450–4,100 m (8,000–13,500 ft).

AM 1972 (Sandling Park), flowers reddish-purple with darker spots (KW7744).

Epithet: From the Tsangpo River.   Hardiness 3.   May–June.

R. tsangpoense var. curvistylum Kingdon Ward ex Cowan & Davidian

Corolla smaller, narrowly tubular-campanulate, deep cerise-coloured.

Flowers very similar to those of *R. prunifolium*.

SE Tibet. 3,650–4,000 m (12,000–13,000 ft).

Epithet: Curved style.   Hardiness 3.   May–June.

An isolated population in the wild, but most plants in cultivation are said to be hybrids.

R. glaucophyllum ssp. glaucophyllum

R. glaucophyllum Rehder
(*R. glaucum* Hook.f.)

Shrub 0.3–1.5 m (1–5 ft) high; **stem** and **branches** with smooth, brown, flaking bark; **branchlets** scaly.

Leaves lanceolate, oblong-lanceolate or oblong, aromatic; **lamina** 3–9 cm long, 1–2.6 cm broad; **upper surface** dark green, shiny, scaly; **underside** markedly glaucous, densely scaly, the smaller pale yellow scales 0.5–2 times their own diameter apart, larger brown scales widely or closely separated.

Inflorescence 4–10 flowers.

Calyx 5-lobed, leafy, 0.5–1 cm long, lobes ovate-lanceolate or lanceolate.

Corolla campanulate, 1.4–2.6 cm long, pink, rose, pinkish-purple or reddish-purple.

Sikkim, Nepal, Bhutan. 2,750–3,650 m (9,000–12,000 ft).

Epithet: With bluish-grey leaf.   Hardiness 3.   April–May.

Very similar in
leaf shape to
R. prunifolium,
though the flowers
are usually paler.

The smaller
scales are further
apart than in
R. pruniflorum
(usually 1–6 times
their own diameter
apart).

Larger brown scales
widely or closely
separated.

Corolla
campanulate.

Corolla reddish-purple,
pinkish-purple, pink or
rarely cerise.

R. charitopes ssp. *tsangpoense* KW5844 at Deer Dell

Usually has pointed
leaves, which are
narrower than those
of *R. charitopes* ssp.
charitopes or
R. charitopes ssp.
tsangpoense.

Leaf underside
also more densely
scaly than that of
R. charitopes ssp.
tsangpoense.

Leaves usually lanceolate
or oblong-lanceolate.

Leaf underside with
larger brown scales
widely or closely
separated.

Calyx leafy, lobes ovate-
lanceolate or lanceolate.

Corolla pink, rose,
pinkish-purple or
reddish-purple.

R. glaucophyllum ssp. *glaucophyllum* 'Deer Dell' at Deer Dell

Rhododendron
(Glauca)

R. glaucophyllum ssp. glaucophyllum var. album
Davidian

R. glaucophyllum var. *album* Davidian

A form was given the cultivar name 'Len Beer' when exhibited in 1977.

Epithet: With white flowers.   Hardiness 3.   April–May.

Of no botanical significance.

R. glaucophyllum ssp. tubiforme (Cowan &
Davidian) D.G.Long

R. tubiforme (Cowan & Davidian) Davidian

Shrub 0.3–2 m, (1–7 ft) high; **stem** and **branches** with smooth, brown, flaking bark; **branchlets** scaly.

Leaves obovate, oblong-obovate, lanceolate or oblanceolate, aromatic; **lamina** 3.2–6 cm long, 1.3–2.5 cm broad; **upper surface** dark green or greyish-green, shiny, scaly; **underside** markedly glaucous, densely scaly, the smaller pale yellow scales 1–3 times their own diameter apart, the larger dark brown scales closely or widely separated.

Inflorescence 3–5 flowers.

Corolla tubular, 2.3–3.8 cm long, pink or deep rose, outside scaly or not scaly.

Assam, E Bhutan, SE Tibet. 2,750–3,650 m (9,000–12,000 ft).

Epithet: With tubular flowers.   Hardiness 3.   April–May.

Cullen suggests that this is a hybrid. In any case, it is better as a subspecies of
R. glaucophyllum *than as a full species as it was previously.*

R. luteiflorum (Davidian) Cullen

Shrub 0.6–1.5 m (2–5 ft) high; **stem** and **branches** with smooth brown flaking bark; **branchlets** scaly.

Leaves lanceolate, oblanceolate or oblong-lanceolate, aromatic; **lamina** 4–6.8 cm long, 1.5–2.6 cm broad; **upper surface** dark green, shiny, sometimes scaly; **underside** markedly glaucous, laxly scaly, pale yellow scales 3–8 times their own diameter apart, larger dark brown scales widely separated.

Inflorescence 3–6 flowers.

Calyx large, leafy, 6–8 mm long.

Corolla campanulate, 2–2.2 cm long, lemon-yellow or greenish-yellow.

Style long, stout, sharply bent.

N Burma. 2,900–3,350 m (9,500–11,000 ft).

AM 1960, FCC 1966 (Brodick) 'Glen Coy' (KW21556).

Epithet: With yellow flowers.   Hardiness 3.   April–May.

Flowers white.

R. glaucophyllum ssp. *glaucophyllum* var. *album* BLM315A 'Len Beer' at Deer Dell

The style is long slender and straight, which is unique in this section.
Corolla tubular; pink or deep rose.
Leaf underside with the smaller pale yellow scales 1–3 times their own diameter apart.

R. glaucophyllum ssp. *tubiforme* L&S2856 at Deer Dell

Corolla lemon-yellow or greenish-yellow.
Leaf underside markedly glaucous, laxly scaly, with pale yellow scales 3–8 times their own diameter apart, and larger dark brown scales widely separated.

R. luteiflorum KW21556 'Glen Coy' at Deer Dell

Rhododendron
(Glauca)

R. pruniflorum Hutch. & Kingdon Ward

Shrub 0.3–1.2 m (1–4 ft) high; **stem** and **branches** with smooth brown flaking bark; **branchlets** scaly.

Leaves obovate, oblong-obovate or oblong, aromatic; **lamina** 1.5–5 cm long, 1–2.6 cm broad; **upper surface** olive-green or dark green, shiny; **underside** usually markedly glaucous, densely scaly, the smaller pale yellow scales contiguous or slightly overlapping or 0.5 times their own diameter apart, the larger brown scales widely or closely separated.

Inflorescence 3–7 flowers.

Calyx 3–5 mm long.

Corolla campanulate, 1–1.6 cm long, nearly crimson, plum-purple, cerise-coloured or lavender-purple.

NE Burma, Assam, S and SE Tibet. 3,350–4,250 m (11,000–14,000 ft).

Epithet: Plum flowers.   Hardiness 3.   May–June.

R. shweliense Balf.f. & Forrest

Indistinguishable from *R. brachyanthum* ssp. *hypolepidotum* out of flower.

Corolla pale pink suffused with yellow, spotted pink.

There is some doubt as to whether the outside of the corolla should be lepidote or not. It is possible that the true species is not in cultivation: those so labelled may be hybrids.

Leaf underside densely scaly: the smaller pale yellow scales contiguous or slightly overlapping or half their own diameter apart.

Corolla nearly crimson, plum-purple, cerise-coloured or lavender-purple.

R. pruniflorum KW8415 at Deer Dell

Rhododendron
(Glauca)

Subgenus *Rhododendron*
Section *Rhododendron*
Subsection *Heliolepida* Sleumer
Heliolepis Series

**Leaves with large conspicuous scales beneath, usually very aromatic.
Corolla funnel-shaped, markedly lepidote outside.
Stamens 10.
Style slender, straight.**

R. bracteatum Rehd. & E.H.Wilson

Shrub 1–3 m (3–10 ft) high; **branchlets** scaly, **leaf-bud scales** persistent.

Leaves elliptic, oblong or oblong-oval; **lamina** 1.3–4.2 cm long, 0.9–1.7 cm broad; **upper surface** dark green, scaly; **underside** scaly, scales very large, yellowish or pale brown, 1–4 times their own diameter apart, without scattered larger darker scales.

Inflorescence 3–6 flowers.

Corolla campanulate, small, 1.2–1.5 cm long, white or white suffused with pink, with or without a crimson blotch, with crimson spots, outside scaly.

W Sichuan. 2,600–3,300 m (8,500–10,800 ft).

Epithet: Furnished with bracts.   Hardiness 3.   June–July.

Very rare in cultivation.

R. rubiginosum Desquamatum Group F24535 (fallen corollas) at Deer Dell

Small leaves and small white to pink campanulate flowers. Leaf-bud scales persistent.

Leaf lamina 1.3–4.2 cm long, 0.9–1.7 cm broad. Scales on the underside very large, yellowish or pale brown, 1–4 times their own diameter apart.

R. bracteatum W4253 at Deer Dell

R. heliolepis var. heliolepis

R. heliolepis Franch.
(*R. oporinum* Balf.f. & Kingdon Ward, *R. plebeium* Balf.f. & W.W.Sm.)
R. fumidum Balf.f. & W.W.Sm.

Shrub or **tree** 0.6–5.5 m (2–18 ft) high; **branchlets** scaly.

Leaves ovate-lanceolate, oblong-lanceolate, lanceolate or ovate, aromatic; **lamina** 4–11.5 cm long, 1.5–4.6 cm broad, apex acute or acuminate; **upper surface** dark green, moderately or densely scaly; **underside** scaly, scales very large, yellowish or brown, nearly contiguous to 4 times their own diameter apart.

Inflorescence 4–9 flowers.

Corolla tubular-campanulate or funnel-campanulate, 1.8–3 cm long, rose, pink, purple, red, lavender-purple, white or white flushed rose, with or without a crimson blotch, and with crimson or greenish spots

Mid-W and W Yunnan, E and NE Upper Burma, E and SE Tibet. 3,000–3,800 m (9,350–12,500 ft).

AM 1954 (Tower Court), flowers white with green and brown spots (F26961).

Epithet: With glistening scales.

Although both var. heliolepis *and var.* brevistylum *are very variable, var.* heliolepis *is distinguished by its* **rounded** *leaf base. Both varieties are hexaploid with 78 chromosomes.*

R. heliolepis var. brevistylum (Franch.) Cullen

R. pholidotum Balf.f. & W.W.Sm.

Shrub 1–2.4 m (3–8 ft) high; **branchlets** scaly.

Leaves ovate-lanceolate, oblong-lanceolate, lanceolate or elliptic-lanceolate, aromatic, 4–7.3 cm long, 1.6–2.8 cm broad, apex acute or acuminate; **upper surface** dark green, shiny, scaly; **underside** scaly, scales very large, yellowish and brown, one-half their own diameter apart, without scattered larger darker scales.

Inflorescence 4–8 flowers.

Calyx 1–3 mm long.

Corolla funnel-campanulate, 2.2–3 cm, long, rose-purple, rose or deep purple, with reddish spots, outside rather densely scaly.

NW Yunnan. 3,050–3,650 m (10,000–12,000 ft).

AM 1933 (Crosfield), flowers pink externally, white inside (KW7108).

Epithet: Scaly.   Hardiness 3.   June–July.

Submerged into R. heliolepsis *by some authorities, var.* brevistylum *can be recognised as having a* **cuneate** *leaf base. Both KW7108 and F26928 have a purple blotch in the centre of the corolla.*

Leaf very aromatic, with bases rounded.

Corolla usually tubular-campanulate or funnel-campanulate.

Leaf underside scaly, **scales very large**, yellowish or brown, nearly contiguous to 4 times their own diameter apart.

Corolla white to pink, rose, red, or purple, or lavender purple.

R. heliolepis var. *heliolepis* **SBEC0285 at Nymans (E. Daniel)**

Leaf underside scaly, scales unequal, very large, yellowish or brown, about half their own diameter apart.

Corolla widely funnel campanulate.

Corolla rose to rose purple or deep purple, slightly zygomorphic.

R. heliolepis var. *brevistylum* **F26928 at Deer Dell**

Rhododendron
(Heliolepida)

R. rubiginosum var. rubiginosum

R. rubiginosum Franch.

Shrub or **tree** 0.6–9 m (2–30 ft) high; **branchlets** scaly.

Leaves ovate-lanceolate, elliptic-lanceolate, lanceolate or oblong-lanceolate, not aromatic; **lamina** 3–9 cm long, 1.3–4.2 cm broad; **upper surface** pale green or dark green, somewhat matt, scaly or not scaly; **underside** densely scaly, scales rust-coloured or brown, overlapping or contiguous, with or without widely scattered larger darker scales.

Inflorescence 4–8 flowers.

Corolla tubular-funnel-shaped, or funnel-shaped, 1.6–3.1 cm long, pink, pale rose, pinkish-purple, lavender-purple or white, with crimson spots, outside scaly.

W and NW Yunnan, SE Tibet, SW Sichuan. 2,450–4,250 m (8,000–14,000 ft).

AM 1960 (Wakehurst) 'Wakehurst', (possibly a hybrid).

Epithet: Reddish-brown.   Hardiness 3.   April–May.

Very common, but variable, in cultivation.

R. rubiginosum Desquamatum Group

R. desquamatum Balf.f. & Forrest

Shrub or **tree** 1.5–8 m (5–26 ft) high; **branchlets** densely scaly.

Leaves oblong-lanceolate, ovate-lanceolate, elliptic-lanceolate or lanceolate; **lamina** 4.5–12.5 cm long, 1.5–4.5 cm broad; **upper surface** olive-green or dark green; **underside** densely scaly scales dark brown or rust-coloured, overlapping or contiguous, with widely or closely scattered larger darker scales.

Inflorescence terminal or terminal and axillary in the uppermost one or two leaves, 3–7 flowers.

Corolla widely funnel-shaped or almost saucer-shaped, 3–4.2 cm long, 3.5–6 cm across, pink, deep rose, purple, lavender, intense bluish-purple or white, with crimson spots, outside scaly.

W and NW Yunnan, NE Upper Burma, SW Sichuan, SE Tibet. 2,500–4,250 m (8,200–14,000 ft).

AM 1938 (Fletcher, Port Talbot) (F24535).

Epithet: Bereft of scales.   Hardiness 3.   April–May.

Not botanically different from R. rubiginosum, *but in cultivation the flower has widely open corollas. Geographically, it is from the western end of the distribution of* R. rubiginosum.

Style longer than the longest stamens.

Leaves less aromatic than the other species in this Subsection.

Leaf underside densely scaly, scales overlapping or contiguous, with or without widely scattered larger darker scales.

Corolla tubular-funnel-shaped or funnel-shaped.

Corolla pink to pinkish-purple, lavender-purple, or rarely white.

R. rubiginosum CNW998 at Deer Dell

Corolla widely funnel-shaped or almost saucer-shaped.

Leaf underside with larger darker scales.

R. rubiginosum Desquamatum Group McLAA19 at Deer Dell

Rhododendron
(Heliolepida)

Subgenus *Rhododendron*
Section *Rhododendron*
Subsection *Lapponica* Sleumer

Lapponicum Series

**Shrubs from moorland or high altitude, small in all their parts
(except *R. cuneatum*).**
Branchlets densely scaly.
Leaves usually less than 2.5 cm long, densely scaly on both surfaces.
Scales on the upper surface almost always prominent.
Inflorescence usually terminal.
Corolla widely funnel-shaped (except *R. intricatum*).
Calyx usually conspicuous.
Stamens 5–10.
Style slender and straight or almost straight, almost always exserted.

Flower colour is mainly at the blue end of the spectrum (lavender, mauve, almost deep blue to purple), but a few species have white, pink or yellow flowers.

Many of the species in Subsection Lapponica *are only found in specialist collections or in colder climates such as Scotland, Scandinavia and Nova Scotia. It is very hard to distinguish between some of them, so only guidelines are given here.*

We have included Davidian's descriptions for **species most likely to be found.** *Others, such as* R. bulu *(Group B2),* R. capitatum *Maxim. not Franch. (Group C2),* R. complexum *(Group A1),* R. nitidulum var. omeiense *(Group B2),* R. polycladum *(Group A2),* R. tapetiforme *(Group A2),* R. thymifolium *(Group A2),* R. websterianum *(Group A2) and* R. yungningense *(Group A2) are Subsection* Lapponica *taxa that are uncommon in cultivation, or of uncertain status. Descriptions of these plants may be found in P.A. and K.N.E. Cox's 'The Encyclopedia of Rhododendron Species' and in Davidian's volume on lepidotes.* **R. setosum** *has been included in Subsection* Fragariiflora.

For ease of identification, the species have been split into three groups, each with two subgroups: once the group and subgroup have been established by studying the scales and flower colour, distinguishing between species common in cultivation is reasonably straightforward. Identifying the rarer species and those in the wild is another matter.

Lapponica Group A: Scales on the leaf underside of one colour

Group A1: Flowers **not** mauve, lavender or blue shades through to deep violet		Group A2: Flowers mauve, lavender, and blue shades through to deep violet	
R. dasypetalum	Rose purple	R. cuneatum	Lavender, lavender-blue to purple
R. flavidum	Yellow	R. fastigiatum	Rose-lavender, lavender blue to deep purple-blue
		R. hippophaeoides var. hippophaeoides	Lavender blue, purplish blue, occasionally white
		R. hippophaeoides var. occidentale	Deep mauve-purple or purple
		R. impeditum	Lavender to purplish blues, violet or blue purple
		R. impeditum Litangense Group	Plum-purple, bluish-purple, reddish-purple or lavender-blue
		R. intricatum	Lavender blue to dark purplish blue
		R. nitidulum var. nitidulum	Rosy lilac to violet purple
		R. polycladum Scintillans Group	Lavender blue to deep blue or blue purple

Lapponica Group B: Scales on the leaf underside of two colours, mainly golden brown with a few scattered darker scales

Group B1: Flowers **not** mauve, lavender, or blue shades through to deep violet		Group B2: Flowers mauve, lavender, and blue shades through to deep violet	
R. orthocladum var. microleucum	White	R. orthocladum var. orthocladum	Pale to deep lavender blue or purple
		R. telmateium	Lavender, rose purple, purplish blue or indigo

Lapponica Group C: Scales on the leaf underside of two colours, with pale and darker scales mixed and almost equal in number

Group C1: Flowers **not** mauve, lavender or blue shades through to deep violet		Group C2: Flowers mauve, lavender, and blue shades through to deep violet	
R. lapponicum	Pink shades (rarely white)	R. russatum	Pale purple, violet, deep rose/violet purple, rarely rose
R. lapponicum Parvifolium Group	Rose or rose-purple		
R. nivale ssp. nivale	Pink shades to magenta to deep purple		
R. nivale ssp. boreale			
R. nivale ssp. australe			
R. rupicola var. rupicola	Deep red purple shades		
R. rupicola var. chryseum	Yellow		

R. dasypetalum Balf.f. & Forrest

Lapponica Gp A1

Rounded **shrub** 30–80 cm (1–2 ft) high; **branchlets** densely scaly.

Leaves oblong, oblong-elliptic or oblong-lanceolate; **lamina** 1–2 cm, shiny, densely scaly; **underside** densely scaly, scales overlapping.

Inflorescence 1–4 flowers.

Corolla widely funnel-shaped, 1.5–2 cm long, bright purplish-rose, outside not scaly, puberulous on the tube and at the base of the lobes.

NW Yunnan. 3,350 m (11,000 ft).

Epithet: With hairy petals.   Hardiness 3–4.   April–May.

Note the epithet. The flowers are unique in the Subsection. Tony Cox considers this to be a hybrid between a Lapponica *and a* Saluenensia *species.*

R. flavidum Franch.

Lapponica Gp A1

(*R. primulinum* Hemsl.)

Shrub 45–90 cm (1.5–3 ft) high; **branchlets** moderately or densely scaly, rather densely puberulous.

Leaves oblong, lanceolate or oblong-oval; **lamina** 0.8–1.8 cm long, 0.4–1 cm broad; **upper surface** dark green, shiny, moderately or densely scaly; **underside** scaly, scales one-half to twice their own diameter apart.

Inflorescence 3–5 flowers.

Calyx 1–4 mm long.

Corolla widely funnel-shaped, 1–1.5 cm long, pale yellow, outside not scaly.

Stamens 10, long-exserted, as long as the corolla or a little longer.

W Sichuan. 3,300–4,300 m (10,800–14,000 ft).

Epithet: Somewhat yellow.   Hardiness 3–4.   April–May.

Rare in cultivation: the taller plants with white flowers commonly found are likely to be hybrids.

Corolla widely funnel-shaped, puberulous on the tube and at the base of the lobes. Scales on the leaf underside dark.

In cultivation, the leaves are usually V- or U-shaped in cross-section.

Leaf under surface densely scaly.

Corolla bright purplish-rose.

R. dasypetalum in W. Berg's garden (M.L.A. Robinson)

Branchlets rather densely puberulous. Leaf upper surface dark green, glossy, the scales being very noticeable against this backgroud. Scales on the leaf underside not overlapping: usually 1–2 times their own diameter apart.

Corolla outside not scaly.

Corolla yellow.

R. flavidum at Deer Dell

R. cuneatum w.w.sm.   Lapponica Gp A2

(*R. ravum* Balf.f. & W.W.Sm.)

Shrub usually large, 0.3–4 m (1–12 ft) high; **branchlets** densely scaly.

Leaves oblong-elliptic, oblong-lanceolate, oval or lanceolate, large; **lamina** 1.4–7 cm long, 0.6–2.8 cm broad; **upper surface** pale green, densely scaly; **underside** densely scaly, scales are characteristic, similar in size and colour, large, dark brown, fawn or creamy-yellow, contiguous or overlapping.

Inflorescence 1–6 flowers.

Calyx large, 0.4–1.2 cm long.

Corolla widely funnel-shaped, large 1.6–3.4 cm long, deep rose, purple, lavender or lavender-blue.

W Yunnan, mid-W and NW Yunnan, SW Sichuan. 2,750–4,300 m (9,000–14,000 ft).

Epithet: Wedge-shaped.   Hardiness 3.   April–May.

Rare in cultivation. Bigger than other members of the Subseries in all its parts. Originally assigned to a separate Subseries.

R. fastigiatum Franch.   Lapponica Gp A2

(*R. capitatum* Franch. not Maxim)

Shrub 0.15–1.2 m (0.5–4 ft) high; **branchlets** densely scaly.

Leaves oblong, elliptic, oblong-elliptic or oblong-oval; **lamina** 0.6–1.8 cm long, 2–8 mm broad; **upper surface** glaucous-grey or pale glaucous-green (in young leaves, markedly glaucous-grey), matt, densely scaly; **underside** densely scaly, scales brown or dark brown, overlapping or nearly contiguous.

Inflorescence 2–5 flowers.

Calyx 3–5 mm long.

Corolla widely funnel-shaped, 1–1.5 cm long, purplish-blue, bright lavender-blue, deep purple-blue or lavender-rose, outside not scaly or sometimes scaly on the lobes.

Mid-W and NW Yunnan. 3,200–4,400 m (10,500–14,500 ft).

AM 1914 (Reuthe), flowers bluish.

AGM 1994.

Epithet: Erect.   Hardiness 3–4.   April–May.

Common in cultivation. The habit is quite variable. Plants (wrongly) labelled R. impeditum *may be this species:* R. impeditum *has non-glaucous young and adult leaves.*

Shrub usually large.
Leaves large, lamina
1.4–7 cm long,
0.6–2.8 cm broad.
Scales similar in size
and colour.
Corolla large
1.6–3.4 mm long.
Unusually large for
this Subsection, and
abberant in it: note
the height.
Leaves usually
almost elliptic, up to
7 cm, matt and
densely scaly above.
Corolla longer than
in any other species
in the Subsection.

Calyx large.

R. cuneatum R11392 at Borde Hill (M.L.A. Robinson)

R. cuneatum (syn. *ravum*) at Deer Dell

Leaf upper surface
glaucous-grey or
pale glaucous-green
(in young leaves,
markedly glaucous-
grey).
Peculiar opaque,
pale (white or
often pinkish)
scales, (eventually
darkening, but some
pale scales should
be found) on the
leaf underside.
The glaucous foliage
is diagnostic,
particularly on the
young growth.

R. fastigiatum at the Rhododendron Species Foundation,
(M.L.A. Robinson)

Rhododendron
(Lapponica Group A2)

R. hippophaeoides var. hippophaeoides
Balf.f. & W.W.Sm.   Lapponica Gp A2
R. hippophaeoides Balf.f. & W.W.Sm.

Shrub 0.2–1.2 m (0.75–4 ft) high; **branchlets** densely scaly.
Leaves oblong, oblong-lanceolate or oblong-oval; **lamina** 1–4 cm long,
    0.4–1.7 cm broad; **upper surface** pale greyish-green, matt, densely scaly;
    underside densely scaly, scales are characteristic, similar in size and colour,
    creamy-yellow, overlapping.
Inflorescence 3–8 flowers.
Corolla widely funnel-shaped, 1–1.3 cm long, lavender-blue, purplish-blue or
    sometimes bright rose.
Stamens 10, short, one-half or two-thirds the length of the corolla.
Style short, shorter than the corolla.
NW Yunnan, SW Sichuan. 2,400–4,250 m (7,900–14,000 ft).
AM 1927 (Bodnant), flowers lavender-blue.
AGM 1993 'Haba Shan'.
Epithet: Resembling sea buckthorn.   Hardiness 3–4.   April–May.

*Common in cultivation; usually upright. In cultivation, the flowers are usually
lavender blue, but a white form has recently been introduced. Inflorescence
generally has fewer flowers than* R. hippophaeodies *var.* occidentale.

R. hippophaeoides var. occidentale Philipson
& Philipson   Lapponica Gp A2
R. fimbriatum Hutch.

Broadly upright **shrub** 0.6–1.2 m (2–4 ft) high; **branchlets** long, erect,
    densely scaly.
Leaves lanceolate, oblong-lanceolate or oblong; **lamina** 1.5–3.5 cm long,
    0.5–1.5 cm broad; **upper surface** pale green, matt, densely scaly, scales
    overlapping; **underside** densely scaly, the scales are characteristic, similar in
    size and colour, fawn or creamy-yellow, overlapping or nearly contiguous.
Inflorescence 3–9 flowers.
Corolla widely funnel-shaped, 1.2–1.6 cm long, deep mauve-purple or purple,
    outside not scaly.
SW Sichuan, NW Yunnan. 3,650 m (12,000 ft).
Epithet: Minutely fringed.   Hardiness 3–4.   April–May.

Forms in cultivation have purple flowers and a narrower leaf. R. tsaii, *which is rare in
cultivation and not separately described here, is very close to this variety.*

Greyish foliage and overlapping yellow scales on the leaf underside,

Leaf upper surface pale greyish-green, matt.

The scales are characteristic, similar in size and colour, creamy-yellow, overlapping.

Corolla usually lavender-blue or purplish-blue.

Stamens one-half or two-thirds the length of the corolla.

R. hippophaeoides var. *hippophaeoides* 'Haba Shan' at Deer Dell

The scales are characteristic, similar in size and colour, fawn or creamy-yellow, overlapping or nearly contiguous.

Inflorescence generally many flowers.

Corolla deep mauve-purple or purple.

The stamens about as long as the corolla and the longer style differentiate this from *R. hippophaeoides* var. *hippophaeoides*.

Leaves lanceolate, oblong-lanceolate or oblong.

R. hippophaeoides var. *occidentale* as (*R. fimbriatum*) F22197A at Deer Dell

Rhododendron
(Lapponica Group A2)

R. impeditum Balf.f. & W.W.Sm.

Lapponica Gp A2

Small very compact, spreading or rounded **shrub** 10–80 cm (0.3–2 ft) high;
    branchlets short, thick, densely scaly.
Leaves elliptic, oblong-oval, oblong or oblong-elliptic; **lamina** 0.5–1.6 cm long,
    3–6 mm broad; **upper surface** dark green in both the adult and young leaves,
    not glaucous, somewhat shiny, scaly; **underside** densely scaly, scales one-half
    their own diameter apart.
Inflorescence 1–3 flowers.
Corolla widely funnel-shaped, 0.9–1.6 cm long, pale or deep purple, intense blue-
    purple, bright violet, lavender or pale or deep rose-purple.
Stamens 10, long-exserted.
Mid-W and NW Yunnan, SW Sichuan. 3,650–4,900 m (12,000–16,000 ft).
AM 1944 (Sunningdale), flowers violet (R59263).
AGM 1993.
Epithet: Tangled.   Hardiness 3–4.   April–May.

Often confused with R. fastigiatum, *which has similar flowers.*

R. impeditum Litangense Group

Lapponica Gp A2

R. litangense Balf.f. ex Hutch.

Habit more upright than *R. impeditum*: resembles a green-foliaged
R. fastigiatum. **Leaf upper surface** dark green, shiny, scaly; **underside**
scaly, scales one-half to twice their own diameter apart. **Corolla** usually
lavender, blue-purple or reddish-purple.

Dark green young and adult foliage and a dense compact habit compared to *R. fastigiatum*

Somewhat shiny scaly upper leaf surface in both the mature and young leaves: lower surface with amber to mid-brown scales which are not overlapping.

Very compact, spreading or rounded shrub.

Leaf under surface densely scaly.

Inflorescence 1–3 flowers.

Corolla usually lavender, blue-purple or violet.

R. impeditum at the Rhododendron Species Foundation (M.L.A. Robinson)

R. impeditum Litangense Group in W. Berg's garden

R. intricatum Franch.   Lapponica Gp A2

(*R. blepharocalyx* Franch.)

Shrub 15–90 cm (0.5–3 ft) high; **branchlets** densely scaly.

Leaves elliptic, oblong-elliptic, oblong-oval or oval; **lamina** 0.5–1.4 cm long,
0.3–1 cm broad; **upper surface** pale greyish-green, matt, densely scaly, scales
overlapping; **underside** densely scaly, the scales are characteristic, similar in size
and colour, large, pale greyish gold, overlapping.

Inflorescence 2–5 flowers.

Calyx minute, 0.5–1 mm long.

Corolla tubular-funnel shaped, small, 0.7–1.3 cm long, lavender-blue, dark
purplish-blue, dark blue or purple-blue, outside not scaly.

Stamens 7–10, included in the corolla-tube, very short, shorter than the corolla.

Style very short, shorter than the stamens.

W and SW Sichuan, NW Yunnan. 3,350–4,600 m (11,000–15,100 ft).

FCC 1907 (Veitch), flowers rosy-red.

Epithet: Entangled.   Hardiness 3–4.   April–May.

Inflorescence often many-flowered in cultivation. **Corolla shape unique in this
Subsection, and showing some resemblance to Section Pogonanthum.**

R. nitidulum var. nitidulum Rehd. & E.H.Wilson

R. nitidulum Rehd. & E.H.Wilson   Lapponica Gp A2

Broadly upright **shrub** 0.6–1.5 m (2–5 ft) high; **branchlets** short thick, densely scaly.

Leaves ovate or elliptic; **lamina** 0.7–1.2 cm long, 5–7 mm broad; **upper
surface** dark green, shiny, densely scaly; **underside** densely scaly, scales
brown, contiguous or overlapping.

Inflorescence 1–2 flowers.

Calyx 1–3 mm long.

Corolla widely funnel-shaped, 1.2–1.3 cm long, violet-purple, outside not scaly.

W Sichuan. 3,300–4,000 m (10,900–13,100 ft).

Epithet: Shining.   Hardiness 3–4.   April–May.

R. nitidulum *var.* omeiense *has a few dark scales on the leaf lower surface: the
distinction from the suggested var.* nitidulum *may be meaningless, and the name
of the former is certainly so, as both are found on Mt Omei.*

Corolla tubular-funnel-shaped.

Style very short, shorter than the stamens.

Matt leaf **upper** surface with overlapping greyish scales, overlapping on the underside.

Inflorescence often many flowered in cultivation.

Leaf under surface densely scaly, the scales similar in size and colour, large, pale greyish gold, overlapping.

Stamens 7–10, included in the corolla-tube.

Corolla shades of blue or purplish blue.

R. intricatum at the Rhododendron Species Foundation (M.L.A. Robinson)

The leaves are small, sharply pointed, usually ovate-elliptic.

In cultivation, the flowers are light to dark blue purple.

Matt leaf upper surface dark with overlapping or almost overlapping scales.

Leaves ovate or elliptic under surface densely scaly, scales contiguous or overlapping.

Inflorescence 1–2 flowers.

Corolla violet-purple.

R. nitidulum var. *nitidulum* (J.C. Birck)

Rhododendron
(Lapponica Group A2)

R. polycladum Franch. Scintillans Group   Lapponica Gp A2

R. scintillans Balf.f. & W.W.Sm.
(*R. compactum* Hutch.)

Shrub 0.15–1 m (0.5–3 ft) high; **branchlets** slender, densely scaly, glabrous.
Leaves lanceolate or oblong; **lamina** 0.8–2.3 cm long, 2–7 mm broad; **upper surface** dark green, shiny, densely scaly; **underside** densely scaly, scales one-half their own diameter apart or nearly contiguous.
Inflorescence 2–5 flowers.
Calyx 0.5–2 mm long.
Corolla widely funnel-shaped, 1–1.5 cm long, lavender-blue, deep blue-purple or pale rose-purple, outside not scaly.

Mid-W and W Yunnan. 3,350–4,000 m (11,000–13,000 ft).

AM 1924 (Bodnant), flowers purplish-rose.

FCC 1934 (Exbury), flowers lavender-blue.

AGM 1993 'Policy'.
Epithet: Sparkling.   Hardiness 3–4.   April–May.

R. polycladum *is almost always (in cultivation) as* Scintillans *Group.*

R. orthocladum var. microleucum Philipson &
Philipson   Lapponica Gp B1

Differs from the R. orthocladum *var.* orthocladum *species type in that it is a compact shrub to 60 cm (2 ft) in height and width. The corolla is white and the leaves vary from oblong to lanceolate. This is the more common form and is an albino form from wild seed.*

Note the epithet: shiny upper leaf surface, lower brown or greyish with dense brown scales. Narrow leaves and flowers of a lovely shade of blue.

R. polycladum Scintillans Group (B. Starling)

R. orthocladum var. *microleucum* at Deer Dell

R. orthocladum var. orthocladum

R. orthocladum Balf.f. & Forrest   **Lapponica Gp B2**

Shrub 0.3–1.2 m (1–4ft) high; **branchlets** densely scaly.

Leaves lanceolate, linear-lanceolate or linear; **lamina** 0.5–2 cm long, 1–4 mm broad; **upper surface** green or pale green, densely scaly; **underside** densely scaly, scales overlapping or nearly contiguous, with or without scattered dark brown scales.

Inflorescence 1–4 flowers.

Calyx minute, 0.5–1 mm long.

Corolla widely funnel-shaped, 0.8–1.3 cm long, lavender, pale purplish-blue, deep blue-purple or purple, outside not scaly.

Style very short, shorter than the stamens.

NW Yunnan, SW Sichuan. 2,750–4,300 m (9,000–14,000 ft).

Epithet: With straight twigs.   Hardiness 3–4.   April–May.

In cultivation, usually upright to 1 m plus.

R. telmateium Balf.f. & W.W.Sm.   **Lapponica Gp B2**

The former R. diacritum, R. drumonium and R. idoneum are now included here.
R. telmateium is a poorly defined species, so the form commonest in cultivation is described below.

Small **shrub** 10–90 cm (0.33–3 ft) high; **branchlets** short, slender, densely scaly.

Leaves lanceolate, oblong-lanceolate or oblanceolate, tiny or small; **lamina** 0.3–1.3 cm long, 2–5 mm broad; **upper surface** green, densely scaly; **underside** densely scaly, scales pale brown or brown, overlapping, with closely or widely scattered dark brown scales.

Inflorescence 1–2 flowers.

Corolla widely funnel-shaped, small 0.8–1.3 cm long, rose-purple, purplish-blue or deep indigo blue, outside scaly on the lobes.

NW Yunnan, SW Sichuan. 2,600–4,600 m (8,500–15,000 ft).

Epithet: From the marshes.   Hardiness 3–4.   April–May.

Narrowly elliptic lanceolate, linear-lanceolate or linear leaves having pale golden or brown scales and scattered darker ones on the underside.

1–4-flowered inflorescence of small lavender blue flowers.

Style very short, shorter than the stamens.

Leaf under surface densely scaly, scales overlapping or nearly contiguous.

Corolla lavender or shades of blue-purple.

R. orthocladum var. *orthocladum* at the Rhododendron Species Foundation (M.L.A. Robinson)

Scales on upper leaf surface overlapping, pale gold.

The leaf usually has a mucronate tip, and the upper surface appears dark green with a little lightening from the dense pale gold scales, which are always present on both surfaces.

In cultivation, usually upright with thin branches and very small elliptic pointed leaves.

Flowers 1–2, small, usually lavender-blue or red purple.

R. telmateium at Deer Dell

R. lapponicum (L.) Wahlenb.

Lapponica Gp C1

(*R. palustre* Turcz.)

Dwarf **shrub** 5–45 cm (0.16–1.5 ft) or sometimes 60–90 cm (2–3 ft) high;
 branchlets short, densely scaly.
Leaves elliptic, oblong, or oblong-obovate; **lamina** 0.5–1.5 cm long, 2–5 mm
 broad; **upper surface** pale green, matt, densely scaly; **underside** densely scaly,
 scales dark brown or brown, overlapping.
Inflorescence 2–5 flowers.
Pedicel densely scaly.
Corolla widely funnel-shaped, small, 0.7–1 cm long, rose-purple, pinkish-purple
 or pink.
Stamens 5–10.

Arctic regions, Lapland, N Sweden, N Norway, Greenland, Labrador, NE USA,
 Canadian Arctic. From sea level up to 1,850 m (6,000 ft).
Epithet: From Lapland.　　Hardiness 3–4.　　March–April.

R. lapponicum Parvifolium Group

Lapponica Gp C1

R. parvifolium Adams.

Shrub 30–90 cm (1–3 ft) high; **branchlets** densely scaly.
Leaves oblong-elliptic, elliptic, oblong-oval, oblong or oblong-lanceolate;
 lamina 0.6–2.5 cm long, 0.3–1 cm broad; **upper surface** pale green,
 matt, densely scaly; **underside** densely scaly, scales overlapping or
 continuous or one-half their own diameter apart.
Inflorescence 2–5 flowers.
Calyx 1–2 mm long.
Corolla widely funnel-shaped, 1–1.3 cm long, deep reddish, bright rose or
 rose-purple, outside not scaly.
Stamens 10.

From Siberia to NE China, N Korea, Sakhalin, the north island of Japan, the
 Aleutians, Alaska.
Epithet: With small leaves.　　Hardiness 3–4.　　January–March.

Usually more upright in cultivation than R. lapponicum *and has larger flowers.*

Falls within the botanical decription of R. lapponicum, *but of sufficient
horticultural merit to be given Group status. The earliest flowering clones
are sometimes referred to as* **'Confertissimum Group'**. *This Group of*
R. lapponicum *makes the best garden plants as* R. lapponicum *itself is very
difficult to grow in cultivation.*

Bicoloured contiguous scales on both surfaces of the leaf: upper and lower surfaces densely scaly; scales dark brown or brown. Stamens 5–10.

Leaf underside densely scaly, scales overlapping.

Corolla small, 0.7–1 cm, pink or purple shades.

R. lapponicum F16299 (Davidian)

R. nivale ssp. nivale
Lapponica Gp C1

R. nivale Hook.f.
(*R. paludosum* Hutch. & Kingdon Ward)

Shrub 8–30 cm (0.25–1 ft) or sometimes 60–90 cm (2–3 ft) or rarely 1.5 m (5 ft)
high; **branchlets** short, densely scaly.

Leaves elliptic, ovate, oval or oblong-oval, tiny or small; **lamina** 0.2–1 cm long;
upper surface pale greyish-green, matt, densely scaly; **underside** densely scaly,
scales brown or pale brown, overlapping or nearly contiguous, with closely or
widely scattered dark brown scales.

Inflorescence 1–2 flowers.

Corolla widely funnel-shaped, small, 0.8–1.1 cm long, deep purple, deep pink,
purple, reddish-purple or lilac.

Nepal, Sikkim, Bhutan, S Tibet. 3,000–5,800 m (10,000–19,000 ft).
Epithet: Snowy.   Hardiness 3–4.   April–May.

R. nivale ssp. boreale Philipson & Philipson
Lapponica Gp C1

(*R. nigropunctatum* Franch., *R. ramosissimum* Franch., *R. alpicola* Rehder & E.H.Wilson,
R. violaceum Rehder & E.H.Wilson, *R. stictophyllum* Balf.f., *R. batangense* Balf.f.)

*Six former species have been submerged here. They are distinguished from
ssp.* nivale *by the calyx, which in ssp.* boreale *is a mere rim. Generally taller
than ssp.* nivale *in cultivation. Rounded leaves, mucronulate, and the two
colours of scales are usually not so strongly contrasting. Some forms have
good blue flowers.*

R. nivale ssp. australe Philipson & Philipson
Lapponica Gp C1

Extremely rare in cultivation. Leaf apex acute. **Calyx lobes 2–4.5 mm,
margin ciliate.**

Rounded small leaves, not mucronulate.

The contrast between the two colours of scales — pale gold and dark brown — is distinctive in all three subspecies of *R. nivale*, but especially marked here.

Calyx lobes 2–4 mm lobed, scaly, not ciliate.

Leaves tiny, lamina 0.2–1 cm long, upper surface pale greyish-green.

Inflorescence 1–2 flowers.

Corolla small, 0.8–1.1 cm long, purple shades or lilac.

R. nivale ssp. *nivale* (as *R. paludosum*) at RBG Edinburgh

Calyx minute or obsolete

R. nivale ssp. *boreale* (as *R. stictophyllum*) (Davidian)

R. rupicola var. rupicola

Lapponica Gp C1

R. rupicola W.W.Sm.
(*R. achroanthum* Balf.f. & W.W.Sm.)

Shrub 0.08–1.2 m (0.25–4 ft) high; **branchlets** densely scaly.

Leaves elliptic, obovate, oblong-oval or oval; **lamina** 0.8–2.5 cm long, 0.4–1.4 cm broad; **upper surface** green, matt, densely scaly, scales overlapping or contiguous; **underside** densely scaly, scales bicolorous, dark brown and yellow, overlapping or nearly contiguous.

Inflorescence 2–8 flowers.

Calyx crimson-purple, outside densely scaly.

Corolla widely funnel-shaped, 1–1.6 cm long, deep plum-crimson or plum-purple, outside scaly.

NW Yunnan, E and NE Upper Burma, SW Sichuan. 3,350–4,300 m (11,000–14,000 ft).

Epithet: Dweller in stony places.   Hardiness 3–4.   April–May.

Merges with R. russatum *in the wild.*

R. rupicola var. chryseum (Balf.f. & Kingdon Ward)

Philipson & Philipson

Lapponica Gp C1

R. chryseum Balf.f. & Kingdon Ward

Shrub 15–90 cm (0.5–3 ft) high; **branchlets** densely scaly.

Leaves oblong, obovate, oval or elliptic; **lamina** 0.8–2 cm long, 0.4–1 cm broad; **upper surface** greyish green or pale green, matt, densely scaly; **underside** densely scaly, scales large bicolorous, dark brown and yellow; or sometimes unicolorous, dark brown, twice their own diameter apart or contiguous or overlapping.

Inflorescence 1–6 flowers.

Corolla widely funnel-shaped, 0.9–1.7 cm long, bright yellow, pale or deep yellow, golden-yellow or greenish, outside scaly on the lobes.

Stamens 5–10.

NW Yunnan, SE Tibet, SW Sichuan. 3,350–4,800 m (10,000–15,650 ft).

Epithet: Golden-yellow.   Hardiness 3–4.   April–May.

Yellow flowers, so in flower can be confused only with R. flavidum, *but* R. rupicola *var.* chryseum *has scales of two colours and a more compact habit.*

R. rupicola *var.* muliense

Rare.

Yellow flowers and a calyx that is both scaly and ciliate.

Corolla deep plum-crimson or plum-purple.

Leaves fairly large, broadly elliptic; upper surface matt.

Bicoloured yellow and dark scales intermingled on both surfaces.

More dark scales than most other species.

R. rupicola var. *rupicola* at the Rhododendron Species Foundation (M.L.A. Robinson)

R. rupicola var. *rupicola* KW7048 at Deer Dell

Corolla bright yellow, pale or deep yellow, golden-yellow or greenish.

Calyx ciliate.
Leaf scales of two colours.

R. rupicola var. *chryseum* at Deer Dell

Rhododendron
(Lapponica Group C1)

R. russatum Balf.f. & Forrest   Lapponica Gp C2
(*R. cantabile* Hutch.)

Shrub 0.35–1.5 m (0.5–5 ft) high; **branchlets** densely scaly.

Leaves oblong-elliptic, oblong-oval, oval, oblong or oblong-lanceolate; **lamina** 0.8–6.5 cm long, 0.4–2.3 cm broad; **upper surface** dark or pale green, densely scaly, scales nearly contiguous or overlapping; **underside** densely scaly, the scales large, bicoloured, dark brown and yellow, or unicolorous, cinnamon-red, dark brown or brown, overlapping or one-half their own diameter apart.

Inflorescence 3–10 or rarely 14 flowers.

Calyx 3–6 mm long.

Corolla widely funnel-shaped, 1.3–2 cm long, deep purple-blue, rose-purple, deep violet-purple, light purple or violet, outside not scaly.

NW and mid-W Yunnan, NE Upper Burma, SW Sichuan. 3,350–4,300 m (11,000–14,000 ft).

AM 1927 (Werrington), flowers intense violet-blue.

FCC 1933 (Exbury), flowers intense purple.

AGM 1993.

Epithet: Reddened.   Hardiness 3–4.   April–May.

Subgenus *Rhododendron*
Section *Rhododendron*
Subsection *Ledum* Kron & Judd

Shrubs to 2 m.

Young shoots both lepidote and tomentose or glandular.

Leaves evergreen, narrow, upper surface recurved.

Underside whitish papillate, scaly with golden rimless scales often setulose; sometimes with a brown tomentum.

Inflorescence as a terminal corymb, many flowers, white, usually relatively small.

Corolla small.

Stamens 7–12.

From cold or sub-Artic northern regions.

Ledum has been assigned to Rhododendron *since Davidian wrote his descriptions: those below have been added by the present authors.*

Larger, wider, leaves than other species in the Subsection, and splendid deep-coloured flowers.

The upper leaf surface, though densely scaly, is not obviously so at first glance. Corolla not scaly, shades of blue, violet or violet purple.

Scales large, bicoloured, dark brown and yellow, or unicolorous, cinnamon-red, or dark brown or brown.

R. russatum F21932 at Deer Dell

R. tomentosum ssp. *tomentosum* at Deer Dell

R. groenlandicum (Oeder) Kron & Judd

(*Ledum groenlandicum* Oeder, *L. latifolium* Jacq., *L. pacificum* Small)

Upright **shrub** 0.7–2m (2–6 ft); **branchlets** rusty tomentose.
Leaves aromatic when bruised, narrowly oblong, 2–6 cm (1–2.5 inches) long,
0.5–1.5 cm (0.25–0.8 inches) wide, margins very revolute; **upper surface** dark
green with a few loose hairs; **underside** tomentose and lepidote, with a thick
indumentum, upper layer felty, ferrugineous, obscuring the scales, and a lower
layer of short white hairs; densely scaly and with reddish glands.
Petiole short, 1–5 mm.
Inflorescence to 5 cm across, many flowers.
Pedicel slender, downy.
Calyx minute.
Corolla white, rotate, 4–8 mm.
Stamens 7–10.
Ovary glandular.
Style glabrous.
Greenland, Canada, Northern USA. 1,000–1,800 m (3300–6,000 ft).
Epithet: From Greenland.   Hardiness 4.   May–June.

This is the species most often seen in cultivation.

R. hypoleucum (Kom.) Harmaja

Ledum hypoleucum Kom., *L. palustre* var. *diversipilosum* Nakai

Erect **shrub** 0.5–1.1 m (1.5–4 ft); **branchlets** rusty tomentose.
Leaves 1.7–8 cm long (1–3.2 inches), 0.5–2cm (0.2–0.8 inches) broad, oblong-
elliptic, apex acuminate, margins revolute, with long brown crisped hairs;
upper surface dark green, somewhat tomentose with ferrugineous hairs;
underside glaucous, densely white-pubescent, somewhat papillate, scales 1–3
times their own diameter apart, midrib only with long crisped rust-coloured
hairs.
Petioles 2–7 mm.
Inflorescence many flowers.
Calyx small.
Corolla small, white, rotate.
Stamens 9–12.
Ovary densely pubescent and scaly.
Style glabrous.
NE Russia, Japan.
Epithet: White beneath.   Hardiness 4.   June–July.

Leaves, narrowly oblong, about twice as wide as the other *Ledum* species, with indumentum.
Leaf underside tomentose and lepidote, with a thick indumentum, upper layer felty, ferrugineous, obscuring the scales.

R. groenlandicum at Deer Dell

Leaf margins revolute, with long brown crisped hairs, undersides white pubescent with long reddish hairs restricted to the midrib.

R. hypoleucum at Deer Dell

Rhododendron
(Ledum)

R. neoglandulosum Harmaja

(*Ledum glandulosum* Nutt., *L. californicum* Kellogg)

Erect **shrub** 0.5–2 m (1.5–6 ft); **branchlets** puberulous and glandular.
Leaves 1.2–5 cm (0.5–2 inches) long by 0.5–2 cm (0.25–0.8 inches) wide,
    broadly elliptic-oval, or ovate apex acuminate, margins flat or slightly revolute;
    upper surface dark green; **underside** glabrous or more or less pubescent,
    with white glistening scales 1–2 times their own diameter apart.
Petioles 0.4–1 cm.
Inflorescence many flowers, about 5 cm across.
Calyx small, margins ciliate.
Corolla small, cupped, white, rotate.
Pedicels 1.5–4 cm, often glandular.
Stamens 8–12.
Ovary densely glandular, scaly.
Style sparsely glandular.

NW USA.
Epithet: Glandular.   Hardiness 4.   May–August.

R. × columbianum

This is a hybrid between R. neoglandulosum *and* R. groenlandicum.

Leaves, slightly recurved and usually with a few rust-coloured hairs on
    the underside.

R. tomentosum ssp. tomentosum

R. tomentosum (Stokes) Harmaja
(*Ledum palustre* L., *Ledum palustre* var. *dilatatum* Wahlenb.)

Dwarf **shrub** of thin habit, 0.3–1.2 m (1–4 ft) high; **branchlets** rusty
    tomentose, glandular.
Leaves 1–5 cm (0.4–1.5 inches) long, 0.2–0.8 cm (0.1–0.3 inches) wide, linear,
    margin strongly revolute; **upper surface** matt dark green, **underside** with a
    bistrate indumentum, upper layer thick woolly, ferrugineous, lower layer of
    short setulose hairs and/or reddish glands, always scaly.
Inflorescence many flowers.
Calyx minute.
Corolla 1.2 cm across white, rotate.
Stamens 7–10.
Ovary glandular.
Style glabrous.

N & Central Europe, Russia (to S Siberia). 0–2,000 m (0–6,000 ft).
Epithet: Tomentose.   Hardiness 4.   June–July.

R. tomentosum ssp. subarcticum (Harmaja) G.Wallace

Ledum minus Hort., *L. palustre* var. *decumbens* Aiton., *L. subarcticum* Harmaja

Indumentum essentially unistrate; underside with few or no setulose hairs.
Leaves 0.6–2 cm (0.2–0.8 inches) long, 0.1–0.3 cm (0.04–0.1 inches) wide.
Arctic regions of Europe, America and Russia, also Japan (Hokkaido) and Korea.

Leaves broadly elliptic-oval or ovate, more or less flat and almost glabrous underneath.

R. neoglandulosum (A. Dome)

Compact growth. Narrower leaves than those of *R. groenlandicum*. Leaf indumentum: upper layer thick, woolly, ferrugineous; lower layer of short setulose hairs.

R. tomentosum ssp. *tomentosum* at Deer Dell

Rhododendron
(Ledum)

Subgenus *Rhododendron*
Section *Rhododendron*
Subsection *Lepidota* Sleumer
Lepidotum Series and Subseries

Small or medium-sized shrubs.
Leaves evergreen or deciduous.
Pedicel slender, longer than the corolla.
Corolla rotate or rotate campanulate.
Style short, stout and sharply bent.

R. cowanianum Davidian

Shrub 0.9–2.4 m (3–8 ft) high; **branchlets** scaly.
Leaves deciduous, obovate or oblong-obovate; **lamina** 2.3–6 cm long, 1.2–2.9 cm broad, base decurrent on the petiole; **upper surface** bright green, shiny, scaly, margin bristly; **underside** scaly, scales yellowish-green or pale brown, 1–4 times their own diameter apart.
Petiole bristly.
Inflorescence 2–5 flowers.
Pedicel slender, 1–1.8 cm long.
Corolla rotate-campanulate, 1.3–1.8 cm long, reddish-purple, crimson-purple or pink.
Central Nepal. 3,050–3,950 m (10,000–13,000 ft).
Epithet: After Dr. J.M. Cowan (1892–1960).   Hardiness 3.   April–May.

Very rare in cultivation.

R. lepidotum (syn. *R. obovatum*) at Deer Dell

Larger in growth and in leaf than *R. lepidotum*.

Leaves deciduous, upper surface shiny, margin and petiole bristly.

Scales on the leaf underside are less dense.

Similar to *R. baileyi* in flower.

Distinctive because the leaves emerge with the flowers, while some flowers drop as others in the truss open.

Corolla rotate-campanulate, usually reddish-purple or crimson purple.

R. cowanianum McB3247 at Glendoick (M.L.A. Robinson)

R. lepidotum Wall. ex. G.Don

R. elaeagnoides Hook.f., *R. obovatum* Hook.f., *R. salignum* Hook.f., *R. sinolepidotum* Balf.f., *R. cremastes* Balf.f. & Farrer

Shrub 5–90 cm (0.16–3 ft) or sometimes up to 1.5 m (5 ft) high; **branchlets** warty, scaly, not bristly.
Leaves evergreen or sometimes deciduous, oblanceolate, lanceolate or oblong-obovate;
lamina 1–3.2 cm long, 0.4–1.1 cm broad; **upper surface** densely scaly; **underside** pale glaucous green, densely scaly, scales overlapping to one-half their own diameter apart.
Inflorescence 1–3 or rarely 4 flowers.
Pedicel slender, longer than the corolla.
Corolla rotate, 1–1.6 cm long, white, pink, purple, rose, scarlet, crimson or yellow, outside scaly.
Style short, stout and sharply bent.

Kashmir, Punjab, Nepal, Sikkim, Bhutan, Assam to S and SE Tibet, NE Upper Burma and NW Yunnan. 2,450–4,900 m (8,000–16,000 ft).
Epithet: Beset with scales.　　Hardiness 2–3.　　April–June.

A very widely occurring and variable species. The species listed above as synonyms may be found in cultivation so labelled, but all have been submerged into R. lepidotum *by botanical authorities. 'Reuthe's Purple' AM may be a hybrid.*

R. lowndesii Davidian

Dwarf **shrub** 5–12 cm (2–5 inches) or rarely 30 cm (1 ft) high; **branchlets** thin, scaly, pubescent, pilose.
Leaves deciduous, obovate, oblong-obovate or oblanceolate; **lamina** chartaceous, 0.9–2.8 cm long, 0.4–1.2 cm broad, base decurrent on the petiole; **upper surface** bright green, scaly or not scaly, puberulous; margin bristly; **underside** pale green, scaly, scales yellowish-green, 2–5 times their own diameter apart.
Petiole bristly.
Inflorescence 1–2 flowers.
Pedicel slender, bristly.
Corolla rotate-campanulate, 1.3–1.7 cm long, pale yellow, with carmine or greenish-yellow spots, outside scaly.

Nepal. 3,050–4,600 m (10,000–15,000 ft).
Epithet: After Col. D.G. Lowndes, who discovered it in 1950.　　Hardiness 3.
May–June.

Extremely rare in gardens: a creeping shrub for the alpine enthusiasts.

Branchlets warty,
scaly, not bristly.
Leaf upper surface
densely scaly;
underside pale
glaucous green,
densely scaly.
Pedicel slender,
longer than the
corolla.
Style short, stout
and sharply bent.
Leaves may be
deciduous.
Note the variation in
flower colour: white,
pink, rose, purple, red,
crimson or yellow.

R. lepidotum (J.C. Birck)

Resembles
R. lepidotum
Eleagnoides Group,
but the branchlets,
leaf margins, and
pedicels have
bristles and the leaf
underside is more
sparsely scaly.
Leaves deciduous.

Leaf lamina chartaceous,
base decurrent on
the petiole.
Scales on the underside
2–5 times their own
diameter apart.
Corolla pale yellow.

R. lowndesii at Glendoick (M.L.A. Robinson)

Rhododendron
(Lepidota)

Subgenus *Rhododendron*
Section *Rhododendron*
Subsection *Maddenia* Sleumer

Maddenii and Ciliatum Series: Ciliicalyx, Maddenii and Megacalyx Subseries

*This large Subsection has been subject to many botanical revisions: the number of species keeps changing, the differences between many related species being small. Some are extremely difficult to identify and, **in cultivation, many cannot be identified out of flower**.*

The Subsection is divided here into groups that assist identification. The division follows the taxonomy of Cullen, but uses the old series where these will help.

Only the species most likely to be seen will be detailed here as many of the species are not hardy and are rarely encountered. A full list may be found on the RHS Rhododendron Group website: www.rhodogroup-rhs.org

The Subsection will be divided into five groups:

Group A. Stamens (15)–17–25; calyx variable.
(*R. maddenii* and its relatives):

> *R. maddenii* (ssp. *maddenii* and ssp. *crassum*).

Group B. Stamens 10; calyx large.
(the former *R. megacalyx* Subseries):

> *R. dalhousiae* (var. *dalhousiae* and var. *rhabdotum*), *R. headfortianum*, *R. liliiflorum*, *R. lindleyi*, *R. megaclayx* and *R. nuttallii*.

Group C. Stamens 8–10; calyx small.
Group C1. The former Ciliatum series and *R. johnstoneanum*.
Branchlets bristly.
Leaf margins usually ciliate.
Petiole and usually ciliate.
Calyx bristly.
Style impressed.

> *R. burmanicum*, *R. ciliatum*, *R. fletcherianum*, *R. johnstoneanum*, *R. valentinianum* (including var. *changii* and var. *oblongilobatum*).

Group C2.
Leaves less rugolose and ciliate than species in Group C1.
Style impressed.

> *R. formosanum* (var. *formosanum* and var. *inequale*), *R. leptocladon*, and *R. pachypodum*.

The other species with an impressed style (four in cultivation) are uncommon and are not listed.

Group C3: The Ciliicalyx aggregate and other species with a tapered style (five species in cultivation).
Style tapered.

> *R. ciliicalyx*, *R. horlickianum*, *R. lyi*, *R. parryae* and *R. veitchianum* (which includes the submerged *R. cubittii*).

Subsection *Maddenia* Sleumer

Group A. *R. maddenii* and its relatives

The old Maddenii Subseries

R. maddenii (ssp. *maddenii* [including the submerged *R. brachysiphon* and *R. polyandrum*] and ssp. *crassum* [including the submerged *R. manipurense*]). All of these taxa could be representatives of a single species.

Leaves dark, usually shiny or somewhat shiny above, principal vein impressed, other veins slightly impressed, so the surface is almost flat or only slightly rugulose.

Underside glaucous in young leaves, the glaucousness often persisting, densely scaly, lateral veins (darker than the lamina) easily visible if not prominent,

Leaf tip somewhat cucullate, leaf base decurrent or slightly decurrent onto the petiole.

Stamens 17–25, or rarely 15.

Calyx small.

Corolla white or pale pink.

Late-flowering.

Leaves of *R. maddenii* and its varieties (M.L.A. Robinson).
Top to bottom, 2 leaves each (left column): *R. maddenii* ssp.
maddenii and *R. maddenii* ssp. *crassum*; (right column):
R. maddenii ssp. *maddenii* as *R. brachysiphon* and as
R. polyandrum. Bar = 5 cm

Rhododendron
(Maddenia)

R. maddenii ssp. maddenii   Maddenia Gp A

R. maddenii Hook.f.
(*R. calophyllum* Nutt., *R. jenkinsii* Nutt., *R. brachysiphon* Balf.f., *R. polyandrum* Hutch.)

Shrub 1–3 m (3–12 ft) high; **stem** and **branches** with smooth, brown, flaking bark; **branchlets** scaly.

Leaves lanceolate, oblong-lanceolate or ovate-lanceolate; **lamina** 5.3–14.5 cm long, 2–5.6 cm broad; **upper surface** dark green or bright green; **underside** densely scaly, scales overlapping or half their own diameter apart.

Inflorescence 2–6 flowers.

Calyx 2–9 mm long.

Corolla tubular-funnel-shaped, 6.5–10 cm long, fragrant, rather densely scaly outside, white, white suffused with rose or pale pink, with or without greenish or pink blotch.

Sikkim, Bhutan. 1,500–2,750 m (5,000–9,000 ft).

AM 1938 (Trengwainton), flowers white, greenish within.

AM 1978 (Sandling Park) 'Ascreavie', flowers white flushed rose (L&S1141).

AGM 1993.

Epithet: After Lt.-Col. E. Madden (d. 1856), traveller in India.   Hardiness 1–2. May–June.

Hardier forms have been grown outside in Southern England for many years.

R. maddenii ssp. crassum (Franch.) Cullen

R. crassum Franch.   **Maddenia Gp A**
R. manipurense Balf.f. & Watt.

Shrub 0.6–4.6 m (2–15 ft), or sometimes a **tree** 3–6.10 m (10–20 ft) high; **stem** and **branches** with rough bark; **branchlets** densely or moderately scaly.

Leaves oblong-lanceolate, lanceolate, elliptic or oblong-obovate; **lamina** 6–18 cm long, 2.3–6.2 cm broad; **upper surface** dark green, shiny; **underside** densely scaly, scales brown or dark brown, their own diameter apart.

Petiole densely scaly.

Inflorescence 3–5 flowers.

Calyx 0.6–2 cm long.

Corolla tubular-funnel-shaped, 5–10 cm long, fragrant, white, creamy-white or white tinged pink, rather densely scaly all over the outside.

Stamens 15–21.

Mid-W, W, NW Yunnan, North Burma, S and SE Tibet. 1,600–4,300 m (5,250–14,000 ft).

AM 1924 (Sunninghill).

AGM 1993.

Epithet: Fleshy.   Hardiness 1–3.   April–August.

Stem and branches with smooth, brown, flaking bark.

R. maddenii ssp. *maddenii* at Deer Dell

Stem and branches with rough bark. Smooth, dark, leaf upper surface dusted with scales.

Corolla white, creamy-white or white tinged pink.

R. maddenii ssp. *crassum* (as *R. manipurense*) at Deer Dell

Rhododendron
(Maddenia Group A)

Subsection *Maddenia* Sleumer
Group B. The former megacalyx Subseries

R. dalhousiae (var. *dalhousiae* and var. *rhabdotum*), *R. headfortianum*,
R. liliiflorum, *R. lindleyi*, *R. megacalyx* and *R. nuttallii*.

Stamens 10.
Calyx large, deeply lobed.

R. dalhousiae var. dalhousiae   Maddenia Gp B
R. dalhousiae Hook.f.

Shrub often epiphytic, 1–3 m (3–10 ft) high; **branchlets** scaly, not bristly or
sometimes bristly.
Leaves oblong, oblong-obovate or oblanceolate; **lamina** 6.5–14.8 cm long,
2.3–6 cm broad; **upper surface** slightly bullate, dark green or bright green;
underside pale glaucous-scaly, scales 1–3 times their own diameter apart.
Inflorescence 2–7 flowers.
Pedicel rather densely scaly, rather densely or moderately pubescent.
Calyx 0.7–1.1 cm long.
Corolla tubular-campanulate, 7–9.5 cm long, fragrant, creamy-white, yellow
or white.
Sikkim, Nepal, Bhutan. 1,850–3,000 m (6,000–9,500 ft).
AM 1930 (Clyne), flowers pale yellow.
AM 1974 (Sandling Park) 'Tom Spring-Smythe', flowers greenish-white.
FCC 1974 (Sandling Park) 'Frank Ludlow', flowers white tinged yellow (LS&T6694).
Epithet: After Lady Dalhousie, wife of the Governor-General of India.
Hardiness 1–2.   April–June.

R. lindleyi (as *R. basfordii*) LS&H19848 at Deer Dell

Red-brown peeling bark.

Young growth bristly.

Pedicel pubescent.

Flowers almost always greenish yellow in bud, fading: scent often not very strong.

Branchlets scaly, not bristly or sometimes bristly.

Leaf upper surface slightly bullate; underside with scales 1–3 times their own diameter apart.

Corolla white, creamy-white or yellow.

R. dalhousiae var. *dalhousiae* LS&T6694 'Frank Ludlow' FCC at Borde Hill

R. dalhousiae var. rhabdotum (Balf.f. & Cooper) Cullen

R. rhabdotum Balf.f. & Cooper

Maddenia Gp B

Shrub or **tree**, often epiphytic, 1–3.5 m (3–12 ft) high; **branchlets** scaly, bristly.
Leaves oblong, obovate or elliptic; **lamina** 5.2–17 cm long, 2.2–7.4 cm broad;
upper surface bright green or dark green, slightly bullate, not scaly;
underside glaucous or glaucous-green, scaly, scales 2–3 times or rarely their
own diameter apart.
Inflorescence 2–5 flowers.
Corolla tubular-campanulate, 7–11 cm long, fragrant, cream, pale yellow or
white, marked with 5 conspicuous red stripes outside, with or without a
golden blotch.
Bhutan, Assam, S Tibet. 1,525–2,800 m (5,000–9,125 ft).
AM 1931 (Bodnant).
FCC 1936 (Exbury).
AGM 1993.
Epithet: Striped.   Hardiness 1–2.   May-July.

R. headfortianum Hutch.   Maddenia Gp B

R. taggianum Hutch.

Small or dwarf, somewhat slender, upright **shrub** often epiphytic, 0.15–1.2 m
(6 inches–4 ft) high; **branchlets** scaly.
Leaves narrowly oblong or oblong-lanceolate; **lamina** 7 -12 cm long, 2–4.5 cm
broad; **upper surface** dark green or bright green; margin recurved; **underside**
glaucous, scaly, scales 2–3 times their own diameter apart.
Inflorescence 1–3 flowers.
Calyx 1.3–1.5 cm long.
Corolla tubular-campanulate, 5.5–7 cm long, creamy-yellow, or creamy slightly
tinged pink.
SE Tibet, Assam. 2,150–2,750 m (7,000–9,000 ft).
AM 1991 (Millais) 'Cliff Hanger'.
Epithet: After the late Marquess of Headfort.   Hardiness 1–2.   May.

Makes a shapely upright shrub more easily than the straggly R. taggianum. *It is
hard to see why this species has been submerged into* R. taggianum, *which closely
resembles* R. lindleyi.

Immediately recognisable in flower: corolla tubular-campanulate, 7–11 cm long, fragrant, cream, pale yellow or white, marked with 5 conspicuous red stripes outside. Not scented.

Late-flowering.

Almost indistinguishable from var. *dalhousiae* out of flower.

Branchlets scaly, always bristly.

Leaf upper surface slightly bullate.

R. dalhousiae var. *rhabdotum* (M.L.A. Robinson)

Narrowish leaves: narrowly oblong or oblong-lanceolate.

Out of flower, could be confused with *R. maddenii* or one of its relatives, but the leaf is paler and the underside is much less densely scaly.

Leaf margin recurved; underside with scales 2–3 times their own diameter apart.

Corolla creamy-yellow or creamy slightly tinged pink.

R. headfortianum 'Cliff Hanger' AM (M.L.A. Robinson)

Rhododendron
(Maddenia Group B)

R. liliiflorum Leveille   Maddenia Gp B

Shrub or **tree** 3–8 m (10–26 ft), upright and straggly; bark reddish, peeling;
   branchlets scaly not bristly.
Leaves oblong-lanceolate or oblong, apex obtuse or rounded, mucronate, 7–16 cm
   long, 2–5 cm broad, margin not bristly; **upper surface** smooth, somewhat matt,
   elepidote; **underside** pale glaucous green or pale brownish green, scales 1–3
   times their own diameter apart.
Inflorescence 2–3 flowers.
Corolla funnel-campanulate, 5.5–9 cm long, white or white flushed pink,
   densely scaly all over the outside, scented.
Calyx lobed more than half its length.
Stamens 10.
Ovary tapered into the style.
Style scaly to the tip.
Guizhou, Guangxi. 600–1,400 m (1,950–4,600 ft).
Epithet: Lily flowers.   Hardiness 1–3?   May–June.

*Davidian's description has been modified to take into account introductions after
1984, some of which are proving to be hardy.*

R. lindleyi T.Moore   Maddenia Gp B

(*R. basfordii* Davidian)

Shrub often epiphytic, 0.75–5 m (2.5–15 ft) high; **branchlets** scaly.
Leaves oblong or oblong-lanceolate; **lamina** 6.5–14 cm, long, 2.1–5.5 cm broad;
   upper surface dark green, reticulate, not bullate, not scaly; **underside** pale
   glaucous-green or glaucous, laxly scaly, scales 2–4 times their own diameter apart.
Inflorescence umbellate, **flower-bud** pale green, 3–6 flowers.
Pedicel 1–2 cm long, scaly.
Calyx 1–2 cm long.
Corolla broadly tubular-campanulate, 7–11.6 cm long, fragrant, white or rarely
   yellow-white, with or without a golden-yellow blotch.
Nepal, Sikkim, Bhutan, Assam, S Tibet. 2,150–3,350 m (7,000–11,000 ft).
AM 1935 (Exbury), flowers white flushed rose-red.
AM 1965 (Sunte House) 'Dame Edith Sitwell', flowers white, suffused with pale
   pink, possibly a hybrid.
FCC 1937 (Clyne), flowers white suffused with pink at the ends of corolla lobes.
AGM 1993.
Epithet: After Dr. John Lindley (1799–1865), botanist and Secretary to the RHS.
Hardiness 1–2.   May–June.

*Some forms are much hardier than stated by Davidian. Growth straggly. Forms
with up to 12 flowers per inflorescence are now in cultivation.*

R. grothausii *Davidian is closely related but has smaller and bullate leaves,* **reddish-
brown flower buds** *which turn crimson-purple, and smaller corollas. It is possibly
hardier than* R. lindleyi *and may be found labelled* R. lindleyi *in cultivation.*

Large lobed calyx;
large capsule.
Corolla white, or
very pale pink,
densely scaly all
over the outside.

Leaves oblong-
lanceolate or oblong,
apex obtuse or
rounded, mucronate.

R. liliiflorum at Glendoick (M.L.A. Robinson)

Distinguishable by
the reticulate upper
leaf surface, the
laxly scaly lower
surface, the large
round flower buds,
and the non-scaly
large calyx covered
with hairs.

Corolla broadly
tubular-campanulate,
7–11.6 cm long,
normally white.

R. lindleyi at Glenarn (Davidian)

Rhododendron
(Maddenia Group B)

R. megacalyx Balf.f. & Kingdon Ward

Maddenia Gp B

Shrub or **tree**, 1.2–7.5 m (4–25 ft) high; **branchlets** scaly, not bristly.
Leaves oblong-oval, oblong-obovate or oblong-elliptic; **lamina** 9.8–16.8 cm long, 3–7.5 cm broad; **upper surface** bright green or olive-green, matt, bullate, not scaly or scaly; **underside** glaucous, densely scaly, scales sunk into pits, 0.5 times to their own diameter apart, primary veins thick, markedly raised.
Petiole densely scaly.
Inflorescence 3–6 flowers.
Pedicel not scaly, not bristly, sometimes glaucous pruinose.
Calyx 1.7–2.6 cm long, not scaly, glabrous, consistently large, greenish or pinkish-green.
Corolla tubular-campanulate, 7.5–11 cm long, fragrant, white, creamy-white or cream suffused with pink, outside scaly.
NE and N Burma, E Tibet, W and mid-W Yunnan. 1,850–4,000 m (6,000–13,000 ft).
AM 1937 (Clyne), flowers pure white.
Epithet: Large calyx.   Hardiness 1–3.   April–June.

Grown outside at Exbury.

R. nuttallii Booth.

Maddenia Gp B

(*R. sinonuttallii* Balf.f. & Forrest)

Shrub or **tree** sometimes epiphytic, 1.2–9 m (4–30 ft) high; **stem** and **branches** with smooth, dark purplish-brown bark; **branchlets** scaly.
Leaves elliptic or oblong-elliptic; **lamina** 13.2–26.5 cm long, 5–13.1 cm broad; **upper surface** dark green (crimson-purple in young growths), strongly bullate, reticulate, somewhat matt; **underside** pale glaucous green, conspicuously wrinkled, scaly, scales 1–2 times their own diameter apart, midrib prominent, primary veins thick, markedly raised, prominently looped and branched.
Inflorescence 3–7 or rarely up to 12 flowers.
Pedicel pubescent.
Calyx 1.6–2.6 cm long, lobes upright.
Corolla tubular-campanulate, 8.7–13.6 cm long, fragrant, white or creamy-white suffused greenish, the petals or corolla tinged pink, or white flushed rose, with yellow or golden-orange blotch, outside scaly.
Bhutan, Assam. SE and E Tibet, Upper Burma, NW Yunnan, N Vietnam. 1,200–4,450 m (4,000–14,500 ft).
AM 1936 (Exbury) as var. *stellatum* (KW6333).
AM 1955 (Sunningdale) as *R. sinonuttallii*, flowers white with pale crimson spots and pale orange blotch (LS&E12117).
FCC 1864 (Victoria Nursery, Highgate).
AGM 1993.
Epithet: After Thomas Nuttall (1786–1859), botanist and traveller.   Hardiness 1.
April–May.

Recent introductions from Vietnam that have larger leaves are much hardier than stated above. **R. goreri** *Davidian is similar but without the glaucous leaf underside; it is rare in cultivation but may sometimes be found labelled* R. nuttallii.

> *Recent introductions claiming to be* **R. excellens** *Hemsl. & E.H.Wilson have all had 12 stamens, and are referable to* R. nuttallii.

Distinct because of the combination of the grooved petiole, the more or less bullate leaves, underside glaucous and the scales sunk into pits.

Unmistakable in flower or fruit because of the 'mega-calyx'. Stamens 10.

Corolla normally white.

R. megacalyx (B. Starling)

Large strongly bullate, wrinkled leaves which could only be confused with an elepidote species.

Corolla lobes often reflexed, usually white.

Stem and branches with smooth, dark purplish-brown bark.

Leaves of young growth crimson-purple.

R. nuttallii leaves showing the variation (M.L.A. Robinson). Bar = 5 cm

R. nuttallii at Borde Hill

Rhododendron
(Maddenia Group B)

Subsection *Maddenia* Sleumer
Group C1. The former Ciliatum Series and *R. johnstoneanum*

R. burmanicum, R. ciliatum, R. fletcherianum, R. johnstoneanum, R. valentinianum (var. *valentinianum*, var. *changii* and var. *oblongilobatum*).

This series of more compact shrubs, which are common in cultivation, is worth separating for identification purposes: R. johnstoneanum has similar characteristics but is a bigger shrub.

Shrubs 0.3–1.8 m, not epiphytic, often compact, and all fairly hardy.
Branchlets bristly.
Leaf margins bristly; main vein above totally impressed.
Petiole bristly.
Pedicel and calyx not pale, pruinose.
Calyx small, margin bristly.
Stamens usually 10 (occasionally 8).
Style impressed, not tapering onto the ovary.

R. burmanicum Hutch.   Maddenia Gp C1

Shrub 1–1.8 m (3–6 ft) high; **branchlets** scaly, slightly or moderately bristly.
Leaves oblanceolate, obovate or oblong-obovate; **lamina** 4–8 cm long, 1.6–4 cm broad; **upper surface** dark or bright green, scaly; **underside** densely scaly, scales brown, slightly overlapping or contiguous or half their own diameter apart, with or without scattered larger scales.
Petiole densely scaly, often bristly.
Inflorescence 4–6 flowers.
Corolla tubular-campanulate, 2.8–5 cm long, yellow, greenish-yellow or greenish-white, outside moderately or rather densely scaly.
Style straight.
SW Burma. 2,750–3,000 m (9,000–9,900 ft).
Epithet: From Burma.   Hardiness 1–2.   March–May.

Now given a hardiness score of 2–3. Many plants labelled R. burmanicum *that are less densely scaly and have smaller leaves and deep yellow flowers are hybrids.*

Leaves larger than those of *R. valentinianum* and without the hairs on the upper surface midrib.

Leaf underside densely scaly with dark scales.

Leaves oblanceolate, obovate or oblong-obovate.

Scales slightly overlapping or contiguous or half their own diameter apart.

Corolla usually yellow or greenish-yellow.

R. burmanicum at Mount Tomah Botanic Gardens, NSW.

Rhododendron
(Maddenia Group C1)

R. ciliatum Hook.f.   Maddenia Gp C1

Shrub 0.3–1.5 m (1–5 ft) or rarely 1.8 m (6 ft) high; **stem** and **branches** with smooth, brown flaking bark; **branchlets** setose.

Leaves evergreen, elliptic, oblong-elliptic or oblong-lanceolate; **lamina** 3.6–8.6 cm long, 1.8–3.6 cm broad; **upper surface** moderately or sparsely setose, margin rather densely to sparsely setose; **underside** scaly, scales 2–3 times their own diameter apart, midrib setose.

Petiole setose.

Inflorescence 2–5 flowers.

Pedicel setose.

Corolla tubular-campanulate, 2.6–5 cm long, white, white suffused with pink or red, or pale pink.

Style slender, straight.

Sikkim, Nepal, Bhutan, S and SE Tibet. 2,450–4,000 m (8,000–13,000 ft).

AM 1953 (Minterne), flowers white suffused with pink.

AGM 1993.

Epithet: Fringed.   Hardiness 3.   March–May.

Now classed as Hardiness 3–4. In flower, only likely to be confused with the hybrid R. × cilpinense which has less hairy, more shiny, leaves.

R. fletcherianum Davidian   Maddenia Gp C1

Shrub 0.6–1.2 m (2–4 ft) high; **branchlets** scaly, bristly.

Leaves elliptic, oblong-elliptic or oblong-lanceolate; **lamina** 2.3–5.6 cm long, 1.1–2.8 cm broad, base decurrent on the petiole; **upper surface** dark green, shiny, often scaly; margin crenulate, bristly; **underside** pale green, scaly, scales 3–6 times their own diameter apart.

Petiole narrowly winged, scaly, bristly.

Inflorescence 2–5 flowers.

Pedicel pilose.

Calyx 0.8–1 cm long, margin pilose.

Corolla widely funnel-shaped, 3.6–4.2 cm long, pale yellow, outside not scaly or sparsely scaly.

SE Tibet. 4,100–4,300 m (13,500–14,000 ft).

AM 1964 (Glendoick) 'Yellow Bunting', flowers primrose yellow (R22302).

Epithet: After H.R. Fletcher, Regius Keeper, Royal Botanic Garden, Edinburgh, 1956–1970.   Hardiness 3.   March–May.

Now given a hardiness of H4.

Leaf, leaf margin
and underside
midrib, petiole and
calyx all hairy (not
bristly).

Stem and branches
with smooth, brown
flaking bark.
Branchlets setose.
Corolla usually white
or white tinged with
pink or red.

R. ciliatum in R. White's garden, Surrey

Bristly branchlets,
petioles and
pedicels.
Leaf base decurrent
on the petiole and
usually with a
pointed apex.
Scales on the leaf
underside 3–6 times
their own diameter
apart.
Petiole narrowly
winged, scaly, bristly.
Corolla pale yellow.

R. fletcherianum at RBG Edinburgh

R. johnstoneanum Watt ex Hutch.   Maddenia Gp C1

Shrub 1.2–3.7 m (4–12 ft); **branchlets** scaly, bristly.

Leaves elliptic, obovate or oblong-elliptic; **lamina** 5–10 cm long, 2–4.6 cm broad; **upper surface** dark green, shiny; margin bristly or not bristly; **underside** pale green, densely scaly, scales contiguous to 0.5 times their own diameter apart.

Petiole densely scaly, moderately or rather densely bristly.

Inflorescence 3–4 flowers.

Pedicel not bristly.

Calyx minute, 1 mm long, margin bristly.

Corolla funnel-shaped, 4.6–6 cm long, fragrant, yellow, pale yellow, creamy-white, white tinged pink or white with pink bands along the middle of the lobes outside, with or without a yellow blotch, and with or without red spots.

Assam. 1,350–3,350 m (6,000–11,000 ft).

AM 1934 (Nymans), flowers pale creamy-white with yellow blotch (KW7732).

AM 1941 (Trengwainton) to clone 'Rubeotinctum', flowers white with deep pink stripes along the middle of the lobes, and a pinkish-yellow blotch (KW7732).

AM 1975 (Leonardslee) to a clone 'Demi-John', flowers white flushed yellow-green at the throat.

AGM 1993.

Epithet: After Mrs. Johnstone, wife of the Political Agent, Manipur, 1882. Hardiness 1–3.   April–June.

Widely grown.
A double form named 'Double Diamond' is quite common in cultivation.

R. valentinianum var. valentinianum

R. valentinianum Forrest ex Hutch.   Maddenia Gp C1

Shrub 0.3–1.2 m (1–4 ft) high; **branchlets** scaly, rather densely brown-setose.

Leaves elliptic, oblong-elliptic, oblong-oval or oval; **lamina** 2–4.6 cm long, 1–2.9 cm broad; **upper surface** dark green, shiny, scaly, not setose or setose; margin recurved, rather densely brown-setose; **underside** pale green, densely scaly, scales contiguous or half their own diameter apart, not setose.

Petiole rather densely brown-setose.

Inflorescence 1–6 flowers.

Pedicel bristly.

Corolla tubular with spreading lobes or tubular-campanulate, 2.6–3.5 cm long, bright yellow or bright sulphur-yellow, pubescent on the tube.

W and mid-W Yunnan, NE Upper Burma. 2,750–3,650 m (9,000–12,000 ft).

AM 1933 (Bodnant).

Epithet: After Pere S.P. Valentin, Tsedjong Mission, China.   Hardiness 1–2. April–May.

Now given a hardiness score of H2–3. Related to R. fletcherianum.

Branchlets bristly.

Almost always has bristly leaf margins and rounded slightly rugulose leaves.

Petiole densely scaly, moderately or rather densely bristly.

Leaf underside densely scaly, scales contiguous to half their own diameter apart.

Corolla slightly fragrant, usually yellow, pale yellow, or creamy-white.

R. johnstoneanum KW7732 at Deer Dell

R. johnstoneanum 'Double Diamond' at Deer Dell

Compact growth, with densely setose branchlets, petioles and pedicels.

Rounded leaves densely scaly on their undersides.

Style sharply bent.

Leaf margin recurved, rather densely brown-setose.

Corolla bright yellow or bright sulphur-yellow.

R. valentinianum var. *valentinianum* F24347 at RBG Edinburgh

Rhododendron
(Maddenia Group C1)

R. valentinianum var. changii W.P.Fang

R. changii W.P.Fang   **Maddenia Gp C1**

Small **shrub**; **branchlets** tinged purple.
Leaves oblong-elliptic, 3–4.5 cm long, 2–3 cm broad; apex rounded, base
    rounded; **upper surface** pale green, not scaly; **underside** densely scaly; lateral
    veins inconspicuous.
Petiole short–(nearly sessile), scaly.
Inflorescence terminal, 2–4 flowers.
Corolla funnel campanulate, deep yellow.
Stamens 10.
Epithet: After Mr. Chang, a collector.   Hardiness 3?   March–April.

*Has recently been introduced but is very rarely seen as yet. Young leaves tinged
purple, but this may be a juvenile characteristic. Said to have a glabrous calyx and
pedicel. Treated as a full species in the* 'Flora of China'.

R. valentinianum var. oblongilobatum

R.C.Fang   **Maddenia Gp C1**

Upright to rounded **shrub** to over 2 m.
Leaves rugose, medium-green, oval, apex obtuse; **lamina** 4.1–6.1 cm long,
    3.1–5.5 cm broad; **upper surface** dark green, moderately shiny, margins
    bristly, **underside** somewhat glaucous.
Inflorescence 2–4 flowers.
Corolla 5.5–6.5 cm long, deep yellow, without markings.
SE Yunnan, 2,750–2,850 m.
Epithet: With oblong lobes, referring to the calyx.   Hardiness 3?   May–June.

Recently introduced. Growth may be lax.

Subsection *Maddenia* Sleumer
Group C2.

R. formosum (var. *formosum* [including the var. *formosum* Iteophyllum Group]
and var. *inequale*), *R. leptocladon* and *R. pachypodum*.

> **Style impressed.**
> **Leaves less rugulose, and with less hairy margins, than those of the
>   species in group C1.**
> **Stamens 10, occasionally 12.**

Pedicel and calyx lobes not bristly.

Leaves elliptic to oblong-elliptic; upper side smoother, not scaly, lateral veins inconspicuous; slightly shiny. Petiole very short.

R. valentinianum var. *changii* (new foliage) at Glendoick (M.L.A. Robinson)

Larger in every sense, and appears to be hardier than *R. valentinianum* var. *valentinianum*.

Pedicel not setose.

Calyx margins glabrous, lobes shorter than in var. *valentinianum*.

Corolla 5.5–6.5 cm long, deep yellow.

R. valentinianum var. *oblongilobatum* at Glendoick (M.L.A. Robinson)

R. formosum var. formosum   Maddenia Gp C2

R. formosum Wallich.

Shrub 1–3 m (3–10 ft) high; **branchlets** scaly, often rather densely bristly.

Leaves oblanceolate, oblong-obovate or lanceolate; **lamina** 2–7.6 cm long, 0.6–2.4 cm broad, base tapered, decurrent or slightly decurrent on the petiole; **upper surface** dark green; margin often bristly; **underside** glaucous or pale green, scaly, scales 1–2 times their own diameter apart.

Petiole with narrow wings or ridges at the margins, scaly, often bristly.

Inflorescence 2–3 flowers.

Calyx minute, 1 mm long, lepidote.

Corolla funnel-shaped, 4–6.5 cm long, white, with or without 5 pale bands outside, with or without a yellow blotch.

Assam, NW Upper Burma. 750–2,200 m (2,500–7,200 ft).

AM 1960 (RBG Edinburgh), flowers white with yellow blotch.

AM 1988 (Glendoick) 'Khasia' (C&H320).

AGM 1993.

Epithet: Beautiful.   Hardiness 1–2.   April–June.

Hardier than stated above.

R. formosum var. formosum Iteophyllum Group

R. iteophyllum Hutch.   **Maddenia Gp C2**

Shrub 1–2.4 m (3–8 ft) high; **branchlets** scaly, bristly.

Leaves linear, linear-lanceolate, lanceolate or oblanceolate; **lamina** 2.6–8 cm long, 0.5–1.5 cm broad, apex acute, base tapered, decurrent on the petiole; **upper surface** dark green; **underside** pale green, scaly, scales 1–2 times their own diameter apart.

Petiole with narrow wings at the margins, scaly, bristly.

Inflorescence 1–4 flowers.

Calyx minute, 0.5–1 mm long.

Corolla funnel-shaped, 4.8–6 cm long, white or white tinged pink, with or without yellow blotch, outside scaly.

Assam. 600–1,850 m (2,000–6,000 ft).

AM 1979 (Mrs Mackenzie) to a clone 'Lucy Elisabeth'.

Epithet: Willow-leaved.   Hardiness 2–3.   February–June.

The hardiest Group of this species.

Branches very bristly.

Leaves often small and somewhat narrow: 3 or more times as long as broad, and even narrower in the Iteophyllum Group.

Leaves often with bristly margins, lamina decurrent on to the petiole.

Petiole narrowly winged, often with bristles.

Small disc-like calyx.

Corolla normally white.

R. formosum var. *formosum* – a pink form in the USA (M.L.A. Robinson)

Leaves distinctively narrower, usually linear, linear-lanceolate, otherwise identical to *R. formosum*.

Corolla normally white.

R. formosum var. *formosum* Iteophyllum Group at Abbotsbury, Dorset (M.L.A. Robinson)

R. formosum var. inequale C.B.Clarke

R. inaequale (C.B.Clarke) Hutch.   **Maddenia Gp C2**

Shrub sometimes epiphytic, 0.9–3 m (3–10 ft) high; **branchlets** scaly, bristly.
Leaves elliptic-lanceolate, oblong-lanceolate or oblong-obovate; **lamina** 6–11 cm
    long, 1.5–4 cm broad; **upper surface** dark green, not scaly, margin not bristly;
    underside scaly, scales 2–3 times their own diameter apart.
Petiole moderately or densely scaly, not bristly.
Inflorescence 1–6 flowers.
Pedicel densely scaly, not bristly.
Corolla widely funnel-shaped, 5.5–7.5 cm long, white with yellow or green blotch.
Assam. 1,200–2,150 m (4,000–7,000 ft).
AM 1947 (Bodnant).
FCC 1981 (Mrs. Mackenzie), Elizabeth Bennet (C&H301).
Epithet: Of unequal size.   Hardiness 1–2.   March–May.

Unfortunately, this lovely variety is not hardy.

R. pachypodum Balf.f. & W.W.Sm.   **Maddenia Gp C2**

(*R. scottianum* Hutch., *R. supranubium* Hutch.)

Shrub 0.3–1.8 m (1–6 ft) high, or sometimes a tree 7.6 m (25 ft) high;
    branchlets scaly.
Leaves oblong-obovate, obovate, elliptic or oblong-lanceolate; **lamina** 5–10 cm
    long, 2–4 cm broad, apex acuminate or acute; **upper surface** dark green, not
    scaly; **underside** pale glaucous-green, densely scaly, scales nearly contiguous to
    0.5 times their own diameter apart.
Inflorescence 1–5 flowers.
Corolla funnel-shaped, 4.6–6.3 cm or rarely 2.8 cm long, fragrant, yellow, white,
    pink or purplish, with or without a yellow blotch, scaly all over the outside.
W Yunnan. 2,150–3,050 m (7,000–10,000 ft).
FCC 1936 (Exbury), flowers white with a pale yellow streak.
Epithet: Thick-footed.   Hardiness 1–2.   March–May.

Plants labelled R. ciliicalyx *may turn out to be this species.*

R. formosum var. *inequale* at Borde Hill

Branches not bristly.

Leaves variable, but usually obovate or oblanceolate with a pointed tip, not bristly, underside pale glaucous green, scales usually almost touching.

Corolla scaly all over the outside.

Calyx small but occasionally with a single long lobe, usually not bristly.

Corolla normally white or pale pink.

R. pachypodum MF942115 at Hindleap Lodge (M.L.A. Robinson)

R. leptocladon Dop.   Maddenia Gp C2

Shrub epiphytic, 0.5–1 m (2–3.5 ft) tall; **branchlets** brown densely scaly, glabrous, eglandular.

Leaves elliptic, oblong-elliptic or ovate, 4–8 cm long, 2–3.5 cm broad; **lamina** leathery, base cuneate to rounded; apex acuminate; **underside** somewhat glaucous, with brown scales touching or up to half their diameter apart, brown; **upper surface** dark green, elepidote, glabrous.

Petiole densely scaly, glabrous.

Inflorescence 3–4 flowers.

Pedicel short densely scaly with pale scales.

Calyx a rim.

Corolla broadly funnel-shaped or tubular funnel-shaped, 2.9-4.3 cm long, lime-yellow or yellow, deeper in the throat, outer surface very sparsely scaly.

Stamens 10, densely pubescent near the base.

Ovary densely scaly.

Style truncate, scaly to about half its length, or not scaly.

S Yunnan, N Vietnam. 1,800–3,000 m (6,000–10,000 ft).

Epithet: With thin twigs.

Introduced about 1992 from N Vietnam. Appears not to be closely related to
R. lyi, *as originally suggested, because of the impressed style.*

Subsection *Maddenia* Sleumer
Group C3.

R. ciliicalyx, R. horlickianum, R. lyi, R. parryae and *R. veitchianum*
(which includes the submerged *R. cubittii*).

Style lepidote and <u>tapered smoothly</u> on to the ovary.
Calyx small.
Stamens 10, rarely 11–14.

Branchlets brown densely scaly, glabrous.

Apex acuminate.

Corolla lime-yellow or yellow, deeper colour in the throat.

Style impressed, scaly to about half its length.

R. leptocladon at Tregrehan, Cornwall (M.L.A. Robinson)

AC357 (as *R. fleuryi*) – a possible new species from N Vietnam at Deer Dell

R. ciliicalyx Franch.

Maddenia Gp C3

Shrub 1–3 m (2.5–10 ft) high; **branchlets** scaly, often bristly.
Leaves oblong-lanceolate, oblong-obovate or elliptic; **lamina** 5–10.3 cm long, 1.6–3.6 cm broad; **upper surface** usually not scaly, margin usually not bristly; **underside** scaly, scales 1–2 times their own diameter apart.
Petiole scaly, usually moderately or sparsely bristly.
Inflorescence 2–4 flowers.
Calyx 1–7 mm long, margin bristly.
Corolla funnel-shaped, 4.6–7.8 cm long, white or rarely white flushed rose, or rarely rose, with a yellow blotch, outside usually not scaly, usually pubescent at the base of the tube.
Yunnan. 1,800–4,000 m (5,900–13,100 ft).
AM 1923 (Oxford Botanics).
AM 1975 (Sunte House) to a clone 'Walter Maynard', flowers white flushed green at base.
Epithet: Fringed calyx.   Hardiness 1–2.   March–May.

Since Davidian's work, many plants in cultivation labelled R. ciliicalyx *have been shown to be* R. pachypodum *or* R. dendricola.

R. horlickianum Davidian

Maddenia Gp C3

Shrub sometimes epiphytic, 1–3 m (3–10 ft) high; **branchlets** scaly, not bristly.
Leaves elliptic-lanceolate, oblong-lanceolate or oblanceolate; **lamina** 5.2–11.5 cm long, 1.8–4.5 cm broad, apex acutely or shortly acuminate, **upper surface** dark green, shiny, not scaly; **underside** pale glaucous green, scaly, scales 1–3 times their own diameter apart.
Inflorescence 2–3 flowers.
Calyx minute, 1 mm long.
Corolla widely funnel-shaped, 6.5–7 cm long, white or creamy-white, with pink bands outside the lobes, and yellow blotch, outside scaly, rather densely or moderately pubescent.
N Burma. 1,220–2,135 m (4,000–7,000 ft).
Epithet: After Sir James Horlick.   Hardiness 1–2.   April–May.

Uncommon in the UK, but more widely cultivated in the western USA and Canada.

Similar to
R. pachypodum
but the calyx is
noticeably bristly at
least at the margins,
and the corolla is
only **sparsely** scaly.

Corolla normally white.

R. ciliicalyx in the Australian Rhododendron Society's Garden,
NSW, Australia

Similar to
R. formosum var.
inequale, but the leaf
has an acuminate tip.

Corolla flushed pink,
especially on the
lobes, with faint
pink lines down the
centre of each.

Scent faint.

Branchlets not bristly.

Leaf underside scales
1–3 times their own
diameter apart.

Outside of the corolla
rather densely or
moderately pubescent.

R. horlickianum in C. Varcoe's garden, Victoria, British Columbia
(M.L.A. Robinson)

Rhododendron
(Maddenia Group C3)

R. lyi Leveille   Maddenia Gp C3

Shrub 1–2.4 m (3–8 ft) high; **branchlets** scaly, bristly.
Leaves oblanceolate, oblong-lanceolate or lanceolate; **lamina** 3.2–9.5 cm long,
     1.3–3.5 cm broad, apex acute or abruptly acute; **upper surface** dark green,
     shiny, margin not bristly or bristly; **underside** densely scaly, scales contiguous
     to their own diameter apart.
Petiole densely scaly, bristly.
Inflorescence 3–6 flowers.
Pedicel densely scaly, not bristly.
Corolla funnel-shaped, 4.6–6 cm long, fragrant, white, with or without yellowish
     or greenish blotch.
Guizhou. 2,000 m (6,600 ft).
Epithet: After J. Ly, a Chinese collector.   Hardiness 1–2.   April–June.

Some forms, at least, are hardier than stated.

R. parryae Hutch.   Maddenia Gp C3

Shrub or **tree**, sometimes epiphytic, 1.5–3 m (5–10 ft) high; **branchlets** scaly.
Leaves elliptic, oblong-elliptic or somewhat rounded; **lamina** 6–14 cm long,
     3–5.8 cm broad; **upper surface** dark green, shiny; **underside** pale glaucous
     green, scaly, scales 1–2 times their own diameter apart.
Inflorescence 3–5 flowers.
Calyx minute, 1 mm long.
Corolla somewhat narrowly funnel-shaped, 5.5–8.2 cm long, fragrant, white
     with a yellow-orange blotch, without or sometimes with greenish spots, laxly
     scaly outside.
Assam. 1,800–2,150 m (5,800–7,000 ft).
AM 1957 (RBG Edinburgh), flowers white with a yellow-orange blotch.
FCC 1973 (Sunte House).
AGM 1993.
Epithet: After Mrs. A.D. Parry, wife of an officer in the Assam Civil Service.
Hardiness 1–2.   April–May.

*The species is not properly defined. The herbarium specimen of the type form
closely resembles* R. johnstoneanum, *and is unlike the species in cultivation. The
plant in cultivation does not fit* R. johnstoneanum, *nor* R. roseatum, *into which*
R. parryae *was submerged recently: Davidian's judgement that* R. parryae *is a
separate species is maintained here.*

Branchlets bristly.

Leaves usually narrowly oblanceolate with a pointed tip, margins sometimes bristly, underside pale, sometimes slightly glaucous, densely scaly.

Petiole almost always bristly.

Corolla normally white.

R. lyi at Hindleap Lodge (M.L.A. Robinson)

Leaf margins bristly.

Leaf underside smooth, initially glaucous, becoming much less so with age.

Flowers are quite large, somewhat narrowly funnel-shaped, blotched with yellow and very fragrant.

R. parryae Fish 140 at Borde Hill

Leaf of R. parryae (M.L.A. Robinson). Bar = 1 cm

Rhododendron (Maddenia Group C3)

R. veitchianum Hook.f.   Maddenia Gp C3

Shrub or small **tree**, sometimes epiphytic, 1–3.7 m (3–12 ft) high; **branchlets** scaly.

Leaves oblong-obovate, obovate, elliptic or elliptic-lanceolate; **lamina** 5–10 cm long, 2–4.6 cm broad, apex abruptly acuminate or abruptly acute; **upper surface** dark green, shiny, not scaly; **underside** very glaucous, scaly, scales 1–3 times their own diameter apart.

Inflorescence 2–5 flowers.

Corolla funnel-shaped, 5.8–7 cm long, margins of lobes usually crinkled, fragrant, white or white tinged pink or green outside, with or without a yellowish-green blotch.

Central and Lower Burma. Thailand, Laos. 1,200–2,300 m (4,000–7,500 ft).

AM 1978 (Sunte House) to 'Margaret Mead'.

AGM 1993.

Epithet: After the famous family Nurserymen.   Hardiness 1.   May–June.

R. veitchianum Hook.f.   Maddenia Gp C3
R. cubittii Hutch.

Shrub 1.5–2.4 m (5–8 ft) high; **stem** and **branches** with smooth, brown, flaking bark; **branchlets** scaly, often bristly.

Leaves oblong-lanceolate; **lamina** 9–10.8 cm long, 2.5–3.7 cm broad; **upper surface** dark green; **underside** pale green or pale glaucous green, scaly, scales brown, 1.5–5 times their own diameter apart.

Petiole scaly, bristly.

Inflorescence 2–3 flowers.

Calyx minute, 1 mm long.

Corolla widely funnel-shaped, 6–8.3 cm long, white or white flushed rose, with or without a reddish band outside, with a yellowish blotch, and brown spots, outside scaly.

N Burma. 1,700 m (5,500 ft).

AM 1935 (Trengwainton), flowers white, flushed rose.

FCC 1962 (Windsor) 'Ashcombe', flowers white, with a yellowish blotch.

AGM 1993.

Epithet: After G.E.S. Cubitt, who collected in North Burma.   Hardiness 1–2.   February–April.

R. cubittii *is known only from one collection in the wild, and its status needs to be confirmed by the presence or absence of a uniform population.* **The obvious difference between R. cubittii and R. veitchianum appears to be the crinkly flowers, which are said to be a feature of R. veitchianum.** *If this is so, then many plants are wrongly labelled. Widely grown in greenhouses.*

R. veitchianum AM at Borde Hill

Stems and branches with smooth, brown, flaking bark.
Branchlets usually bristly.
Leaf underside pale green or pale glaucous green.
Petiole bristly.
Corolla lobes not crinkled.
Calyx margin bristly.
Corolla normally white, or white flushed rose.

Bark of *R. veitchianum* (as *R. cubittii*) at Borde Hill

Rhododendron
(Maddenia Group C3)

Subgenus *Rhododendron*
Section *Rhododendron*
Subsection *Micrantha* Sleumer
Micranthum Series

R. micranthum Turcz.

Shrub 0.6–2.5 m (2–8 ft) high; **branchlets** thin, scaly, rather densely minutely puberulous.
Leaves oblanceolate or lanceolate; **lamina** 1.5–3.8 cm long, 0.4–1.5 cm broad; **upper surface** dark green, shiny; **underside** densely scaly, scales overlapping or contiguous or 5 times their own diameter apart.
Petiole scaly, minutely puberulous.
Inflorescence racemose, 10–28 flowers, **rhachis** 1–2.6 cm long.
Corolla campanulate, small, 4–6 mm long, white, outside scaly.

N and central China, Manchuria, Korea. 1,600–3,000 m (5,200–9,850 ft).
Epithet: Small flowers.   Hardiness 3–4.   May–July.

The flowers are very close to those of Subsection Ledum.

R. micranthum at Deer Dell

Small oblanceolate or lanceolate leaves, densely scaly underneath, glabrous.

Dense terminal and axillary inflorescences of small white flowers, the rhachis being comparatively long.

R. micranthum at Deer Dell

Rhododendron
(Micrantha)

Subgenus *Rhododendron*
Section *Rhododendron*
Subsection *Monantha* Cullen

A recently created Subsection accommodating the autumn-flowering species recently introduced.

Shrubs epiphytic or terrestrial.
Inflorescence 1–3 flowers.
Corolla tubular-campanulate with 5 small lobes.
Stamens and style exserted beyond the corolla.
Style impressed.
Seeds winged and finned.
Autumn flowering.

R. kasoense Hutch. & Kingdon Ward

Shrub 0.3–2.4 m (1–8 ft) high; **branchlets** scaly.
Leaves lanceolate or oblong-lanceolate; **lamina** 3–7.3 cm long, 1–2.8 cm broad, apex acute, acuminate or obtuse, mucronate, base obtuse or tapered; **upper surface** scaly; **underside** scaly, the scales unequal (possibly detersile), large and medium-sized, pale brown, 1.5–3 times their own diameter apart.
Petiole 0.5–1.5 cm long, scaly.
Inflorescence terminal, shortly racemose, 2–3 flowers.
Calyx 5-lobed or a mere rim, minute, 0.5 mm long, lobes triangular, outside densely scaly, margin densely scaly, eciliate.
Corolla tubular funnel-shaped to tubular-campanulate, 1.6–2 cm long, 5-lobed, yellow, outside scaly, glabrous, reflexed.
Stamens 10, unequal, exserted, 0.8–1.9 cm long; filaments densely villous in the lower one-third or one-half their lengths.
Ovary oblong, 3–4 mm long, 5-celled, densely scaly.
Style long, slender, not scaly.
Assam, SE Tibet. 2,135–2,745 m (7,000–9,000 ft).
Epithet: From Kaso Peak, Delei Valley, Assam.   Hardiness 3?   November.

*Recently introduced by P.A. and K.N.E. Cox. In the specimen examined from HECC 10009, the **mature** leaf underside has sparse dark scales, as well as paler yellowish scales about 5 times their own diameter apart.*

R. kasoense HECC10009 flowering in November at Hindleap Lodge (M.L.A. Robinson)

Inflorescence 2–4 flowers.
Corolla tubular-funnel-shaped to tubular-campanulate, yellow.

R. kasoense HECC10009 at Hindleap Lodge (M.L.A. Robinson)

Foliage of R. kasoense HECC10009 at Hindleap Lodge (M.L.A. Robinson). Bar = 2 cm

R. monanthum Balf.f. & W.W.Sm.

Small spreading **shrub** 0.3–1.2 m (1–4 ft) high; **branchlets** scaly, slightly bristly or not bristly.

Leaves elliptic, oblong-elliptic or oblong; **lamina** coriaceous, 2–5 cm long, 1–2.5 cm broad, apex obtuse or sometimes acute, mucronate, base obtuse or narrowed; **upper surface** scaly; **underside** glaucous, densely scaly, the scales varying much in size, mostly large, brown, one-half their own diameter apart.

Petiole 4–8 mm long, scaly.

Inflorescence terminal, 1–2-flowered, usually solitary.

Pedicel curved or straight, 4–6 mm long, densely scaly.

Calyx very small, undulate-lobulate, 5-lobed, 1–2 mm long, outside densely scaly, margin not ciliate.

Corolla tubular-funnel-shaped to tubular-campanulate, 5-lobed, 1.6–2.3 cm long, bright yellow, outside scaly, not hairy, reflexed.

Stamens 10, exserted.

Yunnan, SE Tibet, NE Upper Burma. 2,750–3,430 m (9,000–14,500 ft).

Epithet: One flower.   Hardiness 2–3?   November.

Recently introduced by P.A. and K.N.E. Cox.

Differs from the
other single-
flowered species (of
Subsection *Uniflora*)
in its larger leaves,
in having markedly
larger scales on the
lower surfaces, in
the usually taller
habit of growth, and
in its autumn
flowering.

Corolla bright yellow.

Foliage of *R. monanthum* in Richard Mossakowski's garden,
Vancouver (M.L.A. Robinson)

Rhododendron
(Monantha)

Subgenus *Rhododendron*
Section *Rhododendron*
Subsection *Moupinensia* Sleumer
Moupinense Series

Small shrubs, straggly in the wild if epiphytic.
Branchlets bristly.
Leaves: various variations on oval or elliptic; densely scaly, thick and rigid, margin slightly recurved.
Inflorescence 1–2 flowers.
Corolla widely funnel-shaped.
Stamens 10.

R. dendrocharis Franch.

Shrub 35–70 cm (1.1–2.3 ft) high; **branchlets** scaly, moderately or rather densely bristly.

Leaves evergreen, elliptic or oval; **lamina** coriaceous, thick, rigid, 0.9–1.6 cm long, 0.4–1 cm broad, apex obtuse or rounded, mucronate, base obtuse or rounded; **upper surface** bright green, somewhat matt, not scaly; margin recurved; **underside** densely scaly, the scales small or somewhat medium-sized, unequal, brown or dark brown, one-half their own diameter apart.

Petiole 2–6 mm long, scaly, bristly.

Inflorescence terminal, 1 or rarely 2 flowers.

Flower bud scales persistent or deciduous.

Pedicel 2–3 mm long, scaly, glabrous or rather densely minutely puberulous, bristly.

Calyx 5-lobed, 1–3 mm long.

Corolla widely funnel-shaped, 2–2.5 cm long, 2.3–3 cm across, 5-lobed, bright rosy-red, outside not scaly, glabrous, inside rather densely pubescent in the tube.

Stamens 10, unequal, 1.2–1.5 cm long, shorter than the corolla.

Ovary conoid or ovoid, 2–3 mm long, 5-celled, densely scaly, glabrous.

Style slender, straight, somewhat short, glabrous or rather densely pubescent at the base.

W Sichuan, 2,600–3,000 m (8,500–9,850 ft).

Epithet: Tree-adorning.   Hardiness 4.   March–April.

Of relatively recent introduction, so uncommon as yet.

R. moupinense W4256 at Deer Dell

R. moupinense new foliage at Deer Dell

Out of flower, can be distinguished from *R. moupinense* in that it is smaller in all respects, particularly the leaves. The corolla, however, is large.

Distinguished from *R. petrocharis* by the flower colour and usually by the less densely scaly lower surface of leaves.

Leaves thick, rigid, elliptic or oval.

Leaf underside densely scaly, the scales one-half their own diameter apart.

Corolla bright rosy-red.

R. dendrocharis 'Glendoick Gem' at Glendoick (M.L.A. Robinson)

R. moupinense Franch.

Shrub often epiphytic, 0.6–1.5 m (2–5 ft) high; **branchlets** scaly, bristly.

Leaves elliptic, ovate-elliptic, ovate or oval; **lamina** coriaceous, thick, rigid, 2–4.6 cm long, 1.1–2.3 cm broad; **upper surface** dark green, shiny, not scaly, glabrous; margin recurved; **underside** pale green, densely scaly, scales their own diameter apart.

Petiole scaly, bristly.

Inflorescence 1–2 flowers.

Corolla widely funnel-shaped, 3–4.8 cm long, white, or white tinged pink or red, with purple or crimson spots.

Style long, slender, straight.

W Sichuan. 2,000–3,300 m (6,550–10,800 ft).
AM 1914 (Great Warley), flowers white.
AM 1937 (Bodnant), flowers rose-pink.
Epithet: From Moupin, W China.   Hardiness 3–4.   February–March.

Widely grown.

R. petrocharis Diels.

Small **shrub**; **branchlets** scaly, glabrous or minutely puberulous, bristly.

Leaves evergreen, elliptic or ovate-elliptic; **lamina** coriaceous, thick, rigid, 1–1.3 cm long, 5–7 mm broad; **upper surface** bright green, somewhat matt, sparsely scaly or not scaly; margin recurved; **underside** densely scaly, the scales medium-sized, unequal, brown or dark brown, slightly overlapping or contiguous.

Petiole 2–7 mm long, scaly, moderately or rather densely bristly.

Inflorescence terminal, 1–2 flowers, **flower-bud** scales persistent or deciduous.

Corolla widely funnel-shaped, 2–3 cm long, 3.5 cm across, 5-lobed, white, outside not scaly, glabrous.

Stamens 10, unequal, about 1.5 cm long, shorter than the corolla; filaments hairy above the base.

Ovary conoid, 3 mm long, 5-celled, densely scaly, glabrous.

Style slender, straight, short, 0.6–1 cm long, shorter than the corolla, shorter than the stamens or about the length of the shortest stamen, 3 times as long as the ovary, not scaly, glabrous.

Sichuan. 1,800 m (5,900 ft).
Epithet: Rock-dwelling.   Hardiness 4.   March–April.

Very rare in cultivation as yet. Whether R. petrocharis *justifies species status is uncertain.*

Rigid small, dark, smooth, oval to elliptic leaves.

Persistent flower bud scales.

Corolla white, or white tinged pink or red.

Leaf underside densely scaly, the scales their own diameter apart.

R. moupinense – a red form, at Deer Dell

Very similar to *R. dendrocharis*, but differs in that the leaves are usually larger; the corolla is white. The scales on the leaf underside are slightly overlapping or contiguous.

R. petrocharis at Glendoick (M.L.A. Robinson)

Subgenus *Rhododendron*
Section *Rhododendron*
Subsection *Rhododendron*

Ferrugineum Series

Compact shrubs.
Branchlets densely lepidote.
Leaves densely lepidote.
Flowers tubular with spreading lobes.
Stamens 10.
Style straight.

European species from the Alps. Linnaeus' name for this Subsection takes precedence over the more modern series name.

R. ferrugineum L.

Shrub 0.3–1.2 m (1–4 ft) high; **branchlets** densely scaly, glabrous, not bristly.
Leaves oblanceolate, lanceolate or oblong; **lamina** 1.6–4.3 cm long, 0.6–1.6 cm broad; **upper surface** dark green, shiny, not scaly, not bristly; **underside** densely scaly, scales dark brown or reddish-brown, overlapping.
Petiole not bristly.
Inflorescence 5–16 flowers.
Calyx minute, 0.5–1 mm long.
Corolla tubular, with spreading lobes, 1.3–1.7 cm long, crimson-purple, rose-scarlet or deep rose, rarely white, outside moderately or rather densely scaly.

European Alps, the Pyrenees, Austrian Alps, W Yugoslavia.
Epithet: Rusty-coloured.   Hardiness 3–4.   June–July.

Forms with white and scarlet flowers were formerly known as R. ferrugineum *var.* album *and* R. ferrugineum *var.* atrococcineum, *respectively.*

Commonly known as the 'Alpine Rose'.

R. ferrugineum in the Valley Gardens

Branchlets not bristly.

Leaves usually darker than those of *R. hirsutum*, underside densely scaly, scales dark brown or reddish-brown, overlapping.

Calyx minute.

Corolla crimson-purple, rose-scarlet or deep rose, outside moderately or rather densely scaly.

A white-flowered form of *R. ferrugineum* (Davidian)

R. hirsutum L.

Small **shrub** 0.3–1 m (1–3 ft) high; **branchlets** scaly, often bristly.
Leaves obovate, oblong, oblanceolate or oblong-lanceolate; **lamina**
  0.8–2.6 cm long, 0.4–1.3 cm broad; **upper surface** dark green, shiny;
  margin crenulate, bristly; **underside** pale green, laxly scaly, scales 2–4 times
  their own diameter apart.
Petiole bristly.
Inflorescence 4–12 flowers.
Calyx 2–5 mm long, margin with long hairs.
Corolla tubular with spreading lobes, 1–2 cm long, rose-pink, scarlet or crimson,
  outside scaly, inside pubescent, margin of the lobes hairy.
Central European Alps, Austrian Alps, NW Yugoslavia.
Epithet: Hairy.   Hardiness 3–4.   June–July.

The white-flowered form was previously designated var. albiflorum. *A rare double-
flowered form is in cultivation.*

R. myrtifolium Schott & Kotschy
R. kotschyi Simonkai

Small **shrub** 30–45 cm (1–1.5 ft) high; **branchlets** scaly, not bristly.
Leaves oblong, oblong-obovate or oblanceolate; **lamina** 0.9–2.3 cm, long,
  5–8 mm broad; **upper surface** dark green, shiny, not scaly; margin
  crenulate; **underside** densely scaly, scales contiguous or slightly overlapping
  or one-half their own diameter apart.
Petiole scaly, not bristly.
Inflorescence 3–7 flowers.
Pedicel scaly, rather densely or moderately minutely puberulous.
Corolla tubular, tube slender, with spreading lobes, crimson-purple, scarlet-purple,
  rosy-purple or rarely white, rather densely pubescent on the tube, moderately
  pubescent on the lobes.
Transylvanian and Carpathian mountains and the mountains of Bulgaria and
  Yugoslav Macedonia. 1,500–2,400 m (4,900–7,900 ft).
Epithet: With myrtle-like foliage.   Hardiness 3.   May–July.

The older name R. myrtifolium *is now used.*

Branchlets scaly, often bristly.

Leaf margin crenulate, bristly; underside laxly scaly, scales 2–4 times their own diameter apart.

Calyx 25 mm long, margin with long hairs.

Corolla rose-pink, scarlet or crimson, outside scaly, margin of the lobes hairy.

R. hirsutum: double-flowered form (J.C. Birck)

Smaller leaves and more compact than *R. ferrugineum*.

Corolla tubular, tube slender, rather densely pubescent on the tube, moderately pubescent on the lobes.

Style shorter than that of *R. ferrugineum*.

Leaf margin crenulate; underside densely scaly. Pedicel scaly, rather densely or moderately minutely puberulous.

Corolla crimson-purple, scarlet-purple, rosy-purple or rarely white.

R. myrtifolium (J.C. Birck)

Rhododendron
(Rhododendron)

Subgenus *Rhododendron*
Section *Rhododendron*
Subsection *Rhodorastra* (Maxim) Cullen
Dauricum Series

Shrubs to 1–4 m.
Branchlets thin.
Leaves wholly or partly deciduous, rarely totally evergreen.
Inflorescence terminal, or terminal and axillary, or axillary, 1–2 flowers.
Corolla widely funnel-shaped.
Stamens 10.
Style slender.
Early flowering.

R. dauricum L.

Shrub 1.5–2.4 m (5–8 ft) high; **branchlets** scaly, rather densely minutely puberulous.

Leaves deciduous, elliptic, oblong-oval or oval; **lamina** 1–3.4 cm long, 0.6–2 cm broad; **upper surface** dark green, shiny, scaly; **underside** densely scaly, scales overlapping or 0.5 to 1.5 times their own diameter apart.

Inflorescence terminal, or terminal and axillary in the uppermost 1 or 2 leaves, 1–2 flowers.

Corolla widely funnel-shaped, 1.3–2.3 cm long, precocious, pink, rose-purple, reddish-purple or dark purple, outside pubescent.

AM 1963, FCC 1969, and AGM 1993 (Windsor) to a clone 'Midwinter', flowers red-purple.

Russia, Manchuria, NE China, Korea, Japan.

Epithet: From Dauria, part of SE Siberia, east of Lake Baikal.   Hardiness 3–4. January–March, sometimes in December.

R. dauricum
R. dauricum L. var. *album* De Candolle

Corolla white.
Korea.
Epithet: With white flowers.   Hardiness 3–4.   January–March.

The clone 'Hokkaido' is a prolific white-flowered form.

R. dauricum in the Valley Gardens

Distinguished by the densely scaly leaf underside, the rounded leaf apex, and the smaller corolla when compared to *R. mucronulatum*: intermediate forms can be found.

Leaves deciduous, elliptic, oblong-oval or oval.

Inflorescence 1–2 flowers.

Corolla precocious, pink, rose-purple, reddish-purple or dark purple, outside pubescent.

R. dauricum 'Hokkaido' AM at Deer Dell

Rhododendron
(Rhodorastra)

R. ledebourii <small>Pojark</small>

R. dauricum var. *sempervirens* Sims.

Altai Mountains, Russia.
AM 1990 (Windsor) 'Hiltingbury'.
Epithet: Evergreen.   Hardiness 3–4.   March–April.

Recently raised to specific status as R. ledebourii. *Somewhat later flowering than the deciduous species. Semi-evergeen or evergreen.*

R. sichotense *Pojark, which is similar to* R. ledebourii *and doubtfully distinct, may be in cultivation. The leaves and flowers of* R. sichotense *are said to be usually larger, the pedicels densely scaly, and the flowers pale purplish-pink or rose.*

R. mucronulatum var. mucronulatum <small>Turcz.</small>

(*R. dauricum* var. *mucronulatum* Turcz.)
R. mucronulatum Turcz. var. *acuminatum* Hutch.

Shrub 1–4 m (3–13 ft) high; **branchlets** slender, scaly.
Leaves deciduous, elliptic, elliptic-lanceolate or lanceolate; **lamina** thin, 3.2–8 cm long, 1.3–3.6 cm broad; **upper surface** bright green, scaly; margin crenulate; **underside** pale green, scaly, scales 2–4 times their own diameter apart.
Petiole scaly, not bristly or sometimes bristly.
Inflorescence terminal, or terminal and axillary, 1–3 flowers.
Corolla widely funnel-shaped, 2.3–2.8 cm long, precocious (flowers appearing before the leaves), rose, rose-purple or reddish-purple, outside not scaly, pubescent.

SE Siberia, Manchuria, N China, Korea, Japan.
AM 1924 (RBG Kew), flowers rich purplish-rose.
AM 1935 (RBG Kew) 'Roseum', flowers bright rose.
AM 1965 (Windsor) 'Cornell Pink', flowers pink-red-purple (Group 62B).
AM 1965 (Windsor) 'Winter Brightness', flowers rich purplish-rose.
FCC 1957 (Windsor) 'Winter Brightness'.
AGM 1993 'Cornell Pink'.
AGM 1993 'Winter Brightness'.
Epithet: With a small point.   Hardiness 3–4.   January–March, sometimes in December.

Evergreen or semi-evergreen leaves.

Corolla pink, rose-purple, reddish-purple, or dark purple.

R. ledebourii or *R. sichotense* at Hindleap Lodge (M.L.A. Robinson)

Leaves deciduous, elliptic, elliptic-lanceolate or lanceolate, apex acuminate or pointed, underside with scales 2–4 times their own diameter apart.

Corolla rose-purple or reddish-purple, outside pubescent.

R. mucronulatum 'Cornell Pink' in the Valley Gardens

Rhododendron
(Rhodorastra)

R. mucronulatum var. taquetii (Lev.) Nakai

R. mucronulatum var. *chejuense* Davidian

The variety differs from the species type in that it is a dwarf plant, 10–30 cm (0.33–1 ft) high.

Island of Cheju, Korea. 1,350–1,850 m (4,500–6,000 ft).

Epithet: For the synonym, from Cheju island, Korea.   Hardiness 3.   March.

Worthy of subspecies status.

Subgenus *Rhododendron*
Section *Rhododendron*
Subsection *Saluenensia* Sleumer
Saluenense Series

Small shrubs.
Leaf undersides densely scaly with overlapping crenulate (with scalloped margins) scales.
Calyx usually coloured, rarely green.
Corolla widely funnel-shaped, rotate or saucer-shaped.
Stamens 10.

Work published after Davidian's manuscript has shown that R. calostrotum *var.* calciphilum *and* R. nitens *are part of a large cline in the wild. Cullen named two new subspecies,* R. calostrotum *ssp.* riparium *and* R. calostrotum *ssp.* riparoides. *In cultivation, however, it is usually sufficient to distinguish the two main species,* R. calostrotum *and* R. saluenense.

Group A. Branchlets, petiole and pedicel <u>not</u> bristly

R. calostrotum (ssp. *calostrotum*, ssp. *riparium* [Calciphilum Group], ssp. *riparium* [Nitens Group], ssp. *keleticum* and ssp. *keleticum* [Radicans Group]).

Group B. Branchlets, petiole and pedicel bristly

R. saluenense (ssp. *saluenense* and ssp. *chameunum* [including Prostratum Group and Charidotes Group]).

A geographically
isolated and stable
dwarf form.
Corolla pink to
purple.

R. mucronulatum var. *taquetii* at Deer Dell

R. calostrotum ssp. *calostrotum* at Eckford

Rhododendron
(Rhodorastra/Saluenensia)

R. calostrotum ssp. calostrotum

R. calostrotum Balf.f. & Kingdon Ward   **Saluenensia Gp A**

Shrub 8–90 cm (0.25–3 ft) or rarely 1.2–1.5 m (4–5 ft) high; **branchlets** densely scaly, not bristly.

Leaves elliptic, oval, oblong-oval or nearly orbicular; **lamina** 1.3–2.5 cm or sometimes up to 3.5 cm long, 0.5–1.5 cm broad; **upper surface** pale green or bluish-green, matt, densely scaly, scales overlapping; **underside** densely scaly, scales overlapping, crenulate.

Petiole densely scaly, not bristly.

Inflorescence 1–3 flowers.

Corolla saucer-shaped or rotate, purple, bright rosy-purple, crimson-purple or scarlet, outside pubescent and scaly.

NE and E Upper Burma, NW Yunnan, SE Tibet, Assam. 3,050–4,900 m (10,000–16,000 ft).

AM 1935 (Nymans), flowers deep rosy mauve (F27065 = F27497).

FCC 1971 and AGM 1993 (Glendoick) 'Gigha', flowers red-purple.

Epithet: With a beautiful covering.   Hardiness 4.   May.

In cultivation, R. calostrotum ssp. riparioides (AM 1983 [Glendoick]) is very similar: the differences between it, R. calostrotum ssp. calostrotum and R. calostrotum ssp. riparium being of botanical interest only. R. calostrotum ssp. ripiarum usually has more flowers in the truss.

The leaf underside of ssp. riparioides *appears smoother and felty, as the scales are not in such definite tiers.* R. calostrotum 'Rock Form' is this subspecies.

R. calostrotum ssp. riparium (Kingdon Ward) Cullen

(Calciphilum Group and Nitens Group)   **Saluenensia Gp A**

R. calostrotum var. *calciphilum* (Hutch. & Kingdon Ward) Davidian
(*R. rivulare* Kingdon Ward)
R. nitens Hutch.

Low spreading or broadly upright **shrub** up to 60 cm (2 ft) high, with pale greyish leaves. The variety differs from ssp. *calostrotum* in its smaller leaves; **lamina** 0.5–1.2 cm long.

Corolla pale pinkish or rosy-purple.

Upper Burma, NW Yunnan. 3,700–4,250 m (12,150–14,000 ft).

Epithet: Lime-loving.   Hardiness 3.   May.

Calciphilum Group has greyish leaves, Nitens Group has shiny leaves.

Young leaves more or less glaucous or greyish, densely scaly.

Scales on leaf underside in 3 or 4 distinct tiers.

Pedicel not bristly.

Inflorescence 1–3 flowers.

Corolla saucer-shaped or rotate, purple, bright rosy-purple, crimson-purple or scarlet.

Foliage of *R. calostrotum* ssp. *calostrotum* (J.C. Birck)

R. calostrotum ssp. *riparioides* 'Rock Form' at Deer Dell

Smaller pale greyish leaves (shiny in Nitens Group).

Scales on leaf underside in 3 or 4 distinct tiers.

Inflorescence 2–5 flowers.

Corolla pale pinkish or rosy-purple.

R. calostrotum ssp. *riparium* at Deer Dell

R. calostrotum ssp. keleticum (Balf.f. & Forrest)
Cullen   Saluenensia Gp A
R. keleticum Balf.f. & Forrest.

A very compact, rounded spreading **shrub**, 15–45 cm (0.5–1.5 ft) high; or a semi-prostrate shrub 8–15 cm (3–6 inches) high; **branchlets** densely scaly, not bristly.

Leaves oblong, oblong-elliptic, lanceolate; **lamina** 0.7–2.1 cm long, 3–9 mm broad; **upper surface** dark green, shiny; **underside** densely scaly, scales overlapping.

Petiole scaly, not bristly.

Inflorescence 1–3 flowers.

Calyx 5–8 mm long, lobes pinkish or crimson.

Corolla widely funnel-shaped, 1.6–3 cm long, deep purplish-crimson or deep purplish-rose, with crimson spots, outside rather densely pubescent, scaly along the middle of the lobes.

E and SE Tibet, NW Yunnan. 3,950–4,575 m (13,000–15,000 ft).

AM 1928 (Gill).

Epithet: Charming.   Hardiness 3.   May–June.

R. radicans is identical to subspecies *keleticum* but is slightly more prostrate.

R. saluenense ssp. saluenense
R. saluenense Franch.   Saluenensia Gp B

Shrub 0.6–1.5 m (2–5 ft) or sometimes 30 cm (1 ft) high; **branchlets** densely scaly, moderately or densely bristly.

Leaves oblong-oval, ovate-elliptic, oblong-elliptic or oval, aromatic; **lamina** 1.4–3.6 cm long, 0.8–2.4 cm broad; **upper surface** dark green, shiny, scaly, margin bristly or not bristly; **underside** densely scaly, scales crenulate, overlapping.

Petiole scaly, bristly.

Inflorescence 1–4 or sometimes 5–7 flowers.

Pedicel scaly, often bristly.

Corolla widely funnel-shaped, 2.1–3 cm long, deep purple-crimson, deep purple or deep crimson-rose, with crimson spots, outside pubescent, scaly.

E and SE Tibet, NW Yunnan, SW Sichuan. 3,050–4,275 m (10,000–14,000 ft).

AM 1945 (Exbury), flowers deep purple.

AM 1965 (Exbury).

Epithet: From the Salween River.   Hardiness 3.   April–June.

Prostrate or mounding shrub.

Leaf upper surface not scaly.

Leaves usually more or less pointed.

Leaf underside densely scaly, scales in 3 or 4 distinct tiers.

Corolla deep purplish-crimson, outside rather densely pubescent, scaly along the middle of the lobes.

R. calostrotum ssp. *keleticum* at Deer Dell

Young growth not glaucous.

Leaf upper surface scaly, underside densely scaly, scales overlapping in different tiers but somewhat flattened.

Calyx and ovary not bristly.

Corolla purple-crimson to crimson-rose.

R. saluenense ssp. *saluenense* F12934 at Deer Dell

Rhododendron
(Saluenensia Group A/B)

R. saluenense ssp. chameunum (Balf.f. & Forrest)

Cullen

Saluenensia Gp B

R. chameunum Balf.f. & Forrest
(*R. charidotes* Balf.f. & Forrest)

Small **shrub** 5–60 cm (0.16–2 ft) high; **branchlets** densely scaly, moderately or densely bristly.

Leaves oblong-oval, oval, elliptic or obovate; **lamina** 0.5–2 cm long, 0.4–1.1 cm broad; **upper surface** dark green, shiny, not scaly or sometimes scaly; **underside** densely scaly, scales overlapping crenulate.

Petiole bristly.

Inflorescence 1–6 flowers.

Pedicel bristly.

Corolla widely funnel-shaped, 1.7–2.9 cm long, deep purplish-rose or purple-crimson, with crimson spots.

W and NW Yunnan, NE Upper Burma, SW Sichuan, SE Tibet. 3,350–4,575 m (11,000–15,000 ft).

Epithet: Lying on the ground.   Hardiness 3.   April–May.

This subspecies is more widely distributed in the wild, and the two groups below are now considered to be part of a continuous variation with altitude and geography.

R. saluenense ssp. chameunum

Charidotes Group

Saluenensia Gp B

R. charidotes Balf.f. & Farrer

Prostrate or rounded **shrub** 8–30 cm (0.25–1 ft) high; **branchlets** scaly, moderately or rather densely bristly.

Leaves elliptic, oblong-oval or oval; **lamina** 0.9–1.8 cm long, 0.4–1 cm broad; **upper surface** dark green, shiny; **underside** densely scaly, scales overlapping crenulate.

Petiole scaly, bristly.

Inflorescence 1–3 flowers.

Pedicel rather densely bristly.

Calyx bristly.

Corolla widely funnel-shaped, 1.8–2 cm long, magenta-crimson or purple-crimson, with crimson spots.

Ovary densely scaly, bristly.

Epithet: Giving joy.   Hardiness 3.   April–May.

Prostratum Group. *The high altitude form of ssp.* chameunum. *Smaller leaves and more prostrate than the species type. Corolla crimson to deep purple-rose, though an almost white form has been found on Beima Shan, NW Yunnan.*

Young growth with crimson purple margins.

Leaf upper surface usually not scaly, underside densely scaly, scales overlapping in different tiers but somewhat flattened.

Calyx and ovary not bristly.

Corolla purple crimson to purplish-rose.

R. saluenense ssp. *chameunum* F12968 at the Rhododendron Species Foundation (M.L.A. Robinson)

Differs in its bristly calyx and ovary.

Otherwise seems to be intermediate between the two subspecies.

Corolla magenta-crimson or purple-crimson.

R. saluenense ssp. *chameunum* Charidotes Group in W. Berg's garden (M.L.A. Robinson)

Subgenus *Rhododendron*
Section *Rhododendron*
Subsection *Scabrifolia* Cullen
Scabrifolium Series

Shrubs or small trees, usually straggly shrubs in cultivation.
Branchlets usually densely pubescent and sometimes bristly.
Leaf upper surface densely or moderately pubescent.
Inflorescence axillary except in *R. spinuliferum*.
Flowers shades of pink, rose or white.
Stamens 10 or rarely 8.

Group A. Leaf underside glaucous
R. hemitrichotum and *R. racemosum*.

Group B. Leaf underside not glaucous
R. mollicomum, *R. scabrifolium* (including var. *scabrifolium* and var. *spiciferum*) and *R. spinuliferum*.

R. scabrifolium var. *pauciflorum* KR342 at Deer Dell. Note the flower buds in the leaf axils

R. mollicomum at Deer Dell

R. hemitrichotum Balf.f. & Forrest   Scabrifolia Gp A

Shrub 0.25–2.4 m (1–8 ft) high; **branchlets** not bristly or rarely bristly, rather densely pubescent, moderately or rather scaly.

Leaves lanceolate, oblanceolate, oblong-lanceolate or oblong; **lamina** 1.2–5 cm long, 0.3–1.3 cm broad; **upper surface** not scaly, not bristly or rarely bristly, rather densely or sometimes moderately pubescent, moderately or sparsely scaly; margin recurved, not bristly, not pubescent; **underside** glaucous, not pubescent, midrib not pubescent or sometimes pubescent, scaly, the scales 0.5 times to their own diameter apart.

Petiole 2–4 mm long, not bristly or rarely bristly, rather densely (or rarely moderately) pubescent, scaly.

Inflorescence axillary in the uppermost few leaves, 1–3 flowers.

Corolla widely funnel-shaped, 0.9–1.4 cm long, pale rose, pink, deep pink or white edged with pink, with or without purple spots, outside scaly.

SW Sichuan, W Yunnan. 2,450–3,950 m (8,000–13,000 ft).

Epithet: Half hairy.   Hardiness 3.   April–May.

R. racemosum Franch.   Scabrifolia Gp A

Shrub 0.15–4.6 m (0.5–15 ft) high; **branchlets** glabrous or minutely puberulous, scaly.

Leaves elliptic, oblong-elliptic, obovate, oval, oblong or oblong-lanceolate; **lamina** 1–5.4 cm long, 0.4–2.6 cm broad; **upper surface** glabrous, scaly or not scaly; **underside** glaucous, scaly, scales 0.5–2 times their own diameter apart.

Inflorescence axillary in the uppermost few leaves, flowers usually in several clusters forming a raceme along the branchlet, 1–4 flowers.

Calyx minute, 0.5 mm long.

Corolla widely funnel-shaped, 0.8–2.3 cm long, white, pink, deep rose or reddish-pink, outside scaly.

Mid-W and NW Yunnan, SW Sichuan. 1,850–4,250 m (6,000–14,000 ft).

AM 1970 (Hydon) 'Rock Rose', flowers bright rose (R11265).

AM 1974 (Glendoick) 'White Lace', flowers white.

FCC 1892 (Veitch).

AGM 1993 'Rock Rose'.

Epithet: Flowers in racemes.   Hardiness 3.   March–May.

Very widely grown.

Similar to
R. mollicomum but
has a glaucous leaf
underside.

Branchlets rather
densely pubescent.

Leaf upper surface
rather densely
or sometimes
moderately
pubescent.

Petiole rather
densely pubescent.

Corolla usually white to
pink or pale rose.

R. hemitrichotum F30940 at Deer Dell

Small smooth
rounded glabrous
leaves, glaucous
beneath.

Branchlets glabrous
or minutely
puberulous.

Inflorescence
axillary, flowers
usually in several
clusters forming a
raceme along the
branchlet.

Corolla white,
pink, deep rose or
reddish pink.

R. racemosum R11265 'Rock Rose' AGM at Deer Dell

R. mollicomum Balf.f. & W.W.Sm.   Scabrifolia Gp B

Shrub 0.6–1.8 m (2–6 ft) high; **branchlets** often bristly, rather densely pubescent.
Leaves lanceolate; **lamina** 1.2–3.6 cm long, 0.3–1.3 cm broad; **upper surface**
rather densely pubescent; margin recurved, pubescent; **underside** rather
densely pubescent, scaly, scales 1–3 times their own diameter apart.
Petiole often bristly, rather densely pubescent.
Inflorescence axillary, 1–3 flowers.
Corolla narrowly tubular-funnel-shaped, oblique, 1.7–2.8 cm long, pale or deep
rose, outside scaly.
NW Yunnan, SW Sichuan. 2,450–3,350 m (8,000–11,000 ft).
AM 1931 (Bodnant), flowers bright rose.
Epithet: Soft-haired.   Hardiness 3.   April–May.

Rare in cultivation.

R. scabrifolium var. scabrifolium

R. scabrifolium Franch.   Scabrifolia Gp B

Shrub 0.2–3 m (0.75–10 ft) high; **branchlets** moderately or rather densely bristly,
densely pubescent, scaly.
Leaves lanceolate, oblong-lanceolate or elliptic; **lamina** 2.3–9.5 cm long,
0.6–2.8 cm broad; **upper surface** bullate, scabrid, bristly, pubescent; margin
recurved, bristly; **underside** rather densely or moderately pubescent, scaly,
scales 1–4 times their own diameter apart.
Petiole moderately or rather densely bristly, rather densely pubescent.
Inflorescence axillary, 1–3(–5) flowers.
Corolla widely funnel-shaped, 1–1.7 cm long, rose, pink, white, deep reddish
or crimson.
Yunnan. 1,800–3,350 m (5,900–11,000 ft).
Epithet: With rough leaves.   Hardiness 3.   March–May.

Similar to
R. hemitrichotum,
but the leaf
underside is pale
green.
Branchlets often
bristly, rather
densely pubescent.
Both leaf surfaces
rather densely
pubescent.

Corolla pale or deep
rose.

R. mollicomum at Deer Dell

Longish, hairy
leaves, distinctively
bullate.

Branchlets densely
pubescent.

Leaf upper surface
bristly, pubescent.

Petiole bristly, rather
densely pubescent.

Corolla widely funnel-
shaped, white to pink,
deep reddish or crimson.

R. scabrifolium var. *scabrifolium* AC774 at Deer Dell

R. scabrifolium var. spiciferum (Franch.) Cullen

R. spiciferum Franch.

Scabrifolia Gp B

Shrub 0.15–1.5 m (0.5–5 ft) high; **branchlets** bristly, densely pubescent.

Leaves lanceolate, linear-lanceolate, linear, oblanceolate or oblong-oval; **lamina** 1.2–3.5 cm long, 0.2–1.3 cm broad; **upper surface** scabrid, bristly, rather densely or moderately pubescent, scaly or not scaly; margin recurved, bristly, pubescent; **underside** rather densely or moderately pubescent, scaly, scales 1–3 times their own diameter apart.

Petiole bristly, densely pubescent, scaly.

Inflorescence axillary in the uppermost few leaves, 1–4 flowers.

Corolla widely funnel-shaped, 1–1.5 cm long, pink, deep pink, rose or white, outside scaly.

W and NW Yunnan, Guizhou, SW Sichuan. 1,500–3,200 m (5,000–10,500 ft).

AM 1955 (Windsor) 'Fine Bristles'.

Epithet: Bearing spikes.   Hardiness 3.   April–May.

Cullen names **R. scabrifolium** *var.* **pauciflorum** *which is intermediate between the other two varieties, and may be a hybrid or part of a cline.*

He also seperates **R. pubescens** *as a full species, whereas Davidian treats it as synonymous with var.* spiciferum. *It has smaller, narrower recurved leaves than either variety of* R. scabrifolium, *and comes from a different location, so may be worth separate status. Flowers pink.*

R. spinuliferum Franch.

Scabrifolia Gp B

Shrub or small **tree**, 0.6–4.6 m (2–15 ft) high; **branchlets** moderately or rather densely bristly, pubescent, scaly.

Leaves oblong-lanceolate, oblanceolate, obovate or oblong-elliptic; **lamina** 2.4–9.5 cm long, 0.6–4.5 cm broad; **upper surface** bullate, scaly; margin often bristly; **underside** rather densely pubescent, scaly, scales 1–3 times their own diameter apart.

Petiole bristly, pubescent, scaly.

Inflorescence axillary in the uppermost few leaves, or terminal, 1–5 flowers.

Corolla tubular, contracted at the upper end, 1.4–2.5 cm long, crimson-red, red, pink or yellowish.

S Yunnan. 800–2,450 m (2,600–8,000 ft).

AM 1974 (Chyverton) 'Jack Hext', flowers red.

AM 1977 (Brodick) 'Blackwater', flowers red.

Epithet: Bearing spines.   Hardiness 2–3.   April–May.

Small narrow hairy leaves, margin recurved so as to narrow the leaf.

The corolla is usually narrower than that of var. *scabrifolium*.

Branchlets bristly, densely pubescent.

Upper leaf surface scabrid, bristly, pubescent, underside pubescent.

Petiole bristly, densely pubescent.

Corolla usually shades of pink.

R. scabrifolium var. *spiciferum* at Deer Dell

R. scabrifolium var. *pauciflorum*, showing possible hybridity with *R. spinuliferum*, in the garden of a Mr. Jenkins in Scotland

R. scabrifolium var. *pubescens* KW3953 at Deer Dell

Unmistakable in flower: inflorescence usually terminal and upward pointing, corolla tubular, contracted at the upper end.

Out of flower, *R. spinuliferum* is very similar to *R. scabrifolium* var. *scabrifolium*, but lacks the bristles on the leaf upper surface.

Leaf upper surface bullate, margin (only) often bristly.

Petiole bristly, pubescent.

Corolla crimson-red, red, pink, or yellowish.

R. spinuliferum (Davidian)

Subgenus *Rhododendron*
Section *Rhododendron*
Subsection *Tephropepla* Sleumer
Tephropeplum Series

Small to medium or tall shrubs.
Branchlets scaly.
Leaves usually relatively narrow.
Underside pale green or glaucous.
Calyx lobes conspicuous.
Corolla tubular-campanulate or campanulate.
Style long, slender, and straight.
Stamens 10, declinate (curved downwards).

R. auritum Tagg.

Shrub 1–3 m (3–10 ft) high; **stem** and **branches** coppery-red with brown, smooth, flaking bark; **branchlets** rather densely scaly.

Leaves oblong, lanceolate or oblong-lanceolate; **lamina** 2.7–6.6 cm long, 1–2.7 cm broad; **upper surface** bright green, scaly; **underside** densely scaly, scales small, one-half to their own diameter apart or nearly contiguous, with closely scattered large scales.

Petiole densely scaly.

Inflorescence 4–7 flowers.

Calyx lobes reflexed.

Corolla tubular-campanulate, 1.8–2.5 cm long, sulphur-yellow, yellow sometimes tinged red, or creamy-white tinged pink on the lobes.

SE Tibet. 2,150–2,600 m (7,000–8,500 ft).

AM 1931 (Exbury), flowers sulphur-yellow (under KW6278).

Epithet: With long ears.   Hardiness 2–3.   April–May.

Plants with deep yellow flowers are rare in cultivation.

R. tephropeplum at Howick (Davidian)

Stem and branches coppery-red with brown, smooth, flaking bark.

Leaf underside densely scaly with partly detersile brown scales.

Calyx lobes reflexed.

Corolla sulphur yellow or creamy white.

Leaf underside densely scaly, scales small, one-half to their own diameter apart or nearly contiguous, with closely scattered large scales.

R. auritum KW6278 at Deer Dell

Rhododendron
(Tephropepla)

R. hanceanum Hemsl.

Shrub 0.3–1.5 m (1–5 ft) high; **branchlets** scaly.

Leaves ovate-lanceolate, ovate or obovate; **lamina** rigid, 3.5–12.8 cm long, 1.6–5.5 cm broad, apex acutely acuminate or acute; **upper surface** olive-green or bright green, scaly; **underside** paler, scaly, scales 1–5 times their own diameter apart.

Inflorescence racemose, 5–11 flower; **rhachis** 0.8–2 cm long.

Corolla funnel-campanulate, 1.3 –2.1 cm long, creamy-white, pale yellow or white.

Style long, slender and straight.

SW Sichuan. 1,200–3,000 m (4,000–9,900 ft).

AM in 1957 (Windsor) 'Canton Consul', flowers cream.

Epithet: After H.F. Hance (1827–1886), Consul at Canton.   Hardiness 3. April–May.

Moved from Subsection Triflora *by Cullen.*

R. hanceanum Nanum Group

R. hanceanum Hemsl. 'Nanum'

Dwarf compact **shrub** 8–15 cm (3–6 inches) high, and up to 25 cm (10 inches) wide, with small **leaves** 2–3.5 cm long, 1–1.6 cm broad, and with yellow **flowers**.

Epithet: Dwarf habit.   Hardiness 3.   April–May.

R. longistylum Rehd. & Wils.

Shrub 0.5–2 m (1.5–7 ft) high; **branchlets** not scaly or scaly, minutely puberulous.

Leaves oblanceolate, lanceolate, oblong-lanceolate; **lamina** 1.6–6 cm long, 0.6–1.5 cm broad; **upper surface** dark green, not scaly or scaly; **underside** scaly, scales 2–4 times their own diameter apart.

Petiole minutely puberulous.

Inflorescence 3–10 flowers.

Calyx 2–3 mm long.

Corolla tubular-funnel-shaped, 1.3–2 cm long, white or white tinged pink, not scaly outside.

W Sichuan. 1,000–2,300 m (3,300–7550 ft).

Epithet: Long-styled.   Hardiness 2–3.   April–May.

Small flowers. Very rare in cultivation. Moved from Subsection Triflora *by Cullen.*

Compact habit with bronzy new growth.

Leaves rigid, smooth, and laxly scaly beneath.

Easily recognised in flower by the long rhachis and the many (5–11)-flowered truss.

Leaves ovate-lanceolate, ovate or obovate, underside with scales 1–5 times their own diameter apart.

Corolla pale yellow, creamy-white or white.

R. hanceanum W882 at Borde Hill

Not as common in cultivation as the larger-growing forms, such as 'Canton Consul'.

The true 'Nanum' has smaller leaves and yellow flowers.

R. hanceanum 'Nanum' in R. White's garden (M.L.A. Robinson)

Small glabrous leaves, almost elliptic, fairly smooth above, moderately scaly beneath.

Corolla tubular-funnel-shaped, small, white or white tinged pink, style impressed.

Stamens and style about 1.5 times longer than the corolla.

Leaf underside the scales 2–4 times their own diameter apart.

R. longistylum at Glendoick (M.L.A. Robinson)

Rhododendron (Tephropepla)

R. tephropeplum Balf.f. & Farrer
(*R. deleiense* Hutch. & Kingdon Ward)

Shrub 0.3–1.2 m (1–4 ft) or rarely 1–2.4 m (6–8 ft) high; **branchlets** moderately or rather densely scaly.

Leaves lanceolate, oblong or oblong-obovate; **lamina** 3–13 cm long, 0.8–4 cm broad; **upper surface** dark green, shiny, scaly; margin recurved; **underside** pale glaucous, densely scaly, scales small, black or brown, one-half to their own diameter apart, with or without widely scattered large scales.

Petiole moderately or densely scaly.

Inflorescence 3–9 flowers.

Calyx 4–8 mm long.

Corolla tubular-campanulate, 1.8–3.2 cm long, dark or pale rose, pink, purplish, rosy-crimson, crimson-purple or rarely white, without spots.

Style long, slender and straight.

NE Upper Burma, NW Yunnan, E Tibet, Assam. 2,450–4,300 m (8,000–14,100 ft).

AM 1929 (Bodnant), flowers pale pink.

AM 1935 (Townhill Park), as *R. deleiense*, flowers purplish-pink.

AM 1975 (Sandling Park) 'Butcher Wood' (KW20844).

Epithet: With ash-grey covering.   Hardiness 2–3.   April–May.

A very variable species in growth habit and leaf size. Deleiense *Group has larger leaves and flowers.*

R. xanthostephanum Merr.
(*R. aureum* Franch. non Georgi.)

Shrub 0.3–3 m (1–10 ft) or rarely 4.5 m (15 ft) high; **branchlets** scaly.

Leaves lanceolate or oblong-lanceolate; **lamina** 5–10.4 cm long, 1.3–3.4 cm broad; **upper surface** dark green or bright green, scaly; **underside** glaucous, densely scaly, scales small, black or brown, one-half to their own diameter apart, with or without widely scattered large scales.

Inflorescence 3–5 or rarely 8 flowers.

Calyx 3–6 mm long.

Corolla tubular-campanulate, 2–2.6 cm long, bright yellow, deep lemon-yellow or canary-yellow, outside rather densely or moderately scaly.

Stamens 10.

W and NW Yunnan, NE Upper Burma, Assam, SE and E Tibet. 1,850–4,100 m (6,000–13,500 ft).

AM 1961 (Windsor) 'Yellow Garland' (F22652).

Epithet: Yellow garland.   Hardiness 1–2.   April–May.

Leaf underside with the smaller scales (one-half to their own diameter apart). The smaller and more numerous scales on the leaf underside are sunk into pits, whereas the larger scales are on the surface of the lamina. Not very hardy, so rare in cultivation.

Leaves glaucous below, as with the much rarer *R. xanthostephanum.*

Corolla pink, purplish, rosy-crimson, crimson-purple.

Branchlets moderately or rather densely scaly. Petiole moderately or densely scaly.

R. tephropeplum KW13006 in the Valley Gardens

Similar to *R. tephropeplum,* but has bright yellow, deep lemon-yellow or canary-yellow flowers.

Leaf underside glaucous with brownish scales.

Calyx lobes not reflexed.

Leaf underside with or without scattered large scales.

R. xanthostephanum CCHH8070 at Glendoick (M.L.A. Robinson)

Rhododendron
(Tephropepla)

Subgenus *Rhododendron*
Section *Rhododendron*
Subsection *Trichoclada* (Balf.f.) Cullen
Trichocladum Series

Small to medium-sized shrubs.
Branchlets usually bristly.
Leaves deciduous or evergreen, often hairy, underside with
vesicular scales.
Corolla cream or yellow shades.
Stamens 10 or occasionally 8.

Group A. Leaves evergreen
R. lepidostylum, *R. rubroluteum* and *R. viridescens*

Group B. Leaves wholly or partly deciduous
R. caesium, *R. mekongense* (var. *mekongense* [including the
Melinanthum Group] and var. *rubrolineatum*)
and *R. trichocladum* var. *trichocladum*.

R. lepidostylum Balf.f. & Forrest   Trichoclada Gp A

Compact **shrub** rounded or sometimes broadly upright, 0.3–1.2 m (1–4 ft) high;
branchlets scaly, rather densely bristly, **leaf-bud** scales persistent.
Leaves evergreen, obovate, oblong-obovate, oblong-oval or oval; **lamina** 2–4.3 cm
long, 1–2 cm broad; **upper surface** olive-green or bluish-green, not glaucous
(in young leaves, bluish-green, markedly glaucous), not scaly, not bristly; margin
recurved, bristly; **underside** glaucous, scaly, scales 1–4 times their own diameter
apart, rather densely bristly.
Inflorescence 1–3 flowers.
Pedicel scaly, bristly.
Calyx bristly.
Corolla funnel-campanulate, 2–2.3 cm long, pale yellow or yellow, tube
bristly outside.
Ovary densely scaly, bristly.
W and mid-W Yunnan. 3,050–3,650 m (10,000–12,000 ft).
AM 1969 (Collingwood Ingram), flowers greenish-yellow.
Epithet: With scaly styles.   Hardiness 3.   May–June.

R. lepidostylum at Deer Dell

Easily recognised by the compact habit and the bluish-green leaves (in young leaves, markedly glaucous), hairy at the margins.

Leaves evergreen.

Branchlets rather densely bristly.

Leaf-bud scales persistent.

Leaf underside glaucous, scaly, scales 1–4 times their own diameter apart, rather densely bristly.

Calyx bristly.

Corolla yellow.

New foliage of *R. lepidostylum* at Deer Dell

Rhododendron
(Trichoclada Group A)

R. viridescens Hutch.   Trichoclada Gp A

Shrub 0.3–1.5 m (1–5 ft) high; **branchlets** scaly, moderately or rather densely bristly.
Leaves evergreen, obovate, oblong-obovate, elliptic, oblong-elliptic or oval;
  lamina 2.3–6.7 cm long, 1.3–3 cm broad; **upper surface** olive-green or bright
  green (in young leaves, pale bluish-green, glaucous), not scaly, glabrous;
  underside pale green, not glaucous or slightly glaucous, scaly, scales 1–3 times
  their own diameter apart.
Petiole bristly.
Inflorescence 3–6 flowers.
Pedicel bristly.
Corolla funnel-campanulate, 1.6–2.4 cm long, pale yellowish-green or yellow,
  with greenish spots, outside scaly.
SE Tibet. 2,900–3,350 m (9,500–11,000 ft).
AM 1972 (Glendoick) 'Doshong La', flowers yellow (KW5829).
Epithet: Becoming green.   Hardiness 3.   May–June.

R. rubroluteum Davidian   Trichoclada Gp A

*Evergreen leaves, unequally sized scales and reddish-yellow flowers. Fits
nowhere comfortably. It is rare in cultivation. Collected only once from an
uncertain locality. Late flowering.*

R. caesium Hutch.   Trichoclada Gp B

Shrub 1–1.5 m (3–5 ft) high; **branchlets** moderately or sparsely scaly, not bristly.
Leaves semi-evergreen, oblong, oblong-lanceolate or elliptic; **lamina** 2.3–5 cm
  long, 1.2–3.4 cm broad; **upper surface** olive-green or bright green, not scaly,
  not bristly; **underside** markedly glaucous, scaly, scales 2–5 times their own
  diameter apart.
Inflorescence 3–5 flowers.
Pedicel scaly, not bristly.
Calyx margin slightly or moderately bristly.
Corolla funnel-campanulate, 1.5–2.5 cm long, yellowish-green, with greenish
  spots, outside scaly.
W Yunnan. 2,450–3,050 m (8,000–10,000 ft).
Epithet: Dullish-blue.   Hardiness 3.   May–June.

Extremely rare in cultivation.

Similar to *R. lepidostylum*, but it is from a different area and differs in the leaves being less hairy and having a relative lack of glaucous colouring

Branchlets scaly, moderately or rather densely bristly.

Leaf bud scales deciduous.

Leaves olive-green or bright green; young leaves pale bluish-green, glaucous.

Corolla pale yellowish-green.

Foliage of *R. viridescens* at Hindleap Lodge (M.L.A. Robinson)

Note the epithet – the leaf upper surface of young leaves is pale bluish green.

Leaf underside glaucous.

Scales large, brown, unequal.

Leaves semi-evergreen.

Leaf underside scales 2–5 times their own diameter apart.

Corolla yellowish-green, with greenish spots.

New foliage of *R. caesium* F26798 at Deer Dell

The remaining two species in this Subsection, R. mekongense *and*
R. trichocladum *merge in the wild. These and all the subdivisions of*
R. mekongense *need taxonomic revision.*

R. mekongense var. mekongense

R. mekongense Franch.   Trichoclada Gp B
(*R. chloranthum* Balf.f. & Forrest, *R. semilunatum* Balf.f. & Forrest)

Shrub 0.3–2.4 m (1–8 ft) high; **branchlets** scaly, often bristly.

Leaves deciduous, oblong-obovate, obovate or oblong-elliptic; **lamina** 1.6–4.6 cm
long, 0.8–2.6 cm broad; **upper surface** olive-green or green, not glaucous, not
scaly, not bristly, glabrous; **underside** not glaucous, scaly, scales 1–4 times their
own diameter apart, not bristly, midrib not bristly or bristly.

Petiole bristly.

Inflorescence 2–5 flowers, flowers precocious.

Pedicel often bristly.

Calyx 0.5–7 mm long.

Corolla campanulate, 1.4–2.3 cm long, yellow, pale yellow tinged green, lemon-
yellow or dark greenish-orange, with or without darker greenish spots.

E and SE Tibet, NW Yunnan, NE Burma. 2,750–4,300 m (9,000–14,000 ft).

Epithet: From the Mekong River, in China.   Hardiness 3.   May.

R. mekongense var. mekongense
Melinanthum Group   Trichoclada Gp B

R. melinanthum Balf.f. & Kingdon Ward

Shrub 1.5–2.5 m (5–8 ft) high; **branchlets** scaly, often bristly.

Leaves deciduous or sometimes semi-deciduous, oblong-obovate,
oblanceolate or elliptic; **lamina** coriaceous, 3.2–5.2 cm long, 1.4–2.5 cm
broad; **upper surface** bluish-green or green, slightly glaucous, not scaly,
not bristly, glabrous; **underside** somewhat glaucous, scaly, scales 1–3
times their own diameter apart, not bristly.

Petiole bristly or not bristly.

Calyx 1–2 mm long.

Corolla widely funnel-campanulate, 1.8–2.4 cm long, yellow, with or without
orange spots, outside scaly,

E Upper Burma. 3,650–4,300 m (12,000–14,000 ft).

AM 1979 (Borde Hill) 'Yellow Fellow', as *R. mekongense* (KW406, i.e. the
type number).

Epithet: Honey flowers.   Hardiness 3.   May–June.

Leaves and stems usually less hairy than those of R. trichocladum. Leaves deciduous, upper surface not glaucous, underside not glaucous. Leaf scales 1–4 times their own diameter apart. Calyx glabrous or sparsely hairy.

Corolla yellow, pale yellow tinged green, lemon-yellow or dark greenish-range.

Petiole bristly. Flowers precocious.

R. mekongense var. *mekongense* F13900 (as *R. chloranthum*) at Deer Dell

Leaves deciduous or sometimes semi-deciduous.

Very similar to the type of *R. mekongense* var. *mekongense*, but with somewhat glaucous leaves.

R. mekongense var. *mekongense* (as *R. melinanthum*) (J.C. Birck)

R. mekongense var. rubrolineatum (Balf.f. &

Forrest) Cullen   **Trichoclada Gp B**

R. rubrolineatum Balf.f. & Forrest

Shrub 0.6–1.5 m (2–5 ft) high; **branchlets** scaly, not bristly.
Leaves deciduous, elliptic, oblong-elliptic or oblong; **lamina** 2.4–6.3 cm long,
    1.3–2.6 cm broad; **upper surface** olive-green, not glaucous, scaly, not bristly;
    underside not glaucous, scaly, scales 1–3 times their own diameter apart.
Petiole scaly, not bristly.
Inflorescence terminal and axillary in the uppermost 2 or 3 leaves, 3–5 flowers.
Corolla funnel-campanulate or campanulate, 1.3–1.9 cm long, creamy-yellow or
    creamy-white, lined and flushed rose on the outside, scaly outside.

Mid-W Yunnan. 3,350 m (11,000 ft).
Epithet: Lined with red.   Hardiness 3.   May–June.

*Plants in cultivation vary and those that are evergeen or semi-evergreen are likely
to be hybrids with* R. racemosum.

R. trichocladum var. trichocladum

R. trichocladum Franch.   **Trichoclada Gp B**
(*R. lophogynum* Balf.f. & Forrest, *R. oulotrichum* Balf.f. & Forrest, *R. mekongense* var.
longipilosum Cullen)

Shrub 0.3–1.8 m (1–6 ft) high; **branchlets** scaly, moderately to rather densely bristly.
Leaves deciduous, oblong-obovate, obovate, oblong-oval or oblong; **lamina**
    1.6–5.6 cm long, 0.6–2.6 cm broad; **upper surface** olive-green or dark green,
    often scaly, bristly or not bristly; margin bristly; **underside** scaly, scales 1–6
    times their own diameter apart, midrib bristly.
Petiole scaly, densely or moderately bristly.
Inflorescence 2–5 flowers, flowers precocious.
Pedicel pilose.
Corolla funnel-campanulate, 1.5–2.3 cm long, sulphur-yellow, bright yellow,
    greenish-yellow or pale lemon-yellow, with or without green spots, outside scaly.

W and Central Yunnan, NE Upper Burma, Burma–Tibet Frontier, SE Tibet, Nepal.
    2,000–4,250 m (6,550–14,000 ft).
AM 1971 (Windsor), as *R. lophogynum*.
Epithet: With hairy twigs.   Hardiness 3.   April–July.

R. trichostomum *var.* longipilosum *Cowan, formerly* R. mekongense *var.*
longipilosum *Cullen, reputed to have longer hairs on the upper leaf surface, seems
to differ only in degree from var.* trichocladum.

Leaves deciduous, usually lacking bristles.

Corolla creamy-yellow or creamy-white, lined and flushed rose on the outside.

Leaf upper surface olive-green, not glaucous; underside not glaucous, scales 1–3 times their own diameter apart.

Inflorescence terminal and axillary in the uppermost 2–3 leaves.

R. mekongense var. *rubrolineatum* F19912 at Deer Dell

Leaves and stems bristly.

Leaf underside green occasionally with a bristly indumentum.

Scales pale- or dark-brown of a more or less uniform size.

Long flowering season.

Leaf upper surface not glaucous, underside with scales 1–6 times their own diameter apart.

Petiole bristly.

Flowers often precocious.

Corolla sulphur-yellow, bright yellow, greenish-yellow or pale lemon-yellow.

R. trichocladum var. *trichocladum* (Davidian)

Rhododendron
(Trichoclada Group B)

Subgenus *Rhododendron*
Section *Rhododendron*
Subsection *Triflora* Sleumer
Triflorum Series, Augustinii, Triflorum and
Yunnanense Subseries

Shrubs, dwarf to over 10 m.
Leaf underside with entire scales.
Inflorescence terminal, or terminal and axillary.
Calyx small.
Corolla zygomorphic, usually widely funnel-shaped – 'butterfly shaped'.
Stamens 10, long.
Style slender, elepidote.
Ovary densely lepidote.

*With many Subsection Triflora species, it is not possible to make a definite
identification out of flower. However, some leaf characteristics are readily
observable, such as the glaucous underside, the presence or absence of hairs
and the density of the scales (magnification is advisable for observation), so
leaves will be considered first.*

Group A. Leaf undersides <u>always</u> glaucous (cf. Group D below)

> *R. ambiguum, R. searsiae* and *R. zaleucum* (var. *zaleucum* and
> var. *flaviflorum*).

> *N.B. In lower light conditions, the glaucescence is reduced, and it is
> usually not present in herbarium specimens. Other species in
> Subsection Triflora occasionally, but not reliably, show glaucescence.*

Group B. Hairs on midrib, lamina or margin

Group B1. Hairs on leaf midrib
R. augustinii (ssp. *augustinii*, ssp. *chasmanthum*, ssp. *hardyi* and
ssp. *rubrum*).

Group B2. Hairs on the lamina or leaf margin
R. yunnanense Suberosum Group, *R. trichanthum* and
R. Sp. Nov. PW20.

Group C. Rounded ovate leaf

> *R. oreotrephes.*

R. augustinii at Deer Dell

Group D. The remainder

R. amesiae – leaves often ovate, petioles bristly.

R. concinnum – leaves often ovate.

R. davidsonianum – leaves usually V-shaped in cross-section.

R. keiskei – dwarf plant.

R. lutescens – leaves markedly long acuminate, long leaves, often reddish-tinged with bronzy brown young growth.

R. polylepis – long rugose (deep-veined) leaves.

R. rigidum – leaves bluish-green and sparsely scaly beneath, glaucous when young.

R. siderophyllum – grey-green leaves.

R. tatsienense – a combination of reddish or purplish branchlets and small, often ovate, leaves.

R. triflorum (var. *triflorum*, var. *bauhiniiflorum* and Mahogani Group): leaves usually elliptic. May or may not have a glaucous leaf underside.

R. yunnanense varieties – if a plant fits none of the above and its leaves are sparsely scaly beneath, it is likely to be *R. yunnanense*. This species and its varieties are very variable and can be difficult to identify out of flower.

continued overleaf

Rhododendron
(Triflora)

Group D1 Very widely funnel-shaped flowers

R. amesiae and *R. concinnum* (including the submerged var. *benthamianum* and the Pseudoyanthinum Group).

Group D2. Flowers not very widely funnel-shaped but yellow or green-yellow

R. keiskei, *R. lutescens*, *R. triflorum* (var. *triflorum*, var. *triflorum* Mahogani Group and var. *bauhiniiflorum*).

Group D3. Flowers not very widely funnel-shaped but white, pink or lavender

R. davidsonianum, *R. polylepis*, *R. rigidum*, *R. siderophyllum*, *R. tatsienense* and *R. yunnanense* (including *R. yunnanense* Hormophorum Group).

Hybrids

R. vilmorinianum and *R. wongii*.

New foliage of *R. lutescens* at Deer Dell

Leaves of Groups A and D of Subsect. *Triflora* at Deer Dell.
Top to bottom: left column, *R. davidsonianum* (Gp D3) and
R. triflorum (Gp D2); middle column, *R. polylepis* (Gp D3),
R. concinnum (Gp D1), *R. yunnanense* (Gp D3), *R. siderophyllum*
(Gp D3) and *R. lutescens* (Gp D2); right column, *R. ambiguum*
(Gp A), *R. ambiguum* (as *R. chengshienianum*) (Gp A),
R. zaleucum (Gp A), *R. lutescens* (Gp D2), *R. rigidum* (Gp D3)
and *R. tatsienense* (Gp D3). Bar = 5 cm

Leaves of Groups B1, B2 and C of Subsect. *Triflora* at Deer Dell.
Top to bottom: left column, *R. oreotrephes* (Gp C) (2 forms);
middle column, *R. suberosum* (Gtp B2) (2 leaves) and
R. trichanthum (Gp B2) (2 leaves); right column, *R. augustinii*
ssp. *augustinii* (Gp B1), *R. augustinii* spp. *hardyi* (Gp B1) and
R. augustinii ssp. *rubrum* (Gp B1). Bar = 5 cm

R. ambiguum Hemsl.   Triflora Gp A
R. chengshienianum Fang.

Shrub 0.6–5.8 m (2–19 ft) high; **branchlets** densely or moderately scaly.
Leaves ovate-lanceolate, lanceolate or elliptic; **lamina** 2.3–8 cm long, 1.2–3.2 cm
 broad; **upper surface** scaly; **underside** glaucous, scaly, scales large, unequal
 differently coloured yellowish-brown, dark brown or blackish, contiguous to
 their own diameter apart.
Inflorescence 2–7 flowers.
Corolla widely funnel-shaped, 2–3.4 cm long, yellow or greenish-yellow, with
 green spots, outside scaly.
W Sichuan. 2,300–4,500 m (7,500–14,800 ft).
AM 1976 (Hergest Croft) 'Jane Banks', flowers yellow-green with greenish spots.
Epithet: Doubtful.   Hardiness 3.   April.

Plants under the synonym **R. chengshienianum** *are even more glaucous, and
have paler flowers.*

R. searsiae Rehd. & E.H.Wilson   Triflora Gp A

Shrub 1.5–5 m (5–16 ft) high; **branchlets** scaly.
Leaves lanceolate, oblong-lanceolate or oblanceolate; **lamina** 2.5–8 cm long,
 1–2.6 cm broad, apex acuminate, acutely acuminate or acute; **upper surface**
 dark green or pale green, scaly; **underside** bluish-glaucous, densely scaly, scales
 yellowish or pale brown, one-half their own diameter apart, with larger dark
 brown scattered scales.
Inflorescence 3–8 flowers.
Corolla widely funnel-shaped, 2–3.4 cm long, white or pale rose-purple, with
 light green spots, not scaly outside.
W Sichuan. 2,300–3,000 m (7,550–9,800 ft).
Epithet: After Sarah C. Sears, American artist.   Hardiness 3.   April–May.

R. ambiguum (as
R. chengshienianum)
at Deer Dell

Leaf underside glaucous, scales large, unequal, differently coloured, contiguous to their own diameter apart.

Corolla yellow or greenish-yellow, with green dots.

R. ambiguum at Deer Dell

Leaf underside bluish-glaucous.

Leaf scales one-half their own diameter apart, with larger dark scattered scales: under a little magnification 3 colours of scales can be seen.

Corolla white or pale rose-purple, with light green spots.

R. searsiae W1343 at Deer Dell

Rhododendron
(Triflora Group A)

R. zaleucum var. zaleucum   Triflora Gp A

R. zaleucum Balf.f. & W.W.Sm.

Shrub or **tree**, 0.3–11 m (1–35 ft) high; **branchlets** scaly.
Leaves lanceolate or oblong-lanceolate, rarely elliptic or obovate; **lamina**
    3.2–8.8 cm long, 1–3 cm broad, apex acuminate or acute; **upper surface**
    pale green or olive-green, not scaly; **underside** markedly glaucous, scaly;
    scales large, 1.5–4 times their own diameter apart.
Inflorescence terminal, or terminal and axillary, 3–5 flowers.
Calyx minute, 0.5–1 mm long.
Corolla funnel-shaped, 2.6–4.8 cm long, white flushed rose, pink, rose, lavender-
    rose or purple, with or without crimson spots, outside scaly.
W, mid-W and NW Yunnan, NE and E Upper Burma. 1,800–4,000 m
    (6,000–13,000 ft).
AM 1932 (Borde Hill), flowers deep rose with crimson spots.
Epithet: Very white.   Hardiness 3–4.   April–May.

R. zaleucum var. flaviflorum   Davidian
Triflora Gp A

Larger **leaves**, up to 10 cm long, yellow **flowers**.
North Triangle, North Burma. 2,750 m (9,000 ft).
Epithet: With yellow flowers.   Hardiness 2–3.   April–May.

Rare in cultivation: less hardy.

R. augustinii ssp. augustinii   Hemsl.   Triflora Gp B
(*R. vilmorinianum* Balf.f. type only)

Shrub 1–7.5 m (3–25 ft) high; **branchlets** scaly, usually rather densely pubescent.
Leaves lanceolate, oblong-lanceolate or oblong-elliptic; **lamina** 3.3–12 cm long,
    1.1–4.5 cm broad; **upper surface** pale green, scaly or not scaly, pubescent or
    glabrous; **underside** scaly, scales 1–5 times their own diameter apart, glabrous
    or sometimes pubescent, except midrib hairy one third to its entire length.
Petiole usually pubescent.
Inflorescence 2–6 flowers.
Corolla widely funnel-shaped, 2–4.3 cm long, intense violet, pale or dark
    lavender-blue, lilac-purple, pale lavender-rose, pink, rose or white tinged pink,
    with yellowish-green, olive-green or brownish spots.
Hubei, W Sichuan, E and SE Tibet, NW Yunnan. 1,300–3,400 m (4,250–11,150 ft).
AM 1926 (South Lodge), flowers lilac-mauve with greenish spots.
AGM 1924.
Epithet: After Augustine Henry (1857–1930), Medical Officer in Chinese
    Customs, later Professor of Forestry, Dublin.   Hardiness 2–3.   April–June.

Leaf underside almost white (but not on young foliage), with large widely-spaced dark scales.

Corolla white flushed rose, pink, rose, lavender-rose or purple.

R. zaleucum var. *zaleucum* F27603 at Deer Dell

R. zaleucum var. *flaviforum* KW20837 at Glenarn

Leaf underside midrib hairy from one-third to its entire length.

Corolla intense violet, pale or dark lavender-blue, lilac-purple, pale lavender-rose, pink, rose or white tinged pink, but in cultivation the flowers are usually shades of near blue.

R. augustinii ssp. *augustinii* W4238 at Deer Dell

R. augustinii ssp. chasmanthum (Diels) Cullen

R. chasmanthum Diels

Triflora Gp B1

Shrub up to about 3 m (10 ft) high.

AM 1930, FCC 1932 (Exbury), flowers bluish-purple with ochraceous spots.
Epithet: With gaping flowers.   Hardiness 2–3.   April–May.

R. augustinii ssp. hardyi (Davidian) Cullen

R. hardyi Davidian

Triflora Gp B1

Shrub 1.2–3 m (4–10 ft) high; **branchlets** scaly.
Leaves completely deciduous or almost deciduous, lanceolate or oblong-lanceolate; **lamina** 4.5–8.2 cm long, 1.8–2.9 cm broad, apex acute or acuminate; **upper surface** green, shiny, scaly, glabrous, midrib puberulous; **underside** scaly, scales 3–6 times their own diameter apart, glabrous, midrib hairy one third to its entire length.
Inflorescence 2–4 flowers.
Corolla widely funnel-shaped, white, or rarely greenish-white faintly tinged lavender, with yellowish or greenish spots at the base.
NW Yunnan, SE Tibet. 3,350–3,650 ft. (11,000–12,000 ft).
Epithet: After Major A.E. Hardy, Sandling Park, Kent.   Hardiness 3.
April–May.

R. augustinii ssp. rubrum (Davidian) Cullen

R. bergii Davidian

Triflora Gp B1

Shrub 1.5–2.8 m (5–9 ft) high; **branchlets** scaly.
Leaves oblong-lanceolate or oblong; **lamina** 4–7.8 cm long, 1.6–3.3 cm broad; **upper surface** dark green, shiny; **underside** pale green, scaly, scales 1–3 times their own diameter apart, midrib rather densely or moderately pubescent.
Petiole hairy.
Inflorescence 3–5 flowers.
Corolla widely funnel-shaped, 2.7–3 cm long, red (Purple Group 73A) with deep red spots on the upper lobes, outside scaly.
NW Yunnan. 3,950 m (13,000 ft).
AM 1978 (Borde Hill), to a clone 'Papillon', flowers reddish with deep red spots (F25914).
Epithet: Red.   Hardiness 3.   March–April.

Rare in cultivation. The AM clone 'Papillon' *is reddish pink, rather than pure red.*

Inflorescence more compact, 5–6 flowers. Corolla lobes **markedly reflexed**, lavender-blue or lavender-rose.

R. augustinii ssp. *chasmanthum* F21470 AM FCC at Deer Dell

A white-flowered deciduous or semi-deciduous *R. augustinii*. Corolla white with yellowish or greenish spots.

R. augustinii ssp. *hardyi* R199 at Deer Dell

Corolla reddish with deeper red spots.

R. augustinii ssp. *rubrum* F25914 (as *R. bergii* 'Papillon') at Borde Hill

R. yunnanense Suberosum Group   Triflora Gp B2

R. suberosum Balf.f. & Forrest

Shrub 1–3 m (3–10 ft) high; **branchlets** scaly, bristly or not bristly.

Leaves lanceolate or oblong-lanceolate; **lamina** 3.5–7.3 cm long, 1–2.2 cm broad, apex acuminate or acute; **upper surface** dark green, shiny, margin rough to the touch, rather densely or moderately bristly; **underside** dark green or pale green, scaly, scales 2–5 times their own diameter apart.

Petiole bristly.

Inflorescence terminal and axillary in the uppermost 2–7 leaves, 1–3 flowers.

Corolla funnel-shaped, 1.7–3 cm long, ivory-white or ivory-white flushed rose, with deep rose or green spots, outside scaly.

SW Sichuan, W and mid-W Yunnan, Upper Burma. 3,350–3,950 m (11,000–13,000 ft).

Epithet: Slightly gnawed.   Hardiness 3.   April–May.

Submerged into R. yunnanense *(q.v.) and close to that species but sufficiently distinct to merit varietal status. Diploid with 26 chromosomes whereas* R. yunnanense *is hexaploid.*

R. trichanthum Rehder.   Triflora Gp B2

(*R. villosum* Hemsl. & E.H.Wilson not Roth.)

Shrub 1–6 m (3–20 ft) high; **branchlets** scaly, densely or moderately bristly and often pubescent.

Leaves oblong-lanceolate, lanceolate or ovate-lanceolate; **lamina** 4–11 cm long, 1.5–3.7 cm broad, apex acuminate to acutely acuminate; **upper surface** scaly, bristly and often pubescent; **underside** scaly, scales 1–4 times their own diameter apart, bristly or pubescent.

Petiole bristly and often pubescent.

Inflorescence 3–5 flowers.

Pedicel bristly.

Calyx bristly.

Corolla widely funnel-shaped, 2.8–3.8 cm long, light to dark purple or rose, outside scaly, bristly on the tube.

W and SW Sichuan. 1,600–3,650 m (5,250–12,000 ft).

AM 1971 (Sandling Park) 'Honey Wood', flowers purple-violet.

Epithet: With hairy flowers.   Hardiness 3.   May–June.

Suberosum Group
has **bristly leaf
margins** and the
upper leaf surface is
usually bristly.
Leaf underside scales
2–5 times their own
diameter apart.
Petiole bristly.
Corolla ivory-white
or ivory-white
flushed rose.

R. yunnanense Suberosum Group F24618 at Deer Dell showing
the leaf bristles

Leaf upper surface
scaly, bristly and
often pubescent.
Branchlets scaly,
densely or
moderately bristly
and often pubescent.
Pedicel bristly.
Calyx bristly.
Corolla light to dark
purple or rose,
bristly on the tube.
Later flowering than
R. augustinii.

R. trichanthum W1342 at Deer Dell

Rhododendron
(Triflora Group B2)

Sp. Nov. PW20 Triflora Gp B2

Branchlets scaly and puberulous with a few short hairs.

Leaves narrowly oblanceolate 4–5 cm long by 1.3–1.8 cm broad, apex markedly acuminate, base tapered; margins with setose hairs, some persistent; **upper surface** dark green with scattered bristly (setose) hairs, which are easily rubbed off; **underside** slightly glaucous, glabrous, midrib raised, densely setose along its entire length, scales 1–4 times their own diameter apart.

Petiole sparsely scaly, pubescent with short pale hairs, or glabrous.

Inflorescence 2–3 flowers.

Pedicel scaly, not bristly.

Calyx 2 mm, densely scaly.

Corolla funnel-campanulate, about 1.5–2.5 cm long, pink with paler lobes; sparsely scaly externally; internally pubescent in the throat with setose hairs: not scaly.

Stamens 8–10, slightly longer than the corolla, more or less lepidote, densely pubescent near the base.

Ovary densely scaly with overlapping pale scales.

Style glabrous except for a very few short hairs near the ovary.

Guizhou.   Hardiness 3?   March–April.

Collected in Guizhou by Peter Wharton as Subsection Triflora. *Seems to be related to R. suberosum or R. polylepis. There may be two species under this number, one with hairy leaf upper surfaces, one without, or the hairs may have rubbed off the specimens examined.*

R. oreotrephes W.W.Sm. Triflora Gp C

(*R. artosquameum* Balf.f. & Forrest, *R. timeteum* Balf.f. & Forrest)

Shrub or **tree**, 0.6–7.5 m (2–25 ft) high; **branchlets** not scaly or scaly.

Leaves evergreen or sometimes semi-deciduous, oblong-elliptic, elliptic, oblong, oval or almost orbicular; **lamina** 1.8–8.9 cm long, 1.2–4.2 cm broad; **upper surface** green, grey-green or bluish-green, not scaly or scaly; **underside** pale glaucous-green or glaucous or brown, scaly, scales contiguous to 4 times their own diameter apart.

Inflorescence 3–10 flowers.

Calyx a mere rim or minute, 0.5–1 mm long.

Corolla funnel-shaped or funnel-campanulate, 1.3–4 cm long, rose, deep rose, purple, grey-lavender or lavender-blue, with or without crimson spots.

Mid-W and NW Yunnan, NE Upper Burma, SW Sichuan, SE Tibet. 2,750–4,900 m (9,000–16,000 ft).

AM 1932 (Exbury), as *R. timetum*, flowers rosy-purple.

AM 1935 (Embley Park), as *R. siderophylloides*, flowers bright pinkish-mauve, with darker spots.

AM 1990 (Glendoick), 'Pentland'.

Epithet: Mountain bred.   Hardiness 3.   April–May.

This species links Subsections Triflora *and* Cinnibarina. *Flowers similar to* Triflora, *foliage to* Cinnibarina.

R. oreotrephes 'Exquisetum' *is a selected clone with more glaucous foliage, but of no botanical significance.*

AM 1937 (Exbury), flowers light mauve pink, spotted red (F20489).

Leaf apex markedly acuminate.

Leaf margins with setose hairs; upper surface dark green with scattered bristly (setose) hairs.

Leaf underside midrib densely setose along its entire length, the hairs detersile.

Corolla pink with paler lobes.

A possible new species or subspecies PW20 at Tregrehan (M.L.A. Robinson)

Leaves oblong-elliptic, elliptic, oblong, oval or almost orbicular, upper surface glaucous, fading to green, grey-green or bluish-green.

Leaf underside scales contiguous to 4 times their own diameter apart.

Corolla rose, deep rose, purple, grey-lavender or lavender-blue.

R. oreotrephes KW5790 at Deer Dell

Rhododendron
(Triflora Group B2/C)

R. amesiae Rehd. & E.H.Wilson   Triflora Gp D1

Shrub 2–4 m (6–13 ft) high; **branchlets** scaly, bristly or not bristly.

Leaves ovate, oblong-elliptic or elliptic; **lamina** 2.8–7 cm long, 1.5–3.4 cm broad; **upper surface** dark green, moderately or rather densely scaly; **underside** densely scaly, scales one-half to their own diameter apart.

Petiole scaly, bristly.

Inflorescence 2–5 flowers.

Corolla widely funnel-shaped, 2.8–4 cm long, purple or dark reddish-purple, with or without darker spots, outside scaly.

W Sichuan. 2,300–3,000 m (7,550–9,850 ft).

Epithet: After Mary S. Ames of North Easton, Massachusetts.   Hardiness 3. April–May.

May not merit full species status.

R. concinnum Hemsl.   Triflora Gp D1

(*R. yanthinum* Bur. & Franch)

Shrub or small **tree** 1–4.5 m (3–15 ft) high; **branchlets** scaly.

Leaves oblong-elliptic, oblong-lanceolate, ovate or ovate-lanceolate; **lamina** 2.5–8.5 cm long, 1.2–3.5 cm broad; **upper surface** dark green; **underside** densely scaly, scales yellowish, pale or dark brown or yellowish-brown, one-half their own diameter apart or contiguous.

Petiole densely scaly.

Inflorescence 2–5 flowers.

Corolla widely funnel-shaped, 1.5–3.2 cm long, purple, deep rosy-purple, deep pinkish or white, outside moderately or rather densely scaly.

W Sichuan, W Hubei, Shaanxi. 1,600–4,150 m (5,250–13,600 ft).

Epithet: Neat.   Hardiness 3.   April–May.

Sometimes mislabelled as var. pseudoyanthum.

R. concinnum Hemsl.   Triflora Gp D1

R. concinnum var. benthamianum (Hemsl.) Davidian

W Sichuan.

Epithet: After Bentham.   Hardiness 3.   April–May.

Perhaps worthy of Group status.

Leaves ovate,
oblong-elliptic or
elliptic.
Petiole scaly, bristly.
**Only the bristly
petioles separate this
from R. concinnum.**
Leaf underside
densely scaly, scales
one-half to their own
diameter apart.
Corolla purple or dark
reddish-purple.

R. amesiae at Deer Dell

Leaves oblong-
elliptic, oblong-
lanceolate, ovate
or ovate-lanceolate,
**but more
commonly ovate
in cultivation.**
Corolla purple, deep
rosy-purple.
Leaf underside densely
scaly, scales one-half
their own diameter
apart or contiguous.
Corolla outside
moderately or rather
densely scaly.

R. concinnum in R. White's garden (M.L.A. Robinson)

Scales dissimilar in
colour, being brown and
yellowish.
Flowers pale lavender
purple or lavender-purple.

'*R. concinnum* var. *benthamianum*' W1869 at Deer Dell

R. concinnum Hemsl. Pseudoyanthinum Group Triflora Gp D1

R. concinnum var. *pseudoyanthinum* (Balf.f. ex Hutch.) Davidian

W Sichuan.
AM 1951 (Wisley), as *R. pseudoyanthinum*.
Epithet: Like *R. yanthinum* (*R. concinnum*).   Hardiness 3.   April–May.

R. keiskei var. keiskei Miq.   Triflora Gp D2
R. keiskei Miq.

Low compact or prostrate **shrub** sometimes broadly upright, or rounded, 0.3–1.8 m
   (1–6 ft) or sometimes 5–8 cm (2–3 inches) high; **branchlets** scaly.
Leaves lanceolate, oblong-lanceolate or oblong-oval; **lamina** 2.3–8.5 cm long,
   0.8–3.5 cm broad; **upper surface** olive-green or bright green, midrib
   moderately or rather densely puberulous; **underside** pale green or green or
   sometimes glaucous, scaly, scales 0.5–5 times their own diameter apart.
Inflorescence 3–6 flowers.
Corolla widely funnel-shaped, 1.4–3 cm long, 1.6–4.5 cm across, pale yellow or
   lemon-yellow, outside scaly.
Japan. 600–1,850 m (2,000–6,000 ft).
AM 1929 (Windlesham), flowers pale yellow.
AM 1970 (Barry Starling) to a prostrate clone 'Yaku Fairy', flowers yellow.
Epithet: After Ito Keisuke (1803–1900), a Japanese botanist.
Hardiness 4.   April–May.

R. keiskei *var.* hypoglaucum *Sato & Suzuki has glaucous leaf undersides.*
R. keiskei *var.* ozawae *T. Yamaz is the most dwarf form, and now includes the
clones 'Yaku Fairy' and 'Ebino'.*

*R. keiskei var.
ozawae* 'Yaku
Fairy' AM at
Deer Dell

Deep ruby-red flowers.

Leaves oblong-lanceolate and often larger.

R. concinnum Pseudoyanthanum Group at Deer Dell

Usually a low compact or prostrate shrub in cultivation.

Midrib on the upper side of the leaf pubescent.

The widely funnel-shaped corolla (pale or lemon-yellow) places this species into Subsection *Triflora*.

Leaf lamina 2.3–7.5 cm long, 0.8–2.8 cm broad.

Scales 0.5–5 times their own diameter apart.

R. keiskei var. *ozawae* 'Ebino' in the UBC Botanic Garden (M.L.A. Robinson)

Rhododendron
(Triflora Group D1/D2)

R. lutescens Franch.

Shrub 1–6 m (3–20 ft) high; **stem** and **branches** with smooth, brown, flaking bark; **branchlets** scaly; young growths bronzy-brown.

Leaves lanceolate, oblong-lanceolate or ovate-lanceolate; **lamina** 4.8–9.3 cm long, 1.3–3.7 cm broad, apex acutely acuminate; **upper surface** bright green or olive-green, scaly; **underside** scaly, scales 0.5–5 times their own diameter apart.

Inflorescence terminal and axillary, 1–3 flowers.

Calyx minute, 0.5–1 mm long.

Corolla widely funnel-shaped, 1.3–2.6 cm long, pale yellow (rarely white) with green spots, outside scaly or not scaly, rather densely or sometimes moderately pubescent.

W Sichuan, NW Yunnan. 800–3,000 m (2,600–9,850 ft).

FCC 1938 (Exbury) 'Exbury', flowers clear lemon-yellow.

AM 1953 (Tower Court) 'Bagshot Sands', flowers primrose-yellow with darker spots. AGM 1969.

Epithet: Becoming yellow.   Hardiness 3.   February–April.

R. triflorum var. triflorum

R. triflorum Hook.f.

Shrub 0.45–4.5 m (1.5–15 ft) high; **stem** and **branches** with smooth, reddish-brown flaking bark; **branchlets** scaly.

Leaves lanceolate, oblong-lanceolate, ovate-lanceolate or elliptic; **lamina** 3–7.8 cm long, 1.3–3.3 cm broad; **upper surface** bright green, not scaly; **underside** glaucous, densely scaly, scales very small, one-half to their own diameter apart.

Inflorescence 2–4 flowers.

Corolla funnel-shaped, 2–3.3 cm long, pale yellow, sulphur-yellow, greenish-yellow or yellow, with greenish-yellow spots, outside rather densely scaly.

Nepal, Sikkim, Bhutan, Assam, S and SE Tibet, Burma–Tibet Frontier. 2,150–4,000 m (7,000–13,000 ft).

Epithet: Three flowers.   Hardiness 3.   May–June.

Leaf **acutely** acuminate; young growth bronzy-brown, sometimes remaining bronze-tinged when mature. **Flowers terminal and axillary.**

Stems and branches with smooth, brown, flaking bark.

Leaf underside scales 0.5–5 times their own diameter apart.

Corolla pale yellow, rather densely or sometimes moderately pubescent.

R. lutescens at Deer Dell

Smooth reddish-purple peeling bark.

Leaf underside sometimes glaucous and sometimes pale green, scales very small, one-half to their own diameter apart.

Corolla pale sulphur-yellow, greenish-yellow or yellow, with greenish-yellow spots.

R. triflorum var. *triflorum* at Deer Dell

Bark of *R. triflorum* var. *triflorum* at Inverewe

Rhododendron
(Triflora Group D2)

R. triflorum var. triflorum Mahogani Group

R. triflorum var. *mahogani* Hutch. **Triflora Gp D2**

SE Tibet. 2,750–3,800 m (9,000–12,500 ft).
Epithet: With mahogany-coloured flowers.   Hardiness 3.   May–June.

Peter Cox reports that this grows mixed with the typical R. triflorum *var.*
triflorum *in the wild.*

R. triflorum var. bauhiniiflorum (Watt ex Hutch.)
Cullen **Triflora Gp D2**
R. bauhiniiflorum Watt ex Hutch.

Lax **shrub** 1.5–1.8 m (5–6 ft) high; **stem** and **branches** with smooth, brown,
flaking bark; **branchlets** scaly.
Leaves oblong-lanceolate or ovate-lanceolate; **lamina** 3.8–6 cm long, 1.4–2.6 cm
broad; **upper surface** bluish-green, not scaly; **underside** densely scaly, scales
very small, one-half their own diameter apart.
Inflorescence 2–3 flowers.
Corolla flat, saucer-shaped, 2.2–2.8 cm long, 3.8–4.3 cm across, deep or pale
yellow or greenish-yellow, with or without yellowish-green spots.
Assam. 2,450–3,900 m (8,000–9,500 ft).
Epithet: With bauhinia-like flowers.   Hardiness 3.   April–June.

R. davidsonianum Rehd. & E.H.Wilson   Triflora Gp D3
(*R. charianthum* Hutch.)

Shrub or **tree** 0.6–5 m (2–16 ft) high; **branchlets** scaly.
Leaves lanceolate or rarely oblong-lanceolate; **lamina** 2.3–7.8 cm long, 0.8–2.6 cm
broad, often V-shaped; **upper surface** dark green or bright green; **underside**
scaly, scales nearly contiguous to their own diameter apart.
Inflorescence terminal, or terminal and axillary in the uppermost one or two
leaves, 3–6 flowers.
Corolla widely funnel-shaped, 1.9–3.3 cm, long, white, white tinged pink, pink,
rose, pale lavender or purple, with or without red spots.
W and SW Sichuan, NW Yunnan. 2,000–3,580 m (6,500–11,800 ft).
AM 1935 (Bodnant).
FCC 1953 (Bodnant), flowers pale rose.
AM 1993 (Clulow), 'Ruth Lyons'.
AGM 1969.
Epithet: After Dr. W.H. Davidson, Friends Mission in China.   Hardiness 3.
April–May.

Corolla with a mahogany-coloured blotch and spots, or suffused mahogany, or of mahogany colour.

R. triflorum var. *triflorum* Mahogani Group KW5687A at Deer Dell

Leaf upper surface bluish green, smaller than those of *R. triflorum* var. *triflorum*.
Corolla flat, saucer-shaped.

R. triflorum var. *bauhiniiflorum* KW7731 at Deer Dell

Leaves almost always V-shaped in cross-section.
Leaf underside scales nearly contiguous to their own diameter apart.

Corolla white, white tinged pink, pink, rose, pale lavender or purple, with or without red spots.

R. davidsonianum W1274 'Ruth Lyons' AM at Deer Dell

R. polylepis Franch.   Triflora Gp D3

Shrub or small **tree**; **branchlets** densely scaly with flaky scales.

Leaves oblong-lanceolate, lanceolate or oblanceolate; **lamina** 4.5–10.2 cm long, 1.2–3.7 cm broad, apex acute or shortly acuminate; **upper surface** dark green; **underside** densely scaly, the scales dark brown or brown, dry flaky, overlapping or contiguous, usually with larger scattered flaky scales.

Inflorescence 3–5 flowers.

Corolla widely funnel-shaped, 2.1–3.5 cm long, pale or deep purple or purplish-violet, with or without yellowish spots.

SW and W Sichuan. 2,000–3,450 m (6,550–11,300 ft).

Epithet: With many scales.   Hardiness 3.   April–May.

R. rigidum Franch.   Triflora Gp D3

(*R. caeruleum* Levl., *R. eriandrum* Levl. ex Hutch.)

Shrub or small **tree**, 0.6–10 m (2–33 ft) high; **branchlets** not scaly or sometimes scaly, glabrous.

Leaves elliptic, oblong-elliptic, oblong-lanceolate or oblanceolate; **lamina** 2.5–6.8 cm long, 1–3.2 cm broad; **upper surface** bluish-green, not scaly; **underside** pale glaucous green, scaly, scales 4–8 (rarely 2–3) times their own diameter apart.

Inflorescence 2–6 flowers.

Corolla widely funnel-shaped, 1.8–3.1 cm long, white, pink or deep rose-lavender, with olive-brown or purple spots, outside not scaly.

NW and W Yunnan. 800–3,350 m (2,600–11,000 ft).

AM 1933 (Sunningdale), as *R. eriandrum*, flowers white, slightly flushed pink (Rock11288).

AM 1939 (Exbury), as *R. caeruleum*, flowers white, spotted red (Rock11288).

AM 1975 (Exbury) 'Louvecienne'.

Epithet: Stiff.   Hardiness 3.   April–May.

Long deep-veined leaves, underside densely scaly the scales dark, dry flaky, overlapping or contiguous, usually with larger scattered flaky scales.

Branchlets densely scaly with flaky scales.

Corolla pale or deep purple or purplish-violet, with or without yellowish spots.

R. polylepis Knott313 at Deer Dell

Leaf upper surface noticeably bluish green, underside pale glaucous green.

Leaf scales 4–8 (rarely 2–3) times their own diameter apart.

Corolla white, pink or deep rose-lavender, with olive-brown or purple spots.

R. rigidum R11288 at Deer Dell

Rhododendron
(Triflora Group D3)

R. siderophyllum Franch. Triflora Gp D3

Shrub or **tree** 1–8 m (3–26 ft) high; **branchlets** scaly.

Leaves oblong-lanceolate, ovate-lanceolate, lanceolate, elliptic or oblong-oval; **lamina** 3–9 cm long, 1.5–4.1 cm broad; **upper surface** pale greyish-green, matt, scaly; **underside** scaly, scales almost contiguous to 1.5 times their own diameter apart.

Inflorescence terminal and axillary in the uppermost one or two leaves, 3–6 flowers.

Corolla widely funnel-shaped, 1.5–3 cm long, white, pink, rose, purple or pale lavender-blue, with or without yellow or rose spots.

SW Yunnan, Guizhou, SW Sichuan. 800–3,350 m (2,750–11,000 ft).

AM 1945 (Exbury).

Epithet: Rusty-coated leaves.   Hardiness 3.   May.

R. tatsienense Franch. Triflora Gp D3

(*R. hypophaeum* Balf.f. & Forrest)

Shrub 0.3–2.75 m (1–9 ft) high; **branchlets** deep crimson-purple, scaly.

Leaves elliptic, obovate, ovate or oblong-lanceolate; **lamina** rigid, 1.6–6 cm long, 1–3.1 cm broad; **upper surface** dark green, shiny, scaly; **underside** pale green, scaly, scales one-half to their own diameter apart.

Inflorescence terminal, or terminal and axillary in the uppermost 1–3 leaves, 1–6 flowers.

Calyx minute, 0.5–1 mm long.

Corolla widely funnel-shaped, 1.4–3 cm long, purple, rose, rose-pink or rose-lavender, with or without red spots, outside scaly.

W Sichuan, NW and mid-W Yunnan. 2,100–4,300 m (7,000–14,000 ft).

Epithet: From Tatsienlu, now Kanding, W China.   Hardiness 3.   April–May.

Leaf upper surface **greyish** green and matt.

Leaf underside scales almost contiguous to 1.5 times their own diameter apart.

Corolla white, pink, rose, purple or pale lavender-blue, with or without yellow or rose spots.

R. siderophyllum CNW1046 at Deer Dell

A combination of long deep purple or reddish stems **and** small ovate leaves.

New leaves glaucous.

Leaf underside pale green, scales one-half to their own diameter apart.

Corolla purple, rose, rose-pink or rose lavender, with or without red spots.

R. tatsienense F29331 at Deer Dell

Rhododendron
(Triflora Group D3)

R. yunnanense Franch.　　　　Triflora Gp D3

(*R. aechmophyllum* Balf.f. & Forrest, *R. chartophyllum* Franch.)

Shrub 0.3–4 m (1–12 ft) high; **branchlets** scaly, not bristly or sometimes bristly.

Leaves evergreen or sometimes semi-deciduous, oblanceolate, oblong-lanceolate or lanceolate; **lamina** 2.5–10.4 cm long, 0.8–2.8 cm broad; **upper surface** bright green, often scaly, sparsely bristly or not bristly, margins bristly or not bristly; **underside** pale glaucous green, scaly, scales 2–6 times their own diameter apart.

Petiole often bristly.

Inflorescence terminal or terminal and axillary, 3–5 flowers.

Calyx minute, 0.5–1 mm long.

Corolla widely funnel-shaped, 1.8–3.4 cm long, 2–4 cm across, white, pink, deep rose, pale rose-lavender or lavender, with deep crimson, deep rose or green spots, scaly or not scaly outside.

Mid-W, W and N Yunnan, W Guizhou, NE Upper Burma, SW Sichuan, SE Tibet. 2,000–4,250 m (6,500–14,000 ft).

AM 1903 (Glasnevin), flowers pink with brown spots.

AM 1993 'Openwood'.

Epithet: From Yunnan.　　Hardiness 3.　　May.

Very common in cultivation, this species should be considered first when identifying plants in Subsection Triflora. *Very widely distributed in the wild, and very variable in flower colour and other attributes. Free-flowering.*

Plants labelled **R. bodinieri**, *which have an elongated leaf tip, belong here.*

R. pleistanthum, *from Sichuan and Yunnan, strongly resembles* R. yunnanense *but has young leaves and petioles* **consistently lacking bristles**. *Having been submerged in the past, it is strangely treated as a full species in* 'The Flora of China', 2005.

R. yunnanense Franch.　　　　Triflora Gp D3

R. hormophorum Balf.f. & Forrest

Shrub 0.25–5 m (0.8–16 ft) high; **branchlets** scaly, puberulous.

Leaves completely deciduous, lanceolate or oblanceolate; **lamina** 2.8–7.3 cm long, 1–2.4 cm broad; **upper surface** bright green, scaly, bristly or not bristly, puberulous or glabrous, margins often bristly; **underside** pale glaucous green, laxly scaly, scales 3–6 times their own diameter apart.

Petiole often bristly.

Inflorescence 3–6 flowers, flowers precocious.

Calyx minute, 0.5 mm long.

Corolla widely funnel-shaped, 1.6–3.1 cm long, rose, white, white suffused with pink, rose-lilac or lavender, with or without crimson, olive-green, rose or orange spots.

SW Sichuan, NW Yunnan. 2,450–3,950 m (8,000–13,000 ft).

AM 1943 (Minterne), flowers white with a few brown spots.

Epithet: Bearing a necklace.　　Hardiness 3.　　May.

A delightful sight with its precocious flowers. Submerged into R. yunnanense *but deserves at least group status because of its deciduous leaves.*

Leaf underside laxly scaly, scales 2–6 times their own diameter apart.

Petiole and leaf margins sometimes bristly.

New stems purplish-red on one side in some forms.

Branchlets not bristly or sometimes bristly.

Leaves evergreen or sometimes semi-deciduous.

Corolla white, pink, deep rose, pale rose-lavender or lavender, with deep crimson, deep rose or green spots.

R. yunnanense 'Openwood' AM at Deer Dell

Leaves completely deciduous.

Corolla white, white tinged pink, rose-lilac or lavender.

R. yunnanense (as *R. hormophorum* F5874) at Deer Dell showing new leaves emerging on bare stems

Rhododendron
(Triflora Group D3)

R. vilmorinianum Balf.f.   Triflora hybrid

Shrub 1–1.5 m (3–5 ft) high; **branchlets** deep crimson or deep crimson-purple, scaly.
Leaves lanceolate, oblong-lanceolate, oblong or oblong-obovate; **lamina** 2.4–6.6 cm long, 0.8–2.6 cm broad; **upper surface** dark green, shiny; **underside** scaly, scales one-half to their own diameter apart.
Petiole often bristly.
Inflorescence 2–4 flowers.
Pedicel often pubescent.
Calyx scaly.
Corolla widely funnel-shaped, 1.8–3.5 cm long, yellowish-white or pink, with or without brownish spots, outside scaly.
E Sichuan.
Epithet: After the famous French seedsman.   Hardiness 3.   May.

Plants in cultivation are probably all hybrids of R. augustinii *and* R. yunnanense, *and do not match the herbarium material.*

R. wongii Hemsl. & E.H.Wilson   Triflora hybrid

Shrub 0.6–2 m (2–6 ft) high; **branchlets** scaly.
Leaves elliptic, oblong-elliptic or oblong; **lamina** 1.5–3 cm long, 0.8–1.5 cm broad; **upper surface** dark green or olive-green, scaly; **underside** slightly glaucous or not glaucous, densely scaly, scales nearly contiguous or their own diameter apart.
Petiole scaly.
Inflorescence 2–5 flowers.
Calyx minute, 0.5–1 mm long.
Corolla widely funnel-shaped, 1.6–2.2 cm long, cream-coloured or yellow.
W Sichuan. 3,650 m (12,000 ft).
Epithet: After I.G. Wong, Ichang, friend and helper of E.H. Wilson.   Hardiness 4.   April–May.

Although R. wongii is described as a full species in 'The Flora of China' 2005, plants in cultivation appear to be hybrids of R. ambiguum *and* R. flavidum *(and such hybrids have been observed by P.A. and K.N.F. Cox in the wild).*

R. wongii at Deer Dell

Rhododendron
(Triflora hybrid)

Subgenus *Rhododendron*
Section *Rhododendron*
Subsection *Uniflora* Sleumer
Uniflorum Series

Usually dwarf, prostrate, spreading, compact shrubs, but can be broadly upright.
Inflorescence 1–2 flowers except for *R. pumilum*.
Corolla being densely pubescent outside.
Style slender and straight.

These species are not easy to please in cultivation.

R. imperator Hutch. & Kingdon Ward
R. patulum Kingdon Ward

Dwarf **shrub** completely prostrate or prostrate, 3–10 cm (1.2–4 inches) high, with spreading or creeping branches; **branchlets** scaly.

Leaves lanceolate to oblanceolate; **lamina** 1.3–3.8 cm long, 0.4–1 cm broad, apex acute or sometimes rounded; **upper surface** dark green, scaly or not scaly; **underside** scaly, scales 1–6 times their own diameter apart.

Inflorescence 1–2 flowers.

Corolla widely funnel-shaped, 2.3–3.2 cm long, dark purple, pinkish-purple or purple, outside rather densely pubescent.

Upper Burma. 3,050–3,350 m (10,000–11,000 ft).

AM 1934 (Wisley), flowers rosy purple (KW6884).

Epithet: Emperor.   Hardiness 2–3.   April–May.

Plants under the name R. patulum *are referable to this species.*

R. ludlowii at Bracken Hill (Davidian)

Leaf apex usually acute.

Flowers large for Subsection Uniflora.

Completely prostrate or prostrate shrub.

Leaf underside scaly, scales 1–6 times their own diameter apart.

Corolla dark purple, pinkish-purple or purple, outside rather densely pubescent.

R. imperator at the Charlton Seal Garden, Vancouver (M.L.A. Robinson)

R. ludlowii Cowan

Dwarf **shrub** up to 30 cm (1 ft) high; **branchlets** scaly, glabrous.

Leaves obovate, oval or rounded; **lamina** 1.2–1.4 cm long, 0.8–1 cm broad; **upper surface** dark green, scaly; margin crenulate-undulate, distinctly notched; **underside** scaly, scales large, 2–3 times their own diameter apart.

Inflorescence 1–2 flowers.

Calyx 5–7 mm long.

Corolla broadly campanulate, 1.5–2.5 cm long, yellow with reddish-brown spots inside the tube, outside scaly, rather densely pubescent.

SE Tibet. 4,100 m (13,500 ft).

Epithet: After F. Ludlow, who collected widely in the Himalayas.   Hardiness 2–3. April–May.

Rare in cultivation.

R. pemakoense Kingdon Ward

Dwarf **shrub** 30–60 cm (1–2 ft) high, often stoloniferous; **branchlets** scaly.

Leaves obovate or oblong-obovate; **lamina** 1–3 cm long, 0.5–1.5 cm broad; **upper surface** greyish-green or bright green, scaly; **underside** pale glaucous green, scaly, scales 0.5–1.5 times their own diameter apart.

Inflorescence 1–2 flowers.

Calyx 1–4 mm long.

Corolla tubular-campanulate or tubular-funnel-shaped, 2.5–3.5 cm long, pinkish-purple, purple or pale pink, outside scaly, rather densely pubescent.

SE Tibet.   2,900–3,050 m (9,500–10,000 ft).

AM 1933 (Bulstrode), flowers white tinged mauve (KW6301).

Epithet: From the province of Pemako, E Tibet.   Hardiness 3.   March–April.

This is the most common species of Subsection Uniflora in cultivation, and differs from the others in its broadly funnel-campanulate corolla.

Large yellow corolla compared to the size of the small notched leaves.

Leaves obovate, oval or rounded, margin crenulate-undulate.

Leaf underside scales large, 2–3 times their own diameter apart.

R. ludlowii (Davidian)

Compact dense habit, leaves usually quite dark.

Corolla normally tubular-funnel-shaped, pinkish-purple, purple or pale pink.

Leaves obovate or oblong-obovate.

Leaf underside scales 0.5–1.5 times their own diameter apart.

R. pemakoense at Deer Dell

R. pumilum Hook.f.

Dwarf **shrub** lax, or low compact prostrate, 5–13 cm (2–5 inches) high, or a lax
broadly upright shrub up to 20 cm (8 inches) or sometimes up to 60 cm (2 ft)
high; **branchlets** scaly, minutely puberulous.

Leaves oval, elliptic, almost orbicular or obovate; **lamina** 0.8–1.9 cm long,
0.4–1.1 cm broad; **upper surface** bright green; **underside** pale glaucous
green, scaly, scales 1–3 times their own diameter apart.

Inflorescence 1–3 flowers.

Pedicel 1–2.6 cm long.

Corolla campanulate, 0.8–1.9 cm long, pink, pale pink, rose or pinkish-purple,
outside rather densely pubescent.

Sikkim, Nepal, Bhutan, S and SE Tibet, Assam, NE Upper Burma. 3,500–4,300 m
(11,500–14,000 ft).

AM 1935 (Townhill Park), flowers pinkish-mauve (KW696l).

Epithet: Dwarfish.   Hardiness 2–3.   April–May.

R. uniflorum Kingdon Ward

Dwarf **shrub**, low-growing up to 30 cm (1 ft) high, or broadly upright shrub up to
1 m (3 ft) high; **branchlets** scaly.

Leaves oblong-oval, obovate or oblong-obovate; **lamina** 1.3–2.4 cm long,
0.6–1.2 cm broad; **upper surface** bright green; **underside** pale glaucous
green, laxly scaly, scales 3–6 times their own diameter apart.

Inflorescence 1–2 flowers.

Pedicel 1.3–2.2 cm long, lengthening in fruit up to 3.5 cm long.

Corolla widely funnel-shaped, 2.2–2.8 cm long, purple, outside rather densely
pubescent, scaly.

S Tibet. 3,350–3,650 m (11,000–12,000 ft).

Epithet: One flower.   Hardiness 2–3.   April–May.

Never recollected after its introduction in 1924 and its status is uncertain.

Differs by having small thimble-like flowers, usually pink, and sparse foliage

Corolla campanulate long, pink, pale pink, rose or pinkish-purple.

Low compact prostrate shrub.

Leaf underside scales 1–3 times their own diameter apart.

Inflorescence 1–3 flowers.

R. pumilum at Arduaine

Allied to *R. pemakoense*, but with upright habit of growth and leaves laxly scaly: leaf scales 3–6 times their own diameter.

Larger leaves than *R. imperator*.

Corolla purple, outside rather densely pubescent.

R. uniflorum KW5876 at Deer Dell

Rhododendron
(Uniflora)

Subgenus *Rhododendron*
Section *Rhododendron*
Subsection *Virgata* (Hutch.) Cullen
Virgatum Series

Sprawling shrub.
Single or paired <u>axillary flowers</u> in the upper 1–12 leaves.

R. virgatum Hook.f.

Shrub 0.3–2.4 m (1–8 ft) high with long branches; **branchlets** scaly.

Leaves lanceolate, oblong-lanceolate or oblong-obovate; **upper surface** dark green or pale green, scaly or not scaly; **underside** pale green or pale glaucous green, scaly, scales one-half to twice their own diameter apart, with large peltate scales widely or closely separated.

Inflorescence axillary in the upper 1–12 leaves, 1 or sometimes 2 flowers.

Corolla funnel-shaped or tubular-funnel-shaped, 1.4–3.9 cm long, pink, pale rose, purple or white, outside scaly, rather densely or moderately pubescent.

Nepal, Sikkim, Bhutan, Assam, SE Tibet, NW and mid-W Yunnan. 1,850–3,800 m (6,000–12,500 ft).

AM 1973 (Sandling Park), flowers white.

Epithet: With willowy twigs.   Hardiness 1–3.   April–May.

Cullen has made a division into two subspecies (ssp. virgatum *and ssp.* oleifolium), *which vary in corolla length. As this variation in corolla size is clinal, it can be accommodated in one species.*

R. virgatum (as *R. oleifolium*) 'Penheale Pink' in the Valley Gardens

Inflorescence axillary in the upper 1–12 leaves.

One or sometimes 2 flowers.

Corolla pink, pale rose, purple or white, outside rather densely or moderately pubescent.

R. virgatum KW6279 (as *R. oleifolium*) at Deer Dell

Rhododendron
(Virgata)

Species not, or very rarely, in cultivation, and/or of doubtful provenance

Argyrophylla

R. brevipetiolatum M.Y.Fang
R. ebaniense M.Y.Fang
R. fangchenense P.C.Tam.
R. hunnewellianum ssp. rockii D.F.Chamb.
R. oblancifolium M.Y.Fang
R. shimenense Q.X.Liu & C.M.Zhang
R. simiarum var. versicolor M.Y.Fang
R. simiarum var. deltoideum P.C.Tam.

Campanulata

R. gannanense Z.C.Feng & X.G.Sun

Campylocarpa

R. henanense W.P.Fang
R. longicalyx M.Y.Fang

Fortunea

R. calophytum var. *jinfuense* Fang & W.K.Hu
R. decorum ssp. *parvistigmatis* W.K.Hu
R. faithae Chun.
R. gingongshanicum P.C.Tam.
R. gonggashanense W.K.Hu
R. guihanianum G.Z.Li
R. hemsleyanum var. chengianum W.P.Fang
R. jingangshanicum P.C.Tam.
R. jingshanense Z.H.Yang
R. liboense Z.R.Chen & K.M.Lan.
R. kwangfuense Chun & Fang
R. magniflorum W.K.Hu
R. miyiense W.K.Hu
R. nymphaeiodes W.K.Hu
R. orbiculare ssp. *oblongum* W.K.Hu
R. praeteritum var. *hirsutum* W.K.Hu
R. verricuiferum W.K.Hu
R. wolongense W.K.Hu
R. xiaoxidongense W.K.Hu

Grandia

R. blumei Nuttall
R. oreogenum L.C.Hu

Irrorata

R. excelsum A.Chev
R. gongshanense T.L.Ming

R. guizhouense M.Y.Fang
R. irroratum ssp. *kontumense* D.F.Chamb.
R. korthalsii Miq.
R. laojunshanense T.L.Ming
R. mengtszense Balf.f. & W.W.Sm.
R. ningyuenense Hand.-Mazz.
R. pingbianense M.Y.Fang
R. spanotrichum Balf.f. & W.W.Sm.
R. tanastylum var. *lingzhiense* M.Y.Fang.

Maculifera

R. pachyphyllum W.P.Fang.
R. polytrichum W.P.Fang.
R. pilostylum W.P.Fang.
R. pseudochrysanthum var. *nankotaisanense* (Hayata) T.Yamaz

Neriiflora

R. bijiangense T.L.Ming
R. cloiophorum var. *mannophorum* Davidian
R. erastum Balf.f. & Forrest
R. euchroum Balf.f. & Kingdon Ward

Parishia

R. flavoflorum T.L.Ming
R. huidongense T.L.Ming
R. urophyllum W.P.Fang

Taliensia

R. barkamense D.F.Chamb.
R. codonanthum Balf.f. & Forrest
R. dachangense G.Z.Li
R. danbaense L.C.Hu
R. detersile Franch.
R. heizhugouense M.Y.He & L.C.Hu
R. lulangense L.C.Hu & Y.Tateishi
R. potaninii Balatin
R. pomense Cowan & Davidian
R. principis Bur. & Franch.
R. punctifolium L.C.Hu
R. puguense L.C.Hu
R. roxieoides D.F.Chamb.
R. shanii W.P.Fang
R. torquescens D.F.Chamb.
R. trichogynum L.C.Hu.
R. zhongdianense L.C.Hu

Thomsoni

R. bonvalotii Franch.
R. dasycladoides Hand.-Mazz.
R. megalanthum M.Y.Fang
R. populare Cowan
R. ramipilosum T.L.Ming

Azaleastrum

R. bachii H. Leveille
R. hangzhouense W.P.Fang & M.Y.He
R. medoense W.P.Fang & M.Y.He
R. mitriforme P.C.Tam.
R. ngawchangense Philipson & Philipson
R. ovatum var. *setiferum* M.Y.He
R. sanidodeum P.C.Tam.
R. tianlinense P.C.Tam.
R. uwaense H.Hara & T.Yamanaka
R. xinganense G.Z.Li

Choniastrum

R. cavaleriei Levl.
R. championiae var. *ovalifolium*
  P.C.Tam.
R. dayaoshanense L.M.Gao & D.Z.Li
R. detampulum Chun ex P.C.Tam.
R. esquirolii Levl.
R. feddei Levl.
R. henryi Hance
R. huguangense P.C.Tam.
R. kaliense W.P.Fang & M.Y.He
R. latouchae var. *ionanthum* (W.P.Feng)
  G.Z.Li
R. linericupulare P.C.Tam.
R. longilobum L.M.Gao & D.Z.Li
R. mitriforme P.C.Tam.
R. shiwandashanense P.C.Tam.
R. stamineum var. *gaozhaiense*
  L.M.Gao
R. stamineum var. *lasiocarpum*
  R.C.Feng & C.H.Yang
R. subestipitatum Chun & P.C.Tam.
R. taiense Hutch.
R. taishunense B.Y.Ding & Y.Y.Fang
R. truncatovarium L.M.Gao & D.Z.Li
R. tutcherae Hemsl. & Wils.
R. vaniotii Levl.

Pogonanthum

R. atropunicum H.P.Yang
R. bellissimum D.F.Chamb.
R. fragrans Maxim.
R. heteroclitum H.P.Yang
R. hoi W.P.Fang
R. luhuoense H.P.Yang
R. mainlingense S.H.Huang & R.C.Fang

R. nyingchiense S.H.Huang & R.C.Fang
R. platyphyllum Balf.f. & W.W.Sm.
R. pogonophyllum Cowan & Davidian
R. praeclarum Balf.f. & Farrer
R radendum W.P.Fang
R. rufescens Franch.
R. tubulosum Ching ex W.Y.Wang

Boothia

R. nanjianense K.M.Feng & Z.H.Yang

Cinnabarina

R. laterifolium R.C.Fang & A.L.Zhang
R. igneum Cowan
R. tenuifolium R.C.Fang & S.H.Huang

Heliolepida

R. hirsutipetiolatum R.C.Fang &
  A.L.Zhang
R. invictum Balf.f. & Farrer

Lapponica

In cultivation but not described:
R. bulu Hutch.
R. capitatum Maxim.
R. complexum Balf.f. & W.W.Sm.
R. polycladum Franch.
R. tapetiforme Balf.f. & Kingdon-Ward
R. thymifolium Maxim.
R. websterianum var. *websterianum*
  Rehder & Wils.
R. yungningense Balf.f.

Not in cultivation:
R. amundsenianum Hand.-Mazz.
R. bamaense Z.J.Zhao
R. burjaticum Malyschev
R. dawuense H.P.Yang
R. declivatum Ching & H.P.Yang
R. gologense C.J.Xu & Z.J.Zhao
R. joniense Ching & H.P.Yang
R. labolengense Ching & H.P.Yang
R. lungchiense W.P.Fang
R. maowenense Ching & H.P.Yang
R. minyaense Philipson & Philipson
R. orthocladum var. *longistylum*
  Philipson & Philipson
R. quinghaiense Ching & W.Y.Yang
R. taibaiense Ching & H.P.Yang
R. tsai Fang
R. websterianum var. *yulongense*
  Philipson & Philipson
R. xiguense Ching & H.P.Yang
R. yulingense W.P.Fang

R. yushuense Z.J.Zhao
R. zekoense Y.D.Sun & Z.J.Zhao
R. zheguense Ching & H.P.Yang

Ledum

R. tolmachevii Harmaja

Maddenia

In cultivation but not described:
R. carneum Hutch.
R. coxianum Davidian
R. cuffeanum Hutch.
R. dendricola Hutch.
R. levinei Merr.
R. ludwigianum Hosseus
R. pseudocilipes Cullen
R. roseatum Hutch.
R. scopulorum Hutch.
R. surasianum Balf.f. & Craib.
R. taggianum Hutch.
R. walongense Kingdon-Ward

(for descriptions see www.rhodogroup-
rhs.org)

Not in cultivation:
R. amandum Cowan
R. chunienii W.P.Fang
R. ciliipes Hutch.
R. crenulatum Hutch. ex Sleumer
R. excellens Hemsl. & Wils.
R. fleuryi Pop.
R. kiangsiense W.P.Fang
R. linearilobum R.C.Fang & A.L.Zhang
R. mianningense Z.J.Zhao
R. nemerosum R.C.Fang
R. rhombifolium R.C.Fang
R. rufosquamosum Hutch.
R. wumingense W.P.Fang
R. yaogangxianense Q.X.Liu
R. yizangense Q.X.Liu
R. yungchangense Cullen

Micrantha

R. brevicaudatum R.C.Fang &
S.S.Chang
R. liaoxigensis S.L.Tung & Z.Lu

Monantha

R. flavanantherum Hutch. & Kingdon-
Ward

Scabrifolia

R. fuyuanense Z.H.Yang
R. spinuliferum var. *glabrescens*
K.M.Feng

Tephropepla

R. tsinlingense W.P.Fang ex J.Q.Fu

Trichoclada

R. nanjianense K.M.Feng & Z.H.Yang

Triflora

R. bivelatum Balf.f.
R. brachypodum W.P.Fang & P.S.Liu
R. gemmiferum Philipson & Philipson
R. guangnanense R.C.Fang
R. kangdingense Z.J.Zhao
R. seguinii H.Lev.
R. shaanxiense W.P.Fang & Z.J.Zhao
R. shimianense W.P.Fang & P.S.Liu
R. tatsienense var. *nudatum* R.C.Fang
R. triflorum ssp. *multiflorum* R.C.Fang
R. xichangense Z.J.Zhao
R. zaleucum var. *pubifolium* R.C.Fang

Virgata

R. virgatum ssp. *olieofolium* var.
glabriflorum K.M.Feng

Therorhodion

R. camtschaticum ssp. *glandulosum*
(Standley ex Small) Millais
R. redowskianum Maxim.

Brachycalyx

R. amakusaense T.Yamaz
R. chilanshanense Kurashige
R. daiyunicum P.C.Tam.
R. huadingense B.Y.Ding & Y.Y.Fang
R. hyuganense T.Yamaz
R. osuzuyamemse T.Yamaz
R. tsurugisanense T.Yamaz
R. yakumontanum T.Yamaz

Tsutsusi

Evergreen azaleas not included but
in cultivation:
R. arunachalense D.F.Chamb & S.J.Rae
R. atrovirens Franch.
R. eriocarpum (Hay *et al.*) Nakai
R. flumineum Fang & M.Y.He
R. kanehirai Wilson

R. microphyton Franch.
R. noriakianum Suzuki
R. oldhamii Maxim
R. ripense Makino
R. rubropilosum Hayata
R. saisiuense Nakai
R. saxicolum Sleumer
R. scabrum G.Don
R. simsii var. *simsii* Planch.
R. stenopetalum (Hogg) Mabb.
R. subsessile Rendle
R. tosaense Makino
R. tschonoskyi Maxim.
R. tsusiophyllum Sugimoto

Evergreen azaleas not in cultivation:
R. adenanthum M.Y.He
R. apricum P.C.Tam.
R. bellum
R. bicorniculatum P.C.Tam.
R. boninense Nakai
R. breviperulatum Hayata
R. chaoanense
R. chrysocalyx Levl. & Vaniot
R. chunii W.P.Fang
R. crassimedium P.C.Tam.
R. crassistylum M.Y.He
R. cretaceum P.C.Tam.
R. florulentum P.C.Tam.
R. flosculum W.P.Fang & G.Z.Li
R. fuchsiifolium Levl.
R. fuscipilum M.Y.He
R. gratiosum P.C.Tam.
R. guizhongense W.P.Fang & M.Y.He
R. hainanense Merrill
R. hejiangense M.Y.He
R. huiyangense Fang & M.Y.He
R. hunanense Chun ex P.X.Tan.
R. hypoblemastosum P.C.Tam.
R. jasminoides M.Y.He
R. jinpingense Fang & M.Y.He
R. jinxiuense Fang & M.Y.He
R. kwangsiense Hu ex P.C.Tam.
R. kwangtungense Merrill & Chun.
R. lasiostylum Hayata
R. linguiense G.Z.Li
R. litchiifolium T.C.Wu & P.C.Tam.
R. longifalcatum P.C.Tam.
R. longiperulatum Hayata
R. loniceriflorum P.C.Tam.
R. macrosepalum Maxim.

R. malipoense M.Y.He
R. mariae ssp. *mariae* Hance
R. mariae ssp. *kwangsiense* D.F.Chamb.
R. matsumurai Komatsu
R. meridionale P.C.Tam.
R. minutiflorum Hu
R. myrsinifolium Ching ex Fang & M.Y.He
R. naamkwanense Merrill
R. nanpingense P.C.Tam.
R. octandrum M.Y.He
R. petilum P.C.Tam.
R. pinetorum P.C.Tam.
R. polyraphidoideum P.C.Tam.
R. pulchroides Chun & W.P.Fang
R. qianyangense M.Y.He
R. rhodoanthum M.Y.He
R. rhuyuenense Chun.
R. rivulare Hand-Mazz.
R. rufo-hirtum Hand-Mazz.
R. rufulum P.C.Tam.
R. saxatile B.I.Ding & I.I.Fang
R. seniavinii Maxim.
R. sikayotaisanense Masam.
R. simsii var. *mesembrinum* Rehder
R. spadiceum P.C.Tam.
R. sparsifolium W.P.Fang
R. strigosum R.L.Liu
R. subcerinum P.C.Tam.
R. subenerve P.C.Tam.
R. subflumineum P.C.Tam.
R. taipaoense T.C.Wu & P.C.Tam.
R. tashiroi var. *lasiophyllum* Hatus.
R. tenue Ching ex W.P.Fang & M.Y.He
R. tenuilaminare P.C.Tam.
R. tingwuense P.C.Tam.
R. tsoi Merr.
R. unciferum P.C.Tam.
R. viscidum C.Z.Guo & Z.H.Liu
R. viscigemmatum P.C.Tam.
R. yangmingshanense P.C.Tam.
R. yaoshanicum Fang & M.Y.He

Pseudovireya

Hardy species in cultivation:
R. emarginatum Hemls. & Wils.
R. euonymifolium Levl.
R. kawakamii Hayata
R. rushforthii Argent & D.F.Chamb.
R. santapaui Sastry *et al.*
R. vaccinioides Hook.f.

Glossary

Abaxial	Pointing away from the axis (usually of stem and leaves) so usually the underside of leaf.
Acuminate	Tapering into a point.
Acute	Pointed.
Adaxial	Pointing towards the axis (usually of stem and leaves) so usually the upperside of leaf.
Adpressed	Lying flat.
Affinity (aff.)	Closely related. Often used when a newly collected specimen does not quite fit a known species but is closely related.
Agglutinate	Glued together.
Alternate	Where only one leaf grows from a **node** on the stem (the alternatives being **opposite** and **whorled**).
Alveola	Cavity.
Anther	The part of the **stamen** containing the pollen grains.
Apex	Tip.
Apiculate	A short sharp point on the leaf apex which is not stiff as in **mucronate** or **cuspidate**.
Arboreal	Tree-like.
Aristate	Awned; tipped by a bristle.
Auricle	An ear-like appendage.
Auriculate	Eared.
Axil	The junction of leaf and stem.
Axillary	Arising in the axil.
Base	The parts nearest the **petiole** (of leaf).
Bistrate	With two layers or strata.
Bloom	A waxy covering appearing powdery (of leaf and **corolla** surfaces).
Bullate	Puckered or blistered.
Calyx, calyces	The outermost part of the flower bud, often left as a cover around the seed **capsule** after flowering.
Campanulate	Bell-shaped.
Candelabroid	An upright **inflorescence** produced when the **rhachis** is long and the **pedicels** emerge from it stiffly (characteristic of subsection *Pontica*).
Capitate	Collected in a dense knob-like head or cluster.
Capitellate	Mop-headed (of **indumentum** hairs).
Capsule	The seed vessel at maturity.
Chartaceous	Thin papery or parchment-like texture.
Chromosome	Rod-like portion of the cell nucleus that determines hereditary characteristics.
Ciliate	With fine hairs.
Ciliolate	Minutely **ciliate**.
Cline	A population of plants in the wild showing systematic gradual gradation from one form to another.

Clone	The vegetatively produced progeny of a single individual.
Compacted	Flattened (of **indumentum**, often the lower layer).
Concave	Incurved.
Conoid	Shaped like a cone.
Contiguous	Touching.
Convex	Curved outward – as opposed to **concave**.
Convolute	Rolled up longitudinally.
Cordate	With two round lobes forming a deep recess at the base.
Coriaceous	Leathery.
Corolla	The **tube** and **lobes** of the flower.
Crenate	Toothed (of **lepidote** scales).
Crenulate	Margin notched, with rounded teeth, scalloped.
Crisped	So curved that the tip is close to the point of attachment (of **indumentum** hairs).
Cultivar	A cultivated variety that is distinct from other varieties.
Cuneate	Wedge-shaped.
Cupular	Cup-shaped.
Cuspidate	Tipped with a sharp rigid point to a greater extent than **mucronate**.
Deciduous	Seasonally falling, not persistent.
Declinate	Curved downwards.
Decumbent	Horizontal for most of its length but turning up at the tip.
Decurrent	Extending down the stem or **petiole**.
Deflexed	Bent downwards.
Dendroid	Tree-like, branching from a distinct trunk (of **indumentum** hairs).
Detersile	**Indumentum**, scales or hairs that are eventually shed.
Dichotomous	Divided in pairs.
Dimorphic	Of two distinct forms.
Diploid	Having two complete sets of **chromosomes** (2n=36).
Discoid	Disk-shaped.
Eciliate	Without fine hairs.
Eglandular	Not **glandular**.
Elepidote	Not scaly.
Elliptic	Tapering equally at both ends (of leaves).
Emarginate	With a notch at the end.
Entire	Margin undivided, without teeth.
Epapillate	Without **papillae**.
Epidermis	The surface or outer layer of cells (of leaves, stems etc.).
Epiphyte	A plant growing on another plant without being parasitic.
Erose	Eaten away (having an irregularly toothed margin as if gnawed).
Evanescent	Soon disappearing (of **indumentum**).
Exserted	Projecting beyond (usually of **stamens** or **style** longer than the **corolla**).
Falcate	Sickle-shaped.

Fasciculate	In bundles (of **indumentum** hairs etc.).
Fastigiate	With erect branches, leading to columnar growth.
Felted	Matted with intertwining hairs (of leaf **indumentum**).
Ferruginous	Rust-coloured.
Fimbriate	With a fringed margin (in rhododendron, usually of hairs).
Filament	The stalk of the **stamen** bearing the **anther**.
Flagellate	Whip-like – a long flexible stem that separates into individual hairs (of leaf **indumentum**).
Floccose	With soft woolly hairs.
Folioliferous	Bearing leaves. For **indumentum** hairs this refers to leaf-shaped structures at the ends of at least some of the hairs.
Fulvous	Reddish-yellow, tawny.
Genus	A group of closely related species, having general characteristics in common – see p. 674.
Glabrescent	Becoming **glabrous**. Also used to mean almost **glabrous**.
Glabrous	Without hairs.
Glandular	With **glands**.
Gland	An appendage discharging a sticky secretion. In rhododendron, usually a hair on the end of which is a secreting bulb.
Glaucous	Covered with a white, bluish or greyish waxy **bloom**.
Globose	Spherical.
Hirsute	Covered with stiff long erect hairs.
Hose-in-hose	Two **corollas** one within the other.
Hypercratiform	Goblet-shaped with spreading lobes (of **corollas**).
Imbricate	Overlapping each other at the margins.
Impressed	Where there is an abrupt junction between the **style** and the **ovary** (as though the **style** had been forced into a soft **ovary**), as opposed to **tapered** (a gradual thickening as one merges into the other).
Indumentum	A hairy covering, particularly of the lower surface of the leaves.
Inflorescence	The whole flowering body, **rhachis**, **pedicel** and **corollas**.
Lacerate	Torn or shredded, as with the leaf-scale margins in section *Pogonanthum*.
Lamina	Leaf blade, surface.
Lanate	Woolly, with soft mingled hairs.
Lanceolate	Lance shaped, tapering more gradually towards the tip (of leaves).
Lateral	On or at the side as opposed to the tip.
Lax	Loose. Where the flowers hang downwards on separated long **pedicels** and/or **rhachis**.
Lepidote	Scaly.
Ligulate	Strap-like.
Linear	Narrow with parallel sides (of leaf shapes).
Lingulate	Resembling a tongue.

Lobe	One of the divisions of a single structure into curved or rounded parts (of the **corolla** or **calyx**).
Long-rayed	With the ends divided into several long strands (of leaf **indumentum**).
Midrib	The central rib of a leaf.
Monomorphic	Of one type or shape only.
Monotypic	A genus of a single species.
Mucro	A hard sharp point.
Mucronate	Terminated by a hard sharp point (see also **cuspidate**).
Mucronulate	Minutely **mucronate** (see also **apiculate**).
Nectar pouches	Vessels or swellings at the base of a **corolla** containing a sweet substance.
Nectary	A **gland** through which a solution of sugar is secreted.
Node	A point on the stem from which one or more leaves arise.
Oblanceolate	Reverse **lanceolate**, i.e. tapering more gradually towards the leaf base and more abruptly towards the apex (of leaves).
Oblique	With unequal sides, as in the **corolla** of some of Subsection *Grandia* and some leaf bases in the genus.
Oblong	With parallel sides and blunt ends (but not narrow, which would be **linear**) (of leaf shapes).
Obovate	Reverse egg-shaped, i.e. narrower near the petiole.
Obsolete	Scarcely apparent: usually of a **calyx**.
Orbicular	Circular (of leaves).
Obtuse	Blunt.
Opposite	Where two leaves grow from the same node on opposite sides of a stem (the alternatives being **alternate** and **whorled**).
Oval	More symmetrical than **ovate** or **obovate**, and longer than broad (of leaf shapes).
Ovary	The undeveloped seed vessel, often enclosed by the **calyx**.
Ovate	Egg-shaped – broader near the **petiole** (of leaf shapes).
Ovoid	Egg-shaped solid with oval outline.
Papillate	Pimpled; covered with minute pimples (**papillae**).
Pannose	Having the appearance of cloth.
Pectinate	Toothed.
Pedicel	Flower-stalk.
Pellicle	Lustrous skin-like covering to **indumentum**.
Pellucid	Transparent (usually of a **gland**).
Peltate	Circular, with a stalk in the middle (of scales).
Perulae	Leaf bud scales that cover the leaf before its emergence.
Petaloid	Petal-like (usually of a **stamen**).
Petiole	Leaf-stalk.
Pilose	Covered with soft long hairs.
Pistil	The **ovary**, **style** and **stigma**.
Plastered indumentum	Indumentum skin-like, with a smooth polished surface.
Polyploid	With three or more complete sets of chromosomes.

Precocious	Flowers produced before the leaves appear.
Pruinose	With a waxy covering appearing powdery (of leaf and **corolla** surfaces).
Pseudowhorl	A circle of leaves looking like a **whorl**, but with the leaves emerging from separate **axils**.
Puberulous, puberulent	Minutely **pubescent**.
Pubescent	Hairy, with short soft hairs.
Punctate	Dotted, with depressions or minute **glands** (usually of a lower leaf surface).
Punctulate	Minutely dotted.
Pustule	A blister or pimple large than a **papilla**.
Pyriform	Pear-shaped.
Racemose	Flowers borne on an unbranched main stalk (hence **raceme**).
Radiate	Emerging from a common centre (of **indumentum** hairs).
Ramiform	Branched from different points along their length (of **indumentum** hairs).
Recurved	Curved backwards.
Reflexed	Bent abruptly backwards.
Reticulate	Netted, like a network (of leaf veins).
Retuse	Having a central depression in its rounded apex (of a leaf).
Revolute	Rolled backwards; margin rolled towards the lower side.
Rhachis, rachis	The part of the inflorescence bearing the **pedicels** and flowers.
Rhomboid (rhombic)	Diamond-shaped, the sides having equal lengths.
Rosulate	Collected into a rosette (of **indumentum** hairs).
Rotate	A very short tube with spreading, almost flat petals or **lobes**.
Rufous	Dull red or rust-coloured.
Rugose	Wrinkled.
Rugulose	Somewhat wrinkled.
Salver-shaped	With a long slender tube and flat spreading petals.
Scabrid	Rough to the touch, warty.
Scale	Minute disc-like object found on branchlets, leaves and flowers.
Sericious	Silky, covered in small soft straight hairs.
Serrate	Saw-toothed with forward pointing notches (of leaf margins).
Serrulate	Minutely **serrate**.
Sessile	Without a stalk.
Seta	A bristle.
Setose	Bristly; with stiff hairs.
Setulose	Covered in small bristles (i.e. less than **setose**).
Short-rayed	With the ends of hairs divided into short strands (of **indumentum** hairs).
Spathulate	With a broadly rounded apex gradually tapering into the stalk: spatula-shaped.

Species, subspecies, variety, forma	*See p. 675.*
Stamen	The male organ of the flower bearing the pollen.
Stellate	Star-shaped, usually on a stalk (of **indumentum** hairs).
Stigma	The small pollen-receptive surface at the tip of the **style**.
Stoloniferous	Bearing runners from near the base of the stem, often below the surface of the soil.
Strigillose	Hairs smaller in size than **strigose**.
Strigose	With stiff hairs lying flat.
Stipitate	Elevated on a stalk (of **glands**).
Style	The thread-like part of a gynoecium (**pistil**) between the **ovary** and **stigma**.
Subacute	Not as pointed as **acute**.
Subagglutinate	Somewhat agglutinate; i.e. somewhat glued together.
Subsessile	Almost **sessile**, on a very short **petiole**.
Swarm	A large group of natural hybrids (in the wild).
Synonym	Superceded ('sunk') plant name (in the context of this book).
Tapered	Gradually decreasing in breadth (especially when an **ovary** tapers on to a **style**).
Taxon, taxa	General term for any group with shared characteristics.
Taxonomy	The science of classification, description and identification.
Terete	Smoothly tapering (e.g. of the **ovary** tapering into the **style**).
Terminal	At the end of a shoot.
Tesselate	Mosaic-like (of leaf veins or **indumentum**).
Testa	The hard covering of a seed.
Tetraploid	With four complete sets of **chromosomes**.
Tomentum	Dense hair covering, often of the upper surface of a leaf: hence **tomentose**.
Triploid	With three complete sets of **chromosomes**.
Truncate	Straight across.
Truss	A cluster of flowers on a single stalk (the **rhachis**).
Type	Strictly, the original introduction of a particular species.
Tube	When present, the part of the **corolla** nearest the **pedicel**.
Undulate	Having a wavy margin.
Unistrate	With one layer or stratum (of **indumentum**).
Velutinous	Velvety.
Venation	The arrangement of the veins.
Ventricose	Swollen on one side, as with some **corollas** in subsection *Grandia*.
Verrucose	Having a wart-like or nodular surface.
Vesicle	A small bladder containing fluid.
Vesicular	Bladder-shaped (of **glands** or **scales**).
Villous, villose	With long soft straight hairs.
Viscous, viscid	Sticky.

Whorl	Where three or more leaves or branches grow from the same **node** on a stem, giving rise to a 'circle' of leaves, often near the end of the stem.
Winged	Broad or flattened often tapering into the leaf base (of **petioles**).
Zygomorphic	Flower of irregular shape that can be divided into equal halves along one vertical line only.

Leaf shapes – descriptions

Linear	Narrow, with parallel opposite sides, the ends tapering, at least 10–12 times as long as broad, e.g. *R. roxieanum* var. *oreonastes*.
Lanceolate	Lance-shaped, widest below the middle, the length of the leaf about three times the breadth, e.g. *R. griersonianum* and *R. yunnanense*.
Oblanceolate	Base tapering, apex broad, widest above the middle, the length of the leaf about three times the breadth, e.g. *R. uvarifolium*.
Oblong	The sides more or less parallel, the ends obtuse or somewhat rounded, the length of the leaf is about twice the breadth, e.g. *R. selense* and *R. cerasinum*.
Elliptic	The sides of the leaf are curved tapering equally to tip and base, widest at the middle, the length of the leaf is about twice the breadth, e.g. *R. campanulatum*.
Obovate	The sides are curved, apex rounded, base narrower, widest above the middle, length is about twice the breadth, e.g. *R. chaetomallum*.
Ovate	Egg-shaped: the sides are curved, widest at the base, the tip narrowed, the length is greater than the breadth, e.g. *R. wasonii*.
Oval	The sides are curved, rounded at both ends, widest at the middle, longer than broad, e.g. *R. callimorphum*.
Orbicular	Circular, e.g. *R. orbiculare*.

Taxa

Genus	A group of clearly homogeneous and related species.
Subgenus	A group with similar characteristics within a genus that will not normally hybridise with other subgenera within that genus.
Section	A subgenus may be divisible into Sections whose members share similar characteristics that differ from the characteristics of other Sections. Hybrids between species from different Sections are possible but appear to be rare. It should be noted here that the term Section implies that all rhododendron species fit into well-defined compartments.

Series	The use of the word 'Series', as previously used by H.H. Davidian and others, implies that some rhododendron species are difficult to place in an appropriate Section and 'flow' together as a series.
Subsection	Fewer differences than for Sections. Species from different Subsections within the same Subgenus and Section readily hybridise, often producing fertile offspring.
Species (sp.)	Must differ in at least two independent characteristics and have distributions in the wild distinct from their closest allies.
Subspecies (ssp.)	A subspecies of a species differs in small taxonomic characteristics from other members of the species (i.e. a taxon q.v. within a species). Although subspecies can breed with each other, they are usually found in different locations or in different populations.
Variety (var.)	Is defined where there is a differing but related form throughout much or all of the geographical area occupied by a species.
Forma	A distinct form found in cultivation whose taxonomic status has not been defined.

Flower shapes of lepidote and elepidote Rhododendrons

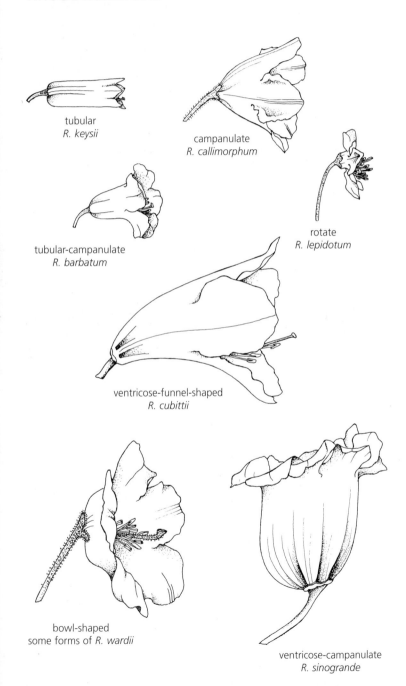

tubular
R. keysii

campanulate
R. callimorphum

rotate
R. lepidotum

tubular-campanulate
R. barbatum

ventricose-funnel-shaped
R. cubittii

bowl-shaped
some forms of *R. wardii*

ventricose-campanulate
R. sinograde

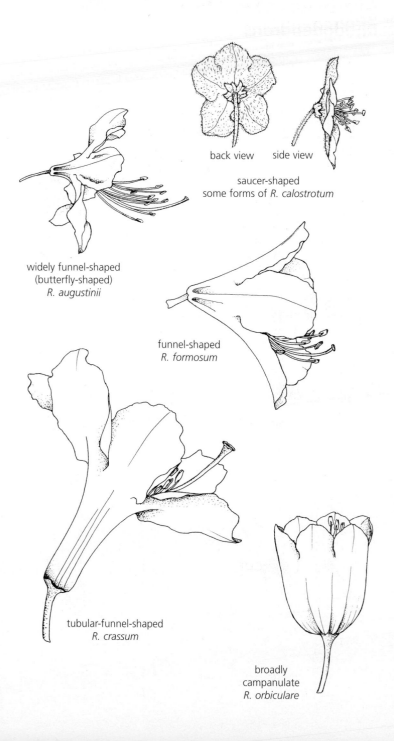

back view   side view

saucer-shaped
some forms of *R. calostrotum*

widely funnel-shaped
(butterfly-shaped)
R. augustinii

funnel-shaped
R. formosum

tubular-funnel-shaped
R. crassum

broadly
campanulate
R. orbiculare

Leaf shapes of lepidote and elepidote Rhododendrons

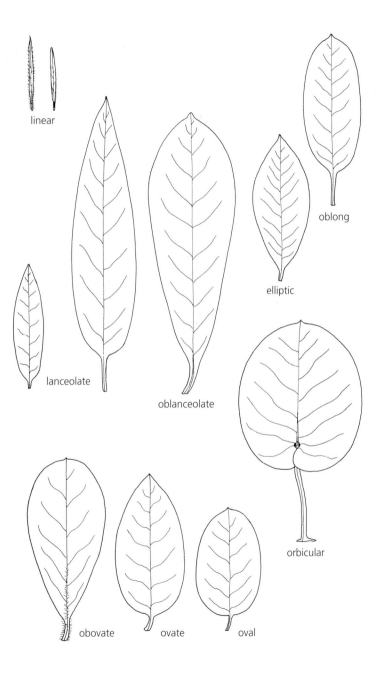

linear

oblong

elliptic

lanceolate

oblanceolate

orbicular

obovate

ovate

oval

Collector abbreviations

AC	Alan Clark
BLM	Beer, Lancaster and Morris
C&H	P. Cox and Hutchison
CCH	Chamberlain, P. Cox and Hutchison
DGEY	Dulong-Galiogong Expedition to Yunnan
EN	Edward Needham
F	George Forrest (also GF)
HECC	Hutchison, Evans, P. Cox and K. Cox
KC	K. Cox
KR	Keith Rushforth
KW	Frank Kingdon Ward
L&S	Ludlow and Sherriff
LS&E	Ludlow, Sherriff and Elliott
LS&H	Ludlow, Sherriff and Hicks
LS&T	Ludlow, Sherriff and Taylor
McB	R. MacBeath
McL	The Hon. J. McLaren
MF	Maurice Foster
PW	Peter Wharton
R	Joseph Rock
RV	Rhododendron Venture (Taiwan)
S&L	Sinclair and Long
TH	Tom Hudson
W	Ernest Wilson

Bibliography

Argent, G. (2006). *Rhododendrons of subgenus vireya*. The Royal Horticultural Society in association with Royal Botanic Gardens Edinburgh

Argent, G., Bond, J., Chamberlain, D.F., Cox, P.A. & Hardy A. (1998). *The Rhododendron Handbook*. The Royal Horticultural Society

Chamberlain, D., Hyam, R., Argent, A., Fairweather, G. & Walter, K.S. (1996). *The Genus Rhododendron, its classification and synonymy*. Royal Botanic Gardens, Edinburgh

Chamberlain, D.F. (1982). *A Revision of Rhododendron 2: Subgenus Hymenanthes*. HMSO

Cowan, J.M. (1950). *The Rhododendron Leaf*. Oliver & Boyd

Cox, P.A. & Cox, K.N.E. (1997). *The Encyclopedia of Rhododendron Species*. Glendoick Publishing

Cullen, J. (1977). Work in progress at Edinburgh on the classification of the Genus Rhododendron. In: *Rhododendrons with Magnolias and Camellias*. Royal Horticultural Society

Cullen, J. (1980). *A Revision of Rhododendron I: Subgenus Rhododendron, sections Rhododendron & Pogonanthum*. HMSO

Davidian, H.H. (1982). *The Rhododendron Species: Vol. I. Lepidotes*. Batsford

Davidian, H.H. (1989). *The Rhododendron Species: Vol. II. Elepidotes, Series Arboreum to Lacteum*. Batsford

Davidian, H.H. (1992). *The Rhododendron Species: Vol. III. Series Neriiflorum to Thomsonii*. Batsford

Davidian, H.H. (1995). *The Rhododendron Species: Vol. IV. Azaleas*. Timber Press

Doleshy, F. (1983). Distribution and classification of certain Japanese Rhododendrons. *American Rhododendron Society Journal*, 37: 81–89

Evans, P.D. (ed) (1998–2005). *Rhododendrons with Camellias and Magnolias*. The Royal Horticultural Society.

Fang, M., Fang, R., He, M., Hu, L., Yang, H. & Chamberlain, D.F. (2005). Rhododendron. *Flora of China Vol. 14*, p. 260–455

Fang, W. (ed). (1986). *Sichuan Rhododendron of China*. Science Press, Beijing

Feng, G. (ed). (1989). *Rhododendrons of China: Vol. I*. Science Press, Beijing

Feng, G. (ed). (1992). *Rhododendrons of China: Vol. II*. Science Press, Beijing

Feng, G. (ed). (1999). *Rhododendrons of China: Vol. III*. Science Press, Beijing

Galle, F.C. (1985). *Azaleas*. Timber Press

Halliday, P. (ed). (2001). *The Illustrated Rhododendron*. Timber Press

Harjma, H. (1991). Taxonomic notes on Rhododendron subsection *Ledum* (*Ledum*, Ericaceae), with a key to its species. *Ann. Bot. Fennici* 28: 171–173

Hayata, Icon. (1913). *Pl. Formosana*, p. 140–141

Judd, W.S. & Kron, K.A. (1995). A revision of Rhododendron, Subgenus *Pentanthera* (sections *Sciadorhodion*, *Rhodora* and *Viscidula*). *Edinburgh J. Bot.* 52: 1

Kneller, M. (1995). *The Book of Rhododendrons*. David & Charles

Kron, K.A. (1993). A revision of Rhododendron section *Pentanthera*. *Edinburgh J. Bot.* 50: 3

Kron, K.A. & Creel, M. (1999). A new species of deciduous Azalea from S. Carolina. *Novon* 9: 377–380

Leach, D.G. (1962). *Rhododendrons of the World*. Allen & Unwin

Leslie, A. (compiler) (2004). *The International Rhododendron Register and Checklist*. The Royal Horticultural Society

Millais, J.G. (1917, 2nd series 1922). *Rhododendrons and the Various Hybrids*. Longmans

Philipson, M.N. & Philipson, W.R. (1975). A revision of *Rhododendron* section *Lapponicum. Notes Roy. Bot. Gard. Edinburgh* 34: 1

Philipson, M.N. & Philipson, W.R. (1986). A revision of *Rhododendron*, Subgenera *Azaleastrum, Mumeazalea, Candidastrum* and *Therorhodion. Notes Roy. Bot. Gard. Edinburgh* 44: 1

Postan, C. (ed.) (1996). *The Rhododendron Story*. The Royal Horticultural Society.

Postan, C. (ed.) (1989–1997). *Rhododendrons with Camellias and Magnolias*. The Royal Horticultual Society.

Stevenson, J.B. (ed). (1947). *The Species of Rhododendron*. The Rhododendron Society

Toothill, E. (1984). *The Penguin Dictionary of Botany*. Penguin

Towe, L.C. (2004). *American Azaleas*. Timber Press

Young, J. & Lu-sheng C. (1980). *Rhododendrons of China*. Binford & Mort

Internet resources

The Flora of China
http://flora.huh.harvard.edu/china/index.html

RHS Rhododendron Camellia and Magnolia Group: for photographs
http://www.rhodogroup-rhs.org/index.htm

RHS: for accepted plant names
http://www.rhs.org.uk/

Glendoick Gardens and Nursery: for new introductions
http://www.glendoick.com/index.shtml

Fraser South Rhododendron Association: for recent taxonomy
http://www.flounder.ca/FraserSouth/Goetsch-Eckert-Hall.asp

The Danish chapter of the American Rhododendron Society: for photographs especially subsection *Lapponica*
http://www.rhododendron.dk/

The Azalea society of America: for photographs, particularly flower buds
http://www.azaleas.org/index.pl

The Rhododendron Species Foundation, Washington USA: for recent introductions
http://www.rhodygarden.org/

Index